GARDEN CENTRES & NURSERIES OF BRITAIN

Published by:
Travel Publishing Ltd
7a Apollo House, Calleva Park
Aldermaston, Berkshire RG7 8TN

EAN 9781904434658

First Published: 2007

Please Note:

All advertisements in this publication have been accepted in good faith by Travel Publishing and they have not necessarily been endorsed by *Country Living* Magazine.

All information is included by the publishers in good faith and is believed to be correct at the time of going to press. No responsibility can be accepted for errors.

Printing by: Ashford Colour Press, Gosport

Location Maps: © Maps in Minutes ™ (2007) © Crown Copyright, Ordnance Survey 2007

Cover Design: Lines & Words, Aldermaston

Foreword

As a mad-keen gardener, wherever I travel in Britain, if there's a nursery or garden centre nearby I can never resist taking a look. The visit almost always yields a plant, gardening tool or book that's on my horticultural shopping list. Often there's a pleasing coffee shop in which to while away a little time and there are always other enthusiasts to engage in conversation. This invaluable new guide from Country Living Magazine and Travel Publishing makes sure that, whenever you're planning a trip, you can check details in advance of the garden centres and nurseries you will find en route. If you're anything like me you'll find it invaluable – a must to keep in the car or on the bedside table.

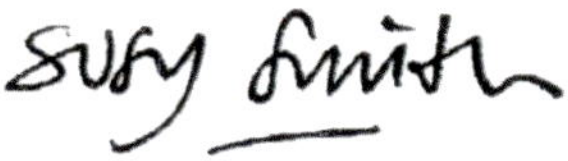

Susy Smith
Editor, Country Living magazine

PS To subscribe to *Country Living Magazine* each month, call 01858 438844

How to Use

Locating a Garden Centre or Nursery

Each garden centre and nursery is numbered and listed alphabetically by county for England and by country for Scotland and Wales.

- If you know the name of the garden centre or nursery in a particular county or country then refer to the contents page at the front of the guide which will direct you to the relevant county/country chapter where the garden centre or nursery can be located in the alphabetic listings.

- If you are seeking a garden centre or nursery within a specific geographical area use the contents page to find the relevant county/country chapter and then refer to the locator map at the beginning of the chapter. Select the numbered locator flag(s) closest to your preferred geographical location and locate the numbered garden centre or nursery within the chapter.

Locating a Garden Centre or Nursery Speciality

Where a garden centre or nursery has identified a speciality in a product or service these have been included under the garden centre/nursery listings. If you are looking for garden centres or nursery offering a particular speciality then refer to the index of specialities to the rear off the guide which will direct you to the relevant listing(s).

Travel Publishing

Contents

BEDFORDSHIRE

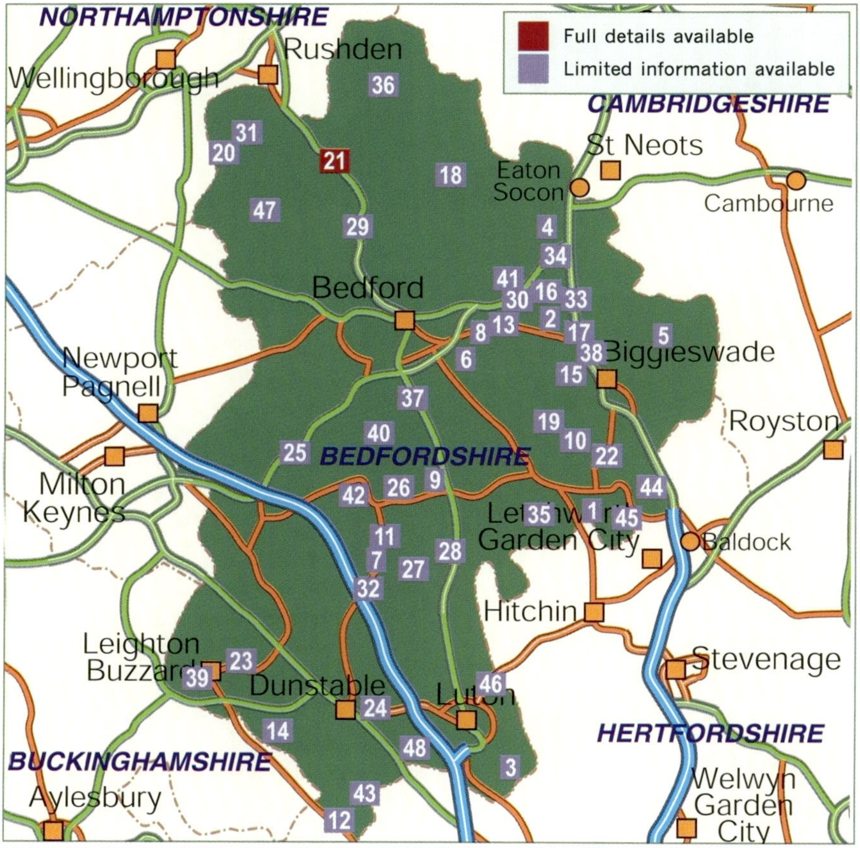

1 Amenity Trees Conifer Nursery

Arlesey, Bedfordshire SG15 6RF
Tel: 01462 731340
Specialities: Conifers, Trees

2 Asterby & Chalkcroft Nurseries

The Ridgeway, Blunham, Bedford, Bedfordshire MK44 3PH
Tel: 01767 640148
Specialities: Annuals/Biennials/Bedding, Shrubs

3 Bridge Valley Nursery

New Mill End, Luton, Bedfordshire LU1 3TS
Tel: 01582 766728
Specialities: Annuals/Biennials/Bedding, Shrubs

4 Cherwood Nursery

Blue Bells, Chawston, Bedford, Bedfordshire MK44 3BL
Tel: 01480 216934
Specialities: Seasonal Plants, Shrubs

5 Churchview Nurseries

5 Sutton Road, Potton, Sandy, Bedfordshire SG19 2DU
Tel: 01767 260400
Specialities: Annuals/Biennials/Bedding, Seasonal Plants

6 Cottage Farm Nurseries

312 Cople Road, Cardington, Bedford, Bedfordshire MK44 3SH
Tel: 01234 838383
Specialities: Hanging Baskets, Landscape Gardening

7 Dingley Dell Nursery

Toddington Road, Westoning, Bedford,
Bedfordshire MK45 5AH
Tel: 01525 874989
Specialities: Herbaceous Perennials, Hanging Baskets

8 Direct Garden Supplies

109 Bedford Road, Willington, Bedford,
Bedfordshire MK44 3PR
Tel: 01234 838070
Specialities: Herbaceous Perennials

9 Farrs

Clophill Road, Maulden, Bedford,
Bedfordshire MK45 2AD
Tel: 01525 861266
Specialities: Herbaceous Perennials

10 Firscreen Nurseries

Broom Road, Stanford, Biggleswade,
Bedfordshire SG18 9JE
Tel: 01462 811264
Specialities: Homegrown, Seasonal Plants

11 Flitvale Garden Centre Limited

Flitwick Road, Westoning, Bedford,
Bedfordshire MK45 5AA
Tel: 01525 712484
Specialities: Garden Products, Garden Furniture

12 Four Acres Nursery

Leighton Buzzard Rd, Dagnall, Berkhamsted,
Bedfordshire HP4 1QR
Tel: 01442 842838
Specialities: Annuals/Biennials/Bedding, Shrubs

13 Frosts at Willington

Sandy Rd, Willington, Bedford,
Bedfordshire MK44 3QP
Tel: 01234 838777

14 Garden Scene Eaton Bray

The Rye, Eaton Bray, Dunstable,
Bedfordshire LU6 2BQ
Tel: 01525 220104
Specialities: Annuals/Biennials/Bedding, Pots, Composts

15 Gardeners Village

Ickwell Road, Upper Caldecote, Biggleswade,
Bedfordshire SG18 9BS
Tel: 01767 627637

16 Grange Nurseres

Grange Road, Blunham, Bedford,
Bedfordshire MK44 3NS
Tel: 01767 640109
Specialities: Seasonal Plants, Annuals/Biennials/Bedding

17 Greensands Plant Centre

37 Girtford Crescent, Sandy,
Bedfordshire SG19 1HR
Tel: 01767 699655
Specialities: Landscape Gardening

18 H.F.R Banks & Son

Myers Hill Farm, Kimbolton Road, Bolnhurst,
Bedford, Bedfordshire MK44 2ES
Tel: 01234 376256
Specialities: Herbaceous Perennials

19 Hasmead Plc

Garden House, Southill Park, Southill, Biggleswade,
Bedfordshire SG18 9LL
Tel: 01462 813390
Specialities: Trees, Shrubs

20 Hinwick Hall Plant Centre

Hinwick, Wellingborough, Bedfordshire NN29 7JD
Tel: 01933 350543
Specialities: Grasses, Herbaceous Perennials, Geranium

21 Ivy Lodge Nurseries

Rushden Road (A6), Sharnbrook,
Bedford MK44 1ND
Tel/Fax: 01234 782108

SEE ADVERTISEMENT PAGE 4

Specialities:

Hanging Baskets; Large Fuchsias; Zonal Geraniums, Californian Dreamers.

Products and Services:

- Seasonal Displays
- Credit Cards Accepted
- Cut Flowers to order
- Floristry
- Fruit Trees
- Garden Furniture
- Giftware
- Information Desk
- Large Car Park
- Plant Guarantee
- Plants
- Seeds and Bulbs
- Trees

Directions:

The nursery is situated on the outskirts of Sharnbrook five miles south of Rushden, eight miles north of Bedford. Keep on the A6 main road. They are located off a small stretch of dual carriageway.

22 Langford Nurseries & Garden Centre

Henlow Road, Langford, Biggleswade,
Bedfordshire SG18 9SD
Tel: 01462 700791
Specialities: Annuals/Biennials/Bedding, Garden Products

23 Leighton Buzzard Garden Centre

Hockliffe Road, Leighton Buzzard,
Bedfordshire LU7 9NX
Tel: 01525 850100

24 Liscombe Road Nurseries

Liscombe Road, Dunstable, Bedfordshire LU5 4PL
Tel: 01582 662957
Specialities: Annuals/Biennials/Bedding, Shrubs

25 Marston Vale Nurseries

Vale Farm, Woburn Road, Lidlington, Bedford,
Bedfordshire MK43 0NL
Tel: 01234 764141
Specialities: Annuals/Biennials/Bedding, Shrubs

26 Mauden Garden Centre

A507 Bypas, Maulden, Bedford,
Bedfordshire MK45 2BJ
Tel: 01525 861696
Specialities: Paving, Garden Products

27 Mill End Nursery

Harlington Road, Sharpenhoe, Bedford,
Bedfordshire MK45 4SJ
Tel: 01525 872385
Specialities: Annuals/Biennials/Bedding, Shrubs

28 Mill Pond Garden Centre

Barton Mill Lane, Barton-le-Clay, Bedford,
Bedfordshire MK45 4RF
Tel: 08454 502012
Specialities: Annuals/Biennials/Bedding, Herbaceous Perennials, Giftware

29 Milton Ernest Garden Centre

Radwell Road, Milton Ernest, Bedford, Bedfordshire
MK44 1SH
Tel: 01234 823033
Specialities: Restaurant/Coffee Shop, Pet Products, Craft Shop

30 Paul Chessum

21 High Street, Great Barford, Bedford, Bedfordshire MK44 3JH
Tel: 01234 870182
Specialities: Trees, Shrubs

31 Podington Garden Centre

High Street, Podington, Wellingborough, Bedfordshire NN29 7HS
Tel: 01933 353656

32 Poplars Nursery & Garden Centre

Harlington Road, Toddington, Dunstable, Bedfordshire LU5 6HE
Tel: 01525 872017

33 The Rockery Centre

21a Tempsford Road, Sandy, Bedfordshire SG19 2AF
Tel: 01767 680371
Specialities: Rockery Plants

34 Roxton Garden Centre

Bedford Road, Roxton, Bedford, Bedfordshire MK44 3DY
Tel: 01480 212701

35 Savin Nurseries

Hillside Rd, Lower Stondon, Henlow, Bedfordshire SG16 6LQ
Tel: 01462 850680
Specialities: Annuals/Biennials/Bedding, Shrubs

36 Scania Nurseries

The Drive, High Street, Riseley, Bedford, Bedfordshire MK44 1BS
Tel: 01234 708233
Specialities: Shrubs, Herbaceous Perennials

37 Seasons Garden Centre

Bedford Road, Wilstead, Bedford, Bedfordshire MK45 3HU
Tel: 01234 741909
Specialities: Restaurant/Coffee Shop, Pet Products, Fencing

38 Seddington Nursery

Great North Street, Seddington, Sandy, Bedfordshire SG19 1NZ
Tel: 01767 680983
Specialities: Annuals/Biennials/Bedding

39 Selections

36 High Street, Leighton Buzzard, Bedfordshire LU7 1EA
Tel: 01525 372204

40 Shears Nursery

Grove Farm, Rectory Lane, Houghton Conquest, Bedford, Bedfordshire MK45 3LD
Tel: 01234 741989
Specialities: Shrubs, Annuals/Biennials/Bedding

41 Southeast Plant Sales

Roxton Road, Great Barford, Bedford, Bedfordshire MK44 3LP
Tel: 01234 871200
Specialities: Annuals/Biennials/Bedding, Herbaceous Perennials

42 Staples Garden Centre Ltd

The Nurseries, Fordfield Road, Millbrook, Bedford, Bedfordshire MK45 2HZ
Tel: 01525 405898
Specialities: Garden Products, Garden Furniture, Annuals/Biennials/Bedding

43 Studham Nurseries

Clements End Road, Studham, Dunstable, Bedfordshire LU6 2NG
Tel: 01582 872958
Specialities: Herbaceous Perennials, Hanging Baskets

44 Taylors Nursery

Taylors Road, Stotfold, Hitchin, Bedfordshire SG5 4AX
Tel: 01462 835288
Specialities: Shrubs, Herbaceous Perennials

45 Thorne's Garden Nursery

Stotfold Road, Letchworth Garden City, Bedfordshire SG6 4JR
Tel: 01462 675767

46 Two Jays Alpines

35 Greenways, Luton, Bedfordshire LU2 8BL
Tel: 01442 864951
Specialities: Herbaceous Perennials

47 Westwood Farm & Nursery

Odell Road, Odell, Bedford, Bedfordshire MK43 7AP
Tel: 01234 721413
Specialities: Annuals/Biennials/Bedding, Herbaceous Perennials

48 Wyevale Garden Centre (Caddington)

Dunstable Road, Caddington, Luton, Bedfordshire LU1 4AN
Tel: 01582 457313

BERKSHIRE

1 Ashridge Manor Garden Centre

Forest Rd, Wokingham, Berkshire RG40 5QY
Tel: 01189 770730
Specialities: Annuals/Biennials/Bedding, Garden Products

2 Big Plant Nursery

Wargrave Road, Twyford, Reading, Berkshire RG10 9PN
Tel: 01189 341757

3 Bone Bros Garden Centre Ltd

Sutton Lane, Slough, Berkshire SL3 8AH
Tel: 01753 544368
Specialities: Garden Products

4 Braywick Heath Nursery Ltd

41 Braywick Road, Maidenhead, Berkshire SL6 1DX
Tel: 01628 622510

5 Brookside Nurseries

Church Road, Swallowfield, Reading, Berkshire RG7 1TH
Tel: 01189 884122
Specialities: Seasonal Plants, Shrubs

6 Chadwell Seeds

81 Parlaunt Road, Slough, Berkshire SL3 8BE
Tel: 01753 542823
Specialities: Seeds

7 Country Homes & Gardens

Ufton Lane, Ufton Nervet, Reading, Berkshire RG7 4HG
Tel: 01189 306611

8 Country Homes & Gardens Plc

656 Reading Road, Winnersh, Wokingham, Berkshire RG41 5HE
Tel: 01189 891222
Specialities: Garden Products, Annuals/Biennials/Bedding, Paving

9 David Good Associates

Cricket View, Oxford Road, Donnington, Newbury, Berkshire RG14 3AA
Tel: 01635 231283
Specialities: Garden Products

10 Dobbies Grovelands Garden Centre

166 Hyde End Road, Shinfield, Reading, Berkshire RG2 9ER
Tel: 01189 884822

11 Edulis

1 Flowers Piece, Ashampstead, Reading, Berkshire RG8 8SG
Tel: 01635 578113
Specialities: Herbaceous Perennials, Landscape Gardening

12 Englefield Garden Centre

The Street, Englefield, Theale, Reading,
BerkshireRG7 5ES
Tel: 0118 930 4898 Fax: 0118 932 3220

Specialities:

65,000 Bedding Plants produced on site; Herbaceous Perennials; Spring Vegetable Plants

Description:

Englefield is part of TACT UK Ltd, a charitable organisation, offering facilities for horticultural therapy sessions and support for people with learning disabilities.

The centre sells a variety of shrubs, bedding and herbaceous plants, and is noted for its colourful hanging baskets.

Open: 9am-5pm, Monday-Saturday; (Sundays: 12-4pm, May-June)

Directions:

The Centre is located ½ mile off the A340 Theale to Pangbourne road, 3 miles from junction 12 of the M4

13 Foxgrove Plants

Foxgrove, Enborne, Newbury, Berkshire RG14 6RE
Tel: 01635 40554
Specialities: Garden Products

14 Garden Art Plus

Barr's Yard, 1 Bath Road, Hungerford,
Berkshire RG17 0HE
Tel: 01488 686811
Specialities: Ornamentals, Garden Products, Giftware

15 Glenvale Nurseries

Hungerford Lane, Southend, Reading,
Berkshire RG7 6JH
Tel: 01189 744006
Specialities: Homegrown, Annuals/Biennials/Bedding

16 Henry Street Garden Centre

Swallowfield Road, Arborfield, Reading,
Berkshire RG2 9JY
Tel: 01189 761223
Specialities: Rose, Restaurant/Coffee Shop, Annuals/Biennials/Bedding

17 Hillier Garden Centre Newbury

Priors Court Road, Hermitage, Thatcham,
Berkshire RG18 9TG
Tel: 01635 200442

18 John Train Plants

Oakley Green Road, Windsor, Berkshire SL4 4PZ
Tel: 01753 751168
Specialities: Hanging Baskets, Information Desk

19 The Laurels Plant Centre

29 Ryeish Green, Hyde End Lane, Ryeish Green,
Reading, Berkshire RG7 1EP
Tel: 01189 883792
Specialities: Annuals/Biennials/Bedding, Shrubs

20 Lillies Shrubs & Climbers

136 Reading Road, Woodley, Reading,
Berkshire RG5 3AA
Tel: 01189 540381
Specialities: Climbers, Shrubs

21 Long Orchard Nurseries

Oakley Green Road, Oakley Green, Windsor,
Berkshire SL4 4QF
Tel: 01628 627731
Specialities: Annuals/Biennials/Bedding, Herbaceous Perennials

22 Moorings Nursery

296 Windsor Road, Maidenhead, Berkshire SL6 2DT
Tel: 01628 622516
Specialities: Shrubs, Homegrown

23 Moss End Garden Centre

Moss End, Warfield, Bracknell, Berkshire RG42 6EJ
Tel: 01344 483488
Specialities: Restaurant/Coffee Shop, Annuals/Biennials/Bedding

24 Nonsuch Nurseries

Church Road, Winkfield, Windsor,
Berkshire SL4 4SF
Tel: 01344 883213
Specialities: Herbaceous Perennials

25 The Nursery

New Road, Ruscombe, Reading,
Berkshire RG10 9LN
Tel: 01189 340011
Specialities: Seasonal Plants, Herbaceous Perennials

26 Oak Tree Nursery

Brock Hill, Warfield, Bracknell, Berkshire RG42 6JX
Tel: 01344 890667
Specialities: Annuals/Biennials/Bedding, Shrubs

27 Primrose Nursery

London Rd, Ruscombe, Reading,
Berkshire RG10 9HW
Tel: 01189 404288
Specialities: Annuals/Biennials/Bedding, Shrubs

28 Pudding Lane Nursery

Reading Road, Arborfield, Reading,
Berkshire RG2 9HP
Tel: 01189 761048
Specialities: Annuals/Biennials/Bedding, Shrubs

29 Scene Easy Garden Centre

Bath Rd, Hare Hatch, Reading, Berkshire RG10 9SB
Tel: 01189 404794

30 Spirals Nursery

Lowesden Works, Lambourn Woodlands,
Hungerford, Berkshire RG17 7RU
Tel: 01488 648744
Specialities: Shrubs, Seasonal Plants

31 Squires Garden Centre (Windsor)

The Willows, Maidenhead Road, Windsor,
Berkshire SL4 5UB
Tel: 01753 865076

32 Stubbings Nursery Ltd

Stubbings Estate, Henley Road, Maidenhead,
Berkshire SL6 6QL
Tel: 01628 825454
Specialities: Annuals/Biennials/Bedding

33 Town & Country Gardening

1 Thorp Close, Binfield, Bracknell,
Berkshire RG42 5LQ
Tel: 01344 484559

34 Valley Nurseries

Whistley Green, Reading, Berkshire RG10 0DU
Tel: 01189 341045
Specialities: Annuals/Biennials/Bedding, Shrubs

35 The Walled Garden Centre @ Dorney Court

Court Lane, Dorney, Windsor, Berkshire SL4 6QP
Tel: 01628 669999

36 White Tower Nursery

Station Road, Aldermaston, Reading,
Berkshire RG7 4LD
Tel: 01189 712123
Specialities: Herbaceous Perennials

37 Wingroves Nursery

Bath Road, Knowl Hill, Reading,
Berkshire RG10 9UR
Tel: 01628 828449
Specialities: Annuals/Biennials/Bedding, Shrubs

38 Wyevale Garden Centre (Binfield)

Forest Road, Binfield, Bracknell,
Berkshire RG42 4HA
Tel: 01344 869456

39 Wyevale Garden Centre (Hare Hatch)

Floral Mile, Bath Road, Hare Hatch, Reading,
Berkshire RG10 9SW
Tel: 01189 403933

40 Wyevale Garden Centre (Hungerford)

4a Bath Road, Hungerford, Berkshire RG17 0HE
Tel: 01488 682916

41 Wyevale Garden Centre (Thatcham)

Bath Road, Thatcham, Berkshire RG18 3AN
Tel: 01635 871760

42 Wyevale Garden Centre (Windsor)

Dedworth Road, Windsor, Berkshire SL4 4LH
Tel: 01753 841791

43 Wyevale Garden Centre (Wokingham)

Heathlands Road, Wokingham, Berkshire RG40 3AS
Tel: 01189 773055

BUCKINGHAMSHIRE

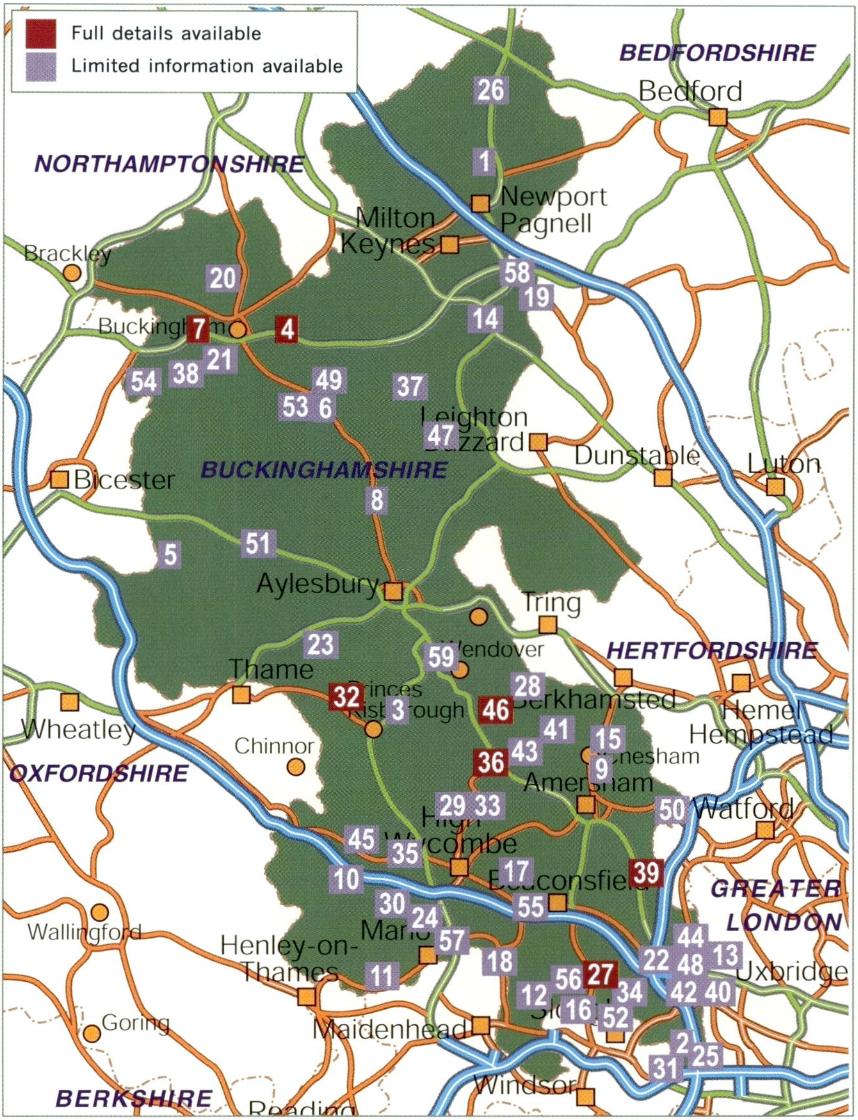

1 Alban Hill Nursery

High Street, Sherington, Newport Pagnell,
Buckinghamshire MK16 9QP
Tel: 01234 711513

2 Alpa Garden Centre

142-144 Swallow Street, Iver Heath,
Buckinghamshire SL0 0HR
Tel: 01753 654101
Specialities: Aquatics, Pet Products

3 Askett Nurseries

Aylesbury Road, Askett, Princes Risborough, Buckinghamshire HP27 9LY
Tel: 01844 274635

4 Barr's Nurseries

Windy Ridge, Bletchley Road, Thornborough, Buckinghamshire MK18 2DZ
Tel: 01280 816855 Fax: 01280 817123
website: barrsnurseries.co.uk

Specialities:

A specialist Nursery growing many unusual plants.

Description:

"We stock a fantastic range of home-produced shrubs and herbaceous perennials, plus trees, roses, herbs, conifers, soft fruit and more. Why not pay us a visit soon?

Directions:

From Buckingham, take the A421 east towards Milton Keynes to the village of Thornborough (3 miles)

5 Bemwode Plants

Kingswood Lane, Ludgershall, Aylesbury, Buckinghamshire HP18 9RB
Tel: 01844 237415
Specialities: Herbaceous Perennials, Shrubs

6 Blooming Marvellous Plants

Aylesbury Road, Shipton, Winslow, Buckinghamshire MK18 3JL
Tel: 07963 747305
Specialities: Herbaceous Perennials, Shrubs, Grasses

7 Buckingham Nurseries & Garden Centre

Tingewick Road, Buckingham, Buckinghamshire MK18 4AE
Tel: 01280 822133 Fax: 01280 815491
e-mail: enquiries@hedging.co.uk
website: www.hedging.co.uk

Specialities:

Hedging; Unusual Plants; Fruit

Description:

The business, founded in 1945 as Byron Hill Nurseries, moved to its present 11-acre site in 1970 when the second generation of the family became involved. The third generation now run a nursery specialising in bare-root hedging, ornamental and forest trees, top and soft fruit, and a wide range of pot-grown shrubs, alpines and perennials. There is an extensive range of gardening requisites, houseplants, giftware and a restaurant.

Directions:

The Nurseries are located about 1 ¼ miles west of Buckingham on the A421, halfway between Buckingham and Tingewick

8 Buckland Landscapes Ltd

The Abbey Tree Nursery, Bushmead Road, Whitchurch, Aylesbury, Buckinghamshire HP22 4LG
Tel: 01296 681988
Specialities: Trees

9 Chessmount Nurseries

41B Chestmount Rise, Chesham, Buckinghamshire HP5 1RB
Tel: 01494 772051
Specialities: Shrubs, Trees

10 Conifers Garden Land

Cadmore End, High Wycombe, Buckinghamshire HP14 3PJ
Tel: 01494 881307

11 Danesfield Gardens

Kingswood Drive, Henley Road, Medmenham, Marlow, Buckinghamshire SL7 2EU
Tel: 01628 484509

12 Davies Bros Nursery

Chalk Pit Lane, Burnham, Slough, Buckinghamshire SL1 8NH
Tel: 01628 666439
Specialities: Lawn, Garden Products, Hanging Baskets

13 Denham Court Nursery

Denham Court Drive, Denham, Uxbridge, Buckinghamshire UB9 5PG
Tel: 01895 832035

14 Dobbies Garden World

Belvedere Lane, Watling Street, Bow Brickhill, Milton Keynes, Buckinghamshire MK17 9JH
Tel: 01908 364890

15 Endeavour Nursery

Cameron Rd, Chesham, Buckinghamshire HP5 3DF
Tel: 01494 793947

16 Farnham Common Nurseries

Crown Lane, Farnham Royal, Slough, Buckinghamshire SL2 3SF
Tel: 01753 647710
Specialities: Hardy Plants, Topiary, Ornamentals

17 Fieldgrove Nurseries & Design Service

Hammersley Lane, Penn, High Wycombe, Buckinghamshire HP10 8HF
Tel: 01494 816754

18 Flowerland

Bourne End Garden Centre, Hedsor Road, Bourne End, Buckinghamshire SL8 5EE
Tel: 01628 523926
Specialities: Garden Products, Restaurant/Coffee Shop

19 Frosts Gardening Centre

Newport Road, Woburn Sands, Milton Keynes, Buckinghamshire MK17 8UE
Tel: 01908 282420

20 Furze Park Nursery

Furze Park Farm, Bycell Road, Maids Moreton, Buckingham, Buckinghamshire MK18 5AA
Tel: 01280 821999
Specialities: Japanese, Hardy Woodland

21 Gawcott Plant Centre & Tree Nurseries

Hillesden Road, Gawcott, Buckingham, Buckinghamshire MK18 4JF
Tel: 01280 821949
Specialities: Garden Products, Shrubs, Trees

22 Greentiles Nursery

Hollybush Lane, Denham, Uxbridge, Buckinghamshire UB9 4HG
Tel: 01895 833759
Specialities: Vegetables, Annuals/Biennials/Bedding, Herbaceous Perennials

23 Haddenham Garden Centre

Stanbridge Road, Haddenham, Aylesbury, Buckinghamshire HP17 8HN
Tel: 01844 290395
Specialities: Pet Products, Giftware, Restaurant/Coffee Shop

24 Heritage Orchids

4 Hazel Close, Marlow, Buckinghamshire SL7 3PW
Tel: 01628 486640
Specialities: Garden Products, Orchids

25 Iver Flowerland

Norwood Lane, Iver Heath, Buckinghamshire SL0 0EW
Tel: 01753 630909
Specialities: Garden Products, Restaurant/Coffee Shop

26 Jeannes Country Garden

98 High Street, Olney, Buckinghamshire MK46 4BE
Tel: 01234 713505
Specialities: Garden Products, Shrubs, Restaurant/Coffee Shop

27 Miles & Miles Nurseries

Beaconsfield Rd, Farnham Royal, Slough, Buckinghamshire SL2 3BY
Tel: 01753 645835
e-mail: eva.miles@tiscali.co.uk
Specialities: Clematis, Climbers

28 Morehavens Camomile Nurseries

Sandpit Hill, Buckland Common, Tring, Buckinghamshire HP23 6NG
Tel: 01494 758642
Specialities: Herbs, Lawn

29 Mr Hildreth's Garden Centre

169 Wycombe Road, Prestwood, Great Missenden, Buckinghamshire HP16 0HJ
Tel: 01494 865576
Specialities: Garden Furniture, Garden Products, Giftware

30 Notcutts Garden Centre

Clay Lane, Booker, Marlow, Buckinghamshire SL7 3DH
Tel: 01494 532532

31 Old Oak Garden Centre

Wood Lane, Iver Heath, Buckinghamshire SL0 0LG
Tel: 01753 653168
Specialities: Garden Furniture, Shrubs, Herbaceous Perennials

32 Pasture Farm Plants

Pasture Farm, Thame Road, Longwick, Princes Risborough, Buckinghamshire HP27 9QX
Tel/Fax: 01 844 343651
e-mail: belgrove@btconnect.com
website: pasturefarmplants.co.uk

Specialities:

Unusual Perennials; Cottage Garden Plants

Description:

Pasture Farm Plants is a small

garden. It offers an ever-changing choice of cottage garden plants and unusual perennials arranged in colour for the customer's selection.

Directions:

From Princes Risborough take the A4129 west towards Thame. Pass the village of Longwick (about 3 miles) and the nursery is ½ mile further, on the right hand side.

33 Penstemons by Colour

Peterley Manor, Peterley Lane, Prestwood,
Great Missenden, Buckinghamshire HP16 0HH
Tel: 01494 866420
Specialities: Penstemon

34 Pinewood Nurseries

Wexham St, Stoke Poges, Slough,
Buckinghamshire SL3 6NB
Tel: 01753 663443
Specialities: Ponsettias, Hanging Baskets, Geranium

35 Plant & Harvest

Chorley Road, High Wycombe,
Buckinghamshire HP14 3AP
Tel: 01494 438635
Specialities: Restaurant/Coffee Shop, Farm Shop

36 The Plant Specialist

Whitefield Lane, Great Missenden,
Buckinghamshire HP16 0BH
Nursery: 01494 866650
e-mail: enquire@theplantspecialist.co.uk
website: www.theplantspecialist.co.uk

SEE ADVERTISEMENT BELOW

Specialities:

Herbaceous perennials, Grasses, Container-grown bulbs, Half hardy perennials

Description:

Established in 2002, the Nursery is located in the village of Great Missenden in the Chilterns and is dedicated to growing a wide range of unusual herbaceous perennials and grasses that also make good, reliable garden plants - many of which can be seen in the display garden. Help and advice is always available from the owners and staff.

Directions:

Approach the village of Great Missenden via the A413. Whitefield Lane is at the end of the High Street opposite the entrance to Missenden Abbey.

37 Potash Nursery

Drayton Parslow, Milton Keynes,
Buckinghamshire MK17 0JE
Tel: 01296 720578

38 Preston Bissett Nurseries

Bushey Lane, Preston Bissett, Buckingham,
Buckinghamshire MK18 4ND
Tel: 01280 848038
Specialities: Herbaceous Perennials

39 Rowan Farm Nurseries

Gorelands Lane, Chalfont St Giles, Bucks HP8 4AB
Tel: 01494 872335 Fax: 01442 891213

Specialities:

Conifers, Fruit Trees, Shrubs, Perennials

Products and Services:

- Pots
- Seasonal Displays
- Credit Cards Accepted
- Fruit Trees
- Garden Products
- Giftware
- Houseplants
- Large Car Park
- Plants
- Seeds and Bulbs
- Trees
- Quality Xmas Trees

Directions:

From the A413, about 4 miles southeast of Amersham, turn left on the B4442 (Vache Lane). Take the 2nd right turn on to Gorelands Lane. Continue along this road until you come to the Chiltern Open Air Museum and Rowan Nursery is a little further, on the right hand side.

Description:

Offering a wide range of plants for all seasons and a wide range of services, the owners of Rowan Farm Nursery are proud to boast that they care as much for their customers as they do for the plants they raise.

The aptly-named Rose family have been here since 1998 and have built up an exceptionally good range of quality plants and both fruit and ornamental trees.

The service too is exceptional – they will even help you load your vehicle if you wish – and the staff are always happy to provide knowledgeable information and advice.

During the Christmas season, the nursery also offers a wide choice of top quality Christmas trees. And all year round, the nursery sells HTA gift vouchers.

40 Shane's Nursery

29 Oxford Road, Denham, Uxbridge,
Buckinghamshire UB9 4DA
Tel: 01895 833289
Specialities: Annuals/Biennials/Bedding, Herbaceous Perennials

41 Shepherd Nursery

Chartridge Lane, Chartridge, Chesham,
Buckinghamshire HP5 2YD
Tel: 01494 775688

42 Smith's Nurseries

95 Oxford Road, Denham, Uxbridge,
Buckinghamshire UB9 4DE
Tel: 01895 233844
Specialities: Water Products and Features, Houseplants, Lawn

43 South Heath Garden Centre & Nursery

Meadow Lane, South Heath,
Buckinghamshire HP16 9SH
Tel: 01494 863269
Specialities: Annuals/Biennials/Bedding, Herbaceous Perennials

44 Springbridge Nurseries

24-26 Oxford Road, Denham, Uxbridge,
Buckinghamshire UB9 4DF
Tel: 08453 701919
Specialities: Conifers, Fruit Trees, Specimen Plants

45 Studley Green Garden Store

Studley Green, High Wycombe,
Buckinghamshire HP14 3XB
Tel: 01494 483761

46 Swan Bottom Nursery

Chesham Lane, Swan Bottom, The Lee,
Buckinghamshire, HP16 9NQ
Tel: 01494 837226

Specialities:

Trees, Roses, Ferns,Shade tolerant plants.

Products and Services:

- Seasonal Displays
- Fruit Trees
- Garden Design
- Information Desk
- Large Car Park
- Plants
- Bulbs
- Trees

Directions:

Take A413 from Aylesbury to Amesham. Turn left after petrol station continue 2 miles along the windy road to crossroads. Go over crossroads into Chesham Lane and Nursery is on the left hand side.

47 Tamarisk Nurseries

Wing Road Farm, Wing Road, Stewkley,
Leighton Buzzard, Buckinghamshire LU7 0JB
Tel: 01525 240747
Specialities: Exotics, Rare and Unusual Plants, Carnivorous Plants

48 Tendercare Nurseries

18 Southlands Road, Denham, Uxbridge,
Buckinghamshire UB9 4HD
Tel: 01895 835544
Specialities: Specimen Trees, Ornamentals, Specimen Plants

49 Tim's Plants

Winslow Nurseries, Little Horwood Road, Winslow,
Buckinghamshire MK18 3JW
Tel: 01296 713820
Specialities: Shrubs, Trees, Herbaceous Perennials

50 The Van Hage Garden Company

Chenies, Rickmansworth,
Buckinghamshire WD3 6EN
Tel: 01494 764545
Specialities: Rose, Grasses, Bamboos

51 Waddesdon Plant Centre

Queen Street, Waddesdon, Aylesbury,
Buckinghamshire HP18 0JW
Tel: 01296 658586
Specialities: Herbaceous Perennials, Houseplants

52 Wexham Nurseries

Wexham Road, Wexham, Slough,
Buckinghamshire SL2 4HE
Tel: 01753 526408
Specialities: Homegrown, Specialist Bedding

53 Winslow Plant Centre

Bell Walk, Winslow, Buckinghamshire MK18 3AH
Tel: 01296 715949

54 Woodfield Nursery Ltd

Radclive Road, Chetwode,
Buckinghamshire MK18 4LB
Tel: 01280 812926

55 Wyevale Garden Centre (Beaconsfield)

London Road, Beaconsfield,
Buckinghamshire HP9 1SH
Tel: 01494 672522

56 Wyevale Garden Centre (Farnham Royal)

Cedar Cottage, Crown Lane, Farnham Royal, Slough,
Buckinghamshire SL2 3SG
Tel: 01753 645627

57 Wyevale Garden Centre (Marlow)

Pump Lane South, Marlow,
Buckinghamshire SL7 3RB
Tel: 01628 482716

58 Wyevale Garden Centre (Woburn Sands)

Newport Road, Woburn Sands, Milton Keynes,
Buckinghamshire MK17 8UE
Tel: 01908 281161

59 Wyevale Garden Centre (Worlds End)

Aylesbury Road, Wendover, Aylesbury,
Buckinghamshire HP22 6BD
Tel: 01296 623116

CAMBRIDGESHIRE

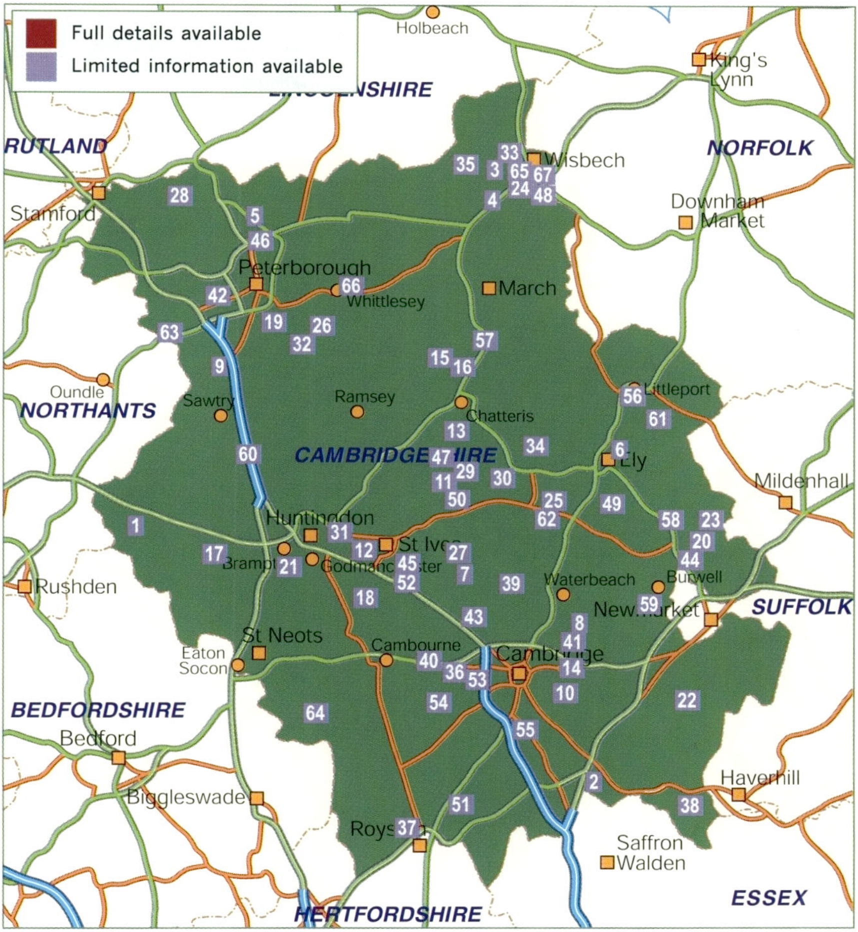

1 A J Button

Station Road, Catworth, Huntingdon, Cambridgeshire PE28 0PE
Tel: 01832 710428

2 All Year Nursery

51 North Road, Abington, Cambridge, Cambridgeshire CB21 6AS
Tel: 01223 893218

3 Allens Nurseries

Barton Road, Wisbech, Cambridgeshire PE13 4TG
Tel: 01945 580433
Specialities: Annuals/Biennials/Bedding, Shrubs

4 Banksia Bonsai

Chapel Lane, South Brink, Wisbech, Cambridgeshire PE14 0RX
Tel: 01945 465176
Specialities: Bonsai

5 Barn Garden & Aquatic Centre

Gunthorpe Road, Peterborough,
Cambridgeshire PE4 7AD
Tel: 01733 320134
Specialities: Aquatics, Garden Products

6 Brands of Ely

42 - 44 Market Street, Ely, Cambridgeshire CB7 4LS
Tel: 01353 663463

7 Brett Nursery

Stanton Mereway, Willingham, Cambridge,
Cambridgeshire CB24 5HJ
Tel: 01954 260134
Specialities: Shrubs, Herbaceous Perennials

8 Cambridge Garden Plants

The Lodge, Clayhithe Road, Horningsea, Cambridge,
Cambridgeshire CB25 9JD
Tel: 01223 861370
Specialities: Herbaceous Perennials, Rose,
Penstemon

9 Cherriash Garden Centre & Pet Supplies

109 North Street, Stilton, Peterborough,
Cambridgeshire PE7 3RR
Tel: 01733 241653

10 Colbrook Nurseries

Fulbourn Road, Teversham, Cambridge,
Cambridgeshire CB1 9AN
Tel: 01223 880496

11 Cranbrook Plants

Colne Road, Somersham, Huntingdon,
Cambridgeshire PE28 3DQ
Tel: 01487 741035
Specialities: Annuals/Biennials/Bedding,
Herbaceous Perennials

12 D J Clifton

St. Ives Road, Hemingford Grey, Huntingdon,
Cambridgeshire PE28 9DX
Tel: 01480 463365

13 D J's Garden Centre

134 London Road, Chatteris,
Cambridgeshire PE16 6SG
Tel: 01354 693937
Specialities: Garden Products, Shrubs

14 Darwin Nurseries & Farm Shop

Quy Water Farm, Newmarket Road, Teversham,
Cambridge, Cambridgeshire CB1 9AT
Tel: 01223 293911
Specialities: Farm Shop

15 Delfland Nurseries Ltd

Benwick Rd, Doddington, March,
Cambridgeshire PE15 0TU
Tel: 01354 740553
Specialities: Organic, Organic Edible Plantrs

16 Doddington Garden Centre & Farm Shop

36 Newgate Street, Doddington, March,
Cambridgeshire PE15 0SR
Tel: 01354 741122
Specialities: Farm Shop, Shrubs

17 Ellington Thorpe Nurseries

Grafham Road, Ellington Thorpe, Huntingdon,
Cambridgeshire PE28 0AP
Tel: 01480 811330
Specialities: Shrubs, Hedging, Topiary

18 Elsworth Herbs

Avenue Farm Cottage, 31 Smith Street, Elsworth,
Cambridge, Cambridgeshire CB23 4HY
Tel: 01954 267414
Specialities: Herbs

19 Farcet Nurseries

Straight Drove, Farcet, Peterborough,
Cambridgeshire PE7 3DJ
Tel: 01733 243827
Specialities: Shrubs

20 Fordham Nursery & Garden Centre

Market Street, Fordham, Ely,
Cambridgeshire CB7 5LQ
Tel: 01638 720455
Specialities: Garden Products

21 Frosts Brampton Garden Centre

Buckden Rd, Brampton, Huntingdon,
Cambridgeshire PE28 4NF
Tel: 01480 453048

22 Fulbrooke Nursery

Home Farm, Westley Waterless, Newmarket,
Cambridgeshire CB8 0RG
Tel: 01638 507124
Specialities: Bamboos

23 G Caffarelli

Floral Farm, Fordham Road, Isleham, Ely,
Cambridgeshire CB7 5QY
Tel: 01638 720622
Specialities: Floristry

24 G W Harrington

173 North Brink, Wisbech,
Cambridgeshire PE13 4TA
Tel: 01945 584758

25 Glen Nurseries

18A High St, Wilburton, Ely,
Cambridgeshire CB6 3RB
Tel: 01353 740252

26 Growing Hand

Ramsey Road, Farcet, Peterborough,
Cambridgeshire PE7 3DR
Tel: 01733 844479

27 Hayden & Son

3 Mill Road, Willingham, Cambridge,
Cambridgeshire CB24 5LA
Tel: 01954 260363

28 Helpston Garden Centre

West Street, Helpston, Peterborough,
Cambridgeshire PE6 7DU
Tel: 01733 252548
Specialities: Garden Furniture, Garden Products, Seasonal Plants

29 Holwood Nurseries

Chatteris Road, Somersham, Huntingdon,
Cambridgeshire PE28 3DN
Tel: 01487 740936

30 Honeysome Aquatic Nursery

The Row, Sutton, Ely, Cambridgeshire CB6 2PB
Tel: 01353 778889
Specialities: Annuals/Biennials/Bedding, Shrubs, Aquatics

31 Huntingdon Garden & Leisure

Banks End, Wyton, Huntingdon,
Cambridgeshire PE28 2AA
Tel: 01480 410750

32 Kew Lodge Nursery

Straight Drove, Farcet, Peterborough,
Cambridgeshire PE7 3DL
Tel: 01733 844468
Specialities: Seasonal Plants, Shrubs

33 L Esser & Son

Grange Farm Nursery, Barton Road, Wisbech,
Cambridgeshire PE13 4TH
Tel: 01945 582262
Specialities: Shrubs, Herbaceous Perennials

34 Lane Farm Nursery

Hale Fen Lane, Wardy Hill, Ely,
Cambridgeshire CB6 2DS
Tel: 01353 778069
Specialities: Annuals/Biennials/Bedding, Shrubs

35 Long Acre Garden Centre Ltd

High Side, Wisbech St. Mary, Wisbech,
Cambridgeshire PE13 4SA
Tel: 01945 410013

36 Madingley Mulch

Madingley Road, Coton, Cambridge,
Cambridgeshire CB23 7PH
Tel: 01954 212144

37 Marden Cottage Nursery

1 Marden Cottage, Old North Road, Bassingbourn,
Royston, Cambridgeshire SG8 5JU
Tel: 01763 244457
Specialities: Shrubs, Cottage Garden Plants

38 Mill Green Meadow Nursery

Mill Green, Horseheath, Cambridge,
Cambridgeshire CB21 4QZ
Tel: 01799 584942
Specialities: Annuals/Biennials/Bedding, Herbaceous Perennials

39 Monksilver Nursery

Oakington Road, Cottenham, Cambridge,
Cambridgeshire CB24 8TW
Tel: 01954 251555
Specialities: Annuals/Biennials/Bedding, Herbaceous Perennials

40 Natureworks Bonsai

3 St. Neots Road, Hardwick, Cambridge,
Cambridgeshire CB23 7QH
Tel: 01954 210194
Specialities: Bonsai

41 Notcutts Garden Centre

High Street, Horningsea, Cambridge,
Cambridgeshire CB25 9JG
Tel: 01223 860320

42 Notcutts Garden Centre

Oundle Road, Orton Waterville, Peterborough,
Cambridgeshire PE2 5UU
Tel: 01733 234600

43 Oakington Garden Centre

Dry Drayton Road, Oakington, Cambridge,
Cambridgeshire CB24 3BD
Tel: 01223 234818
Specialities: Garden Products, Annuals/Biennials/Bedding

44 P C & S L Simpson

42 Station Road, Fordham, Ely,
Cambridgeshire CB7 5LW
Tel: 01638 720194
Specialities: Fruit Trees, Fruit Bushes

45 P T Pinner & Son

Thimbleset Nurseries, Mill Road, Fen Drayton,
Cambridge, Cambridgeshire CB24 4ST
Tel: 01954 232699
Specialities: Seasonal Plants, Herbaceous Perennials

46 Pamwell Plants

152 Fulbridge Rd, Peterborough,
Cambridgeshire PE1 3LF
Tel: 01733 572688
Specialities: Seasonal Plants, Herbaceous Perennials

47 Parkhall Nurseries & Garden Centre

Parkhall Road, Somersham, Huntingdon,
Cambridgeshire PE28 3HW
Tel: 01487 840397
Specialities: Garden Products, Annuals/Biennials/Bedding, Herbaceous Perennials

48 Parkside Nurseries

19 Cromwell Road, Wisbech,
Cambridgeshire PE14 0SF
Tel: 01945 582448

49 Pembroke Farm Nursery

Pembroke Farm, Barway, Ely,
Cambridgeshire CB7 5UB
Tel: 01353 722903
Specialities: Annuals/Biennials/Bedding, Shrubs

50 Pennycross Plants

Earith Road, Colne, Huntingdon,
Cambridgeshire PE28 3NL
Tel: 01487 841520
Specialities: Seasonal Plants, Shrubs

51 Phillimore Garden Centre

Cambridge Road, Melbourn, Royston,
Cambridgeshire SG8 6EY
Tel: 01763 260537

52 The Plant Place

2 Mill Road, Fen Drayton, Cambridge,
Cambridgeshire CB24 4ST
Tel: 01954 231665
Specialities: Herbaceous Perennials, Shrubs

53 Polhill Garden Centre

Coton Orchard, Cambridge Road, Coton, Cambridge,
Cambridgeshire CB23 7PJ
Tel: 01954 210234
Specialities: Garden Products, Annuals/Biennials/Bedding

54 Saunders Nurseries

West Street, Comberton, Cambridge,
Cambridgeshire CB23 7EN
Tel: 01223 262268
Specialities: Annuals/Biennials/Bedding, Shrubs

55 Scotsdale Garden Centre

120 Cambridge Road, Great Shelford, Cambridge,
Cambridgeshire CB22 5JT
Tel: 01223 842777

56 Sharman's Garden Centre

Camel Road, Littleport, Ely,
Cambridgeshire CB6 1EW
Tel: 01353 862298
Specialities: Annuals/Biennials/Bedding, Herbaceous Perennials

57 Skylark Country Store and Garden Centre

Skylark Farm, Manea Road, Wimblington, March,
Cambridgeshire PE15 0PA
Tel: 01354 741212

58 Soham Nurseries

67 East Fen Common, Soham, Ely,
Cambridgeshire CB7 5JJ
Tel: 01353 720848
Specialities: Herbaceous Perennials, Shrubs

59 Spring View Nursery

10 Spring Close, Burwell, Cambridge,
Cambridgeshire CB25 0HF
Tel: 01638 742993
Specialities: Annuals/Biennials/Bedding, Shrubs

60 St Andrews Nurseries

Old Great North Road, Sawtry, Huntingdon, Cambridgeshire PE28 5XN
Tel: 01487 830482

61 Tabbs

The Nursery, Mile End Road, Prickwillow, Ely, Cambridgeshire CB7 4SJ
Tel: 01353 688437
Specialities: Shrubs, Annuals/Biennials/Bedding

62 Twenty Pence Garden Centre

Twenty Pence Road, Wilburton, Ely, Cambridgeshire CB6 3RN
Tel: 01353 741024
Specialities: Fencing, Paving

63 The Walled Garden Centre

Elton Hall, Elton, Peterborough, Cambridgeshire PE8 6SH
Tel: 01832 280058

64 Waresley Park Garden Centre

Gamlingay Road, Waresley, Sandy, Cambridgeshire SG19 3DB
Tel: 01767 650249

65 Waverley Nurseriues

Barton Road, Wisbech, Cambridgeshire PE13 1LE
Tel: 01945 583959

66 Westfield Nurseries

Pinewood Avenue, Whittlesey, Peterborough, Cambridgeshire PE7 1EX
Tel: 01733 206688
Specialities: Annuals/Biennials/Bedding, Shrubs

67 Wisbech Plant Co Ltd

North Brink Nursery, North Brink, Wisbech, Cambridgeshire PE13 1LL
Tel: 01945 582588

CHESHIRE

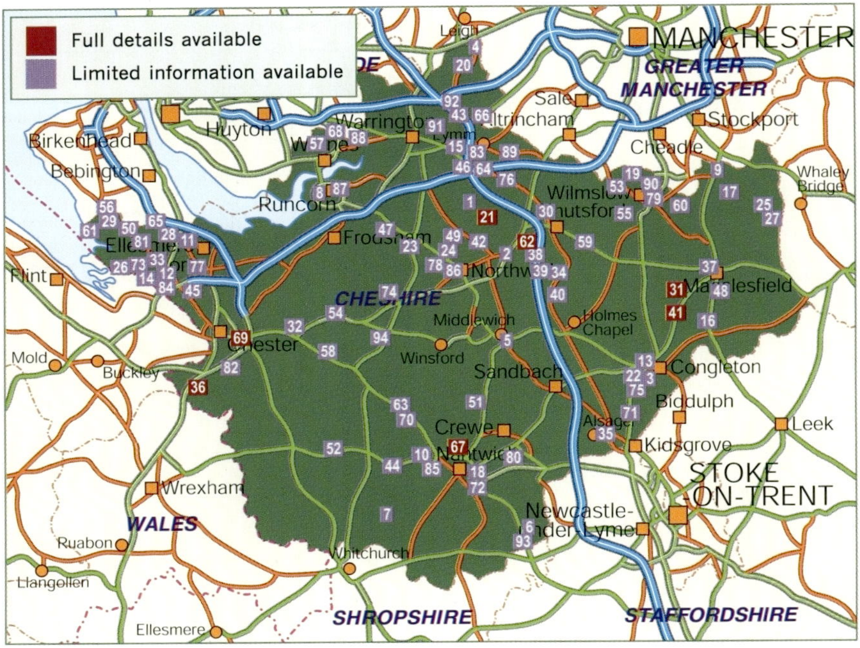

1 Arley Hall Nursery

Arley, Northwich, Cheshire CW9 6NA
Tel: 01565 777479

2 Ascol Drive Nurseries

Ascol Drive, Plumley, Knutsford,
Cheshire WA16 0UD
Tel: 01606 47694

3 Astbury Meadow Garden Centre

Newcastle Road, Astbury, Congleton,
Cheshire CW12 4RL
Tel: 01260 276466

4 Bents Garden Centre

Warrington Road, Glazebury, Warrington,
Cheshire WA3 5NT
Tel: 01942 266300

5 Boosey's Garden Centre

Newton Bank, Middlewich, Cheshire CW10 9EX
Tel: 01606 832324

6 Bridgemere Garden World

Bridgemere, Nr Nantwich, Cheshire CW5 7QB
Tel: 01270 521100

7 Bromac Nursery

Little Porters Hill Farm, Frith Lane, Nantwich,
Cheshire CW5 8HQ
Tel: 01270 780319

8 Brookfield Garden Centre

2 Weston Road, Runcorn, Cheshire WA7 4JT
Tel: 01928 576754

9 Brookside Garden Centre

Macclesfield Road, Poynton, Stockport,
Cheshire SK12 1BY
Tel: 01625 872919

10 Buriand Nurseries

Buriand, Bank Lane, Burland, Nantwich,
Cheshire CW5 8LS
Tel: 01270 524230

11 Burleydam Garden Centre

Chester Road, Childer Thornton, Ellesmere Port, Cheshire CH66 1QW
Tel: 01513 393195

12 Burton Garden Nurseries

Chester High Road, Burton, Neston, Cheshire CH64 8TF
Tel: 01513 362275

13 C Dale

Fol Hollow, Astbury, Congleton, Cheshire CW12 4HT
Tel: 01260 275992
Specialities: Shrubs, Annuals/Biennials/Bedding, Herbaceous Perennials

14 C T Harding

Puddington Lane, Puddington, Neston, Cheshire CH64 5SG
Tel: 01513 362320

15 Caddick's Clematis Nursery

Lymm Road, Thelwall, Warrington, Cheshire WA4 2TG
Tel: 01925 757196
Specialities: Clematis

16 Conquest Plants Nursery

Leek Road, Bosley, Macclesfield, Cheshire SK11 0PP
Tel: 01260 223793
Specialities: Landscape Gardening

17 Coppice Road Nursery

The Coppice, Poynton, Stockport, Cheshire SK12 1SR
Tel: 01625 871173

18 Crewe Road Nurseries

209 Crewe Road, Crewe Road, Willaston, Nantwich, Cheshire CW5 6NE
Tel: 01270 624245

19 Croftland Nursery

Altrincham Road, Styal, Wilmslow, Cheshire SK9 4JE
Tel: 01625 539616
Specialities: Annuals/Biennials/Bedding, Herbaceous Perennials, Cut Flowers

20 Culcheth Garden Centre

Church Lane, Culcheth, Warrington, Cheshire WA3 5DL
Tel: 01925 765500

21 Curbishleys Roses

Bate Heath Nurseries, Budworth Road, Aston by Budworth, nr Northwich, Cheshire CW9 6LT
Tel: 01565 733286

Specialities:

Roses; Shrubs; Clematis

Description:

Run by a brother and sister team, Curbishleys Roses has been established in this peaceful corner of rural Cheshire for more than 75 years. In addition to over 300 varieties of roses, the nursery stocks some 60 types of clematis and a good selection of shrubs.

Directions:

Just 5 minutes from Junction 19 of the M6. Take the B5391 towards Pickmere and then follow the signs for Arley Hall. You will see Curbishleys Roses on the left hand side.

22 D.J.M Nursery

Newcastle Rd, Astbury, Congleton, Cheshire CW12 4RL
Tel: 01260 275032

23 Davenports Florists

Bridge Farm, Warrington Road, Bartington, Northwich, Cheshire CW8 4QU
Tel: 01606 853241
Specialities: Exotics, Floristry, Farm Shop

24 Dave's Woodcrafts

Daisy Bank Lane, Soot Hill, Anderton, Northwich, Cheshire CW9 6AA
Tel: 01606 784270

25 David G Ross

Paddock Lane, Brook Bottom, Kettleshulme, High Peak, Cheshire SK23 7RD
Tel: 01663 732555

26 Dovecote Nurseries

Station Road, Burton, Neston, Cheshire CH64 5SB
Tel: 01513 365748
Specialities: Herbaceous Perennials, Annuals/Biennials/Bedding, Shrubs

27 Dunge Valley Gardens and Hardy Plant Nursery

Windgather Rocks, Kettleshulme, High Peak, Cheshire SK23 7RF
Tel: 01663 733787
Specialities: Hardy Plants, Trees, Rhododendron (Inc Azalea)

28 Follyfield Nurseries

Benty Heath Lane, Hooton, Ellesmere Port, Cheshire CH66 6AG
Tel: 01513 274999
Specialities: Annuals/Biennials/Bedding, Herbaceous Perennials

29 Forest Farm Nurseries

Chester High Road, Neston, Cheshire CH64 3TH
Tel: 01513 364346

30 Fryers Garden Centre

Manchester Road, Mere, Knutsford, Cheshire WA16 0SX
Tel: 01565 755455
Specialities: Rose, Garden Furniture, Barbeques

31 Gawsworth Nurseries

Marton Lane, Gawsworth, nr Macclesfield, Cheshire SK11 9EU
Tel/Fax: 01260 224553

Specialities:

Herbaceous Perennials; Alpines; Bedding Plants

Description:

For more than 20 years, Margaret Stevens has been growing a wide range of plants – bedding, herbaceous perennials and bulbs in bowls for indoor displays. She has supplied plants and hanging baskets to Macclesfield Borough Council which has won awards with her displays at Tatton Park.

Directions:

From the A536 about 3 miles south of Macclesfield, turn left into Gawsworth village. Continue through the village and then turn right, proceeding south into Marton Lane.

32 Goldenfields Nursery

Barrow Lane, Tarvin Sands Tarvin, Chester, Cheshire CH3 8JF
Tel: 01829 741616

33 Gordale Nurseries & Garden Centre

Chester High Road, Burton, Cheshire CH64 8TF
Tel: 01513 362116
Specialities: Garden Furniture, Giftware, Restaurant/Coffee Shop

34 Grasslands Nursery

Free Green Lane, Over Peover, Knutsford, Cheshire WA16 9QY
Tel: 01565 723211

35 Greenacres

Liverpool Road West, Church Lawton, Stoke-on-Trent, Cheshire ST7 3DL
Tel: 01270 883881

36 Grosvenor Garden Centre

Wrexham Road, Belgrave, nr Chester, Cheshire CH4 9EB
Tel: 01244 625200 Fax: 01244 625210
e-mail: info@grosvenorgardencentre.co.uk
website: www.grosvenorgardencentre.co.uk

SEE ADVERTISEMENT PAGE 23

Specialities:

Plants; Garden Furniture; Gifts & Food.

Description:

Voted Top Garden Centre in the Northwest, 2007, Grosvenor Garden Centre is adjacent to Eaton Hall, family home of the Duke of Westminster. You'll find absolutely everything here for the garden, with a special emphasis on garden furniture. The site also contains an excellent café/restaurant.

Directions:

Exit the A55 signposted Wrexham and Chester A483. At the roundabout take the first exit B5445. From the A5 the A483 to Chester. At the roundabout before the A55 interchange turn right on to the B5445.The Garden Centre is about ½ mile on your left.

37 Hadfield's

89b Churchill Way, Macclesfield, Cheshire SK11 6AT
Tel: 01625 423259

38 Heathfield Nursery

Ullard Hall Lane, Plumley, Knutsford, Cheshire WA16 9RN
Tel: 01565 634361

39 The Hidden Nursery

Plumley Moor Road, Plumley, Knutsford, Cheshire WA16 9SD
Tel: 01565 722315
Specialities: Shrubs, Conifers

40 Hills Garden Centre

London Road, Allostock, Knutsford, Cheshire WA16 9LU
Tel: 01565 722567
Specialities: Garden Furniture, Garden Machinery, Garden Products

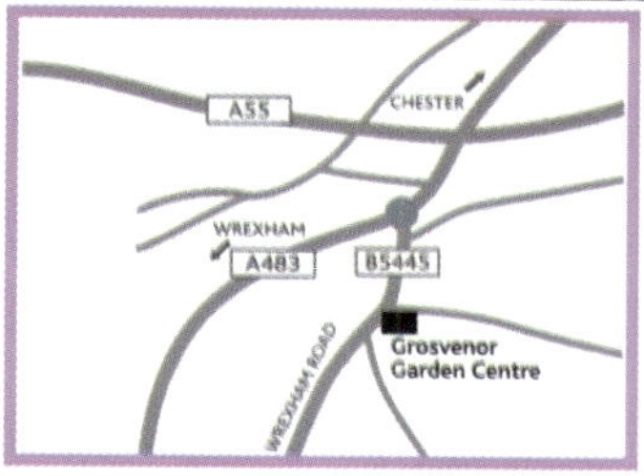

41 Hollands Nurseries & Garden Centre

Gawsworth, nr Congleton, Cheshire SK11 9JB
Tel: 01260 223362
website: www.hollandsnurseries.co.uk

SEE ADVERTISEMENT PAGE 24

Specialities:

Bedding Plants; Farm Shop; Coffee Shop

Products and Services:

- Credit Cards Accepted
- Farm Shop
- Fruit Trees
- Garden Furniture
- Garden Products
- Hot Tubs
- Houseplants
- Information Desk
- Large Car Park
- Paving
- Plants
- Restaurant/Coffee Shop
- Seeds and Bulbs
- Trees

Directions:

Hollands Nurseries are on the A536, 4 miles north of Congleton, 6 miles south of Macclesfield

42 The Hollies Plant Centre

The Hollies, Earles Lane, Marston, Northwich, Cheshire CW9 6DX
Tel: 01565 733503

43 JC Nurseries

Woodlands, Nicol Avenue, Woolston, Warrington, Cheshire WA3 6DL
Tel: 01925 813935

44 K J Nurseries

Wrexham Road, Faddiley, Nantwich, Cheshire CW5 8JE
Tel: 01270 524495
Specialities: Hanging Baskets

45 K's Nursery

Coalpit Lane, Mollington, Chester, Cheshire CH1 6LJ
Tel: 01244 880468

46 Lane End Nursery

Old Cherry Lane, Lymm, Cheshire WA13 0TA
Tel: 01925 752618
Specialities: Annuals/Biennials/Bedding, Herbaceous Perennials, Trees

47 Lodge Lane Nursery & Gardens

Lodge Lane, Dutton, Warrington, Cheshire WA4 4HP
Tel: 01928 713718
Specialities: Herbaceous Perennials, Wild Flowers

Conveniently located on the A536, 4 miles north of Congleton, Hollands Nurseries was established in 1996 and has built up a reputation for stocking healthy quality plants and for providing friendly and knowledgeable service. Hanging baskets are a speciality but the site also provides a full range of garden items such as fertilizers, paving stones and garden tools. The recently opened farm shop features local produce and the excellent coffee shop serves a good choice of meals and snacks. Hollands also has the franchise for Sundance Spas and stocks a wide choice of tubs.

- Credit Cards Accepted
- Farm Shop
- Fruit Trees
- Garden Furniture
- Garden Products
- Hot Tubs
- Houseplants
- Information Desk
- Large Car Park
- Paving
- Plants
- Restaurant/Coffee Shop
- Seeds and Bulbs
- Trees

Bedding Plants, Farm Shop, Coffee Shop.....

Hollands Nurseries & Garden Centre

Gawsworth, nr Congleton, Cheshire SK11 9JB
Tel: 01260 223362
Website: www.hollandsnurseries.co.uk

48 Macclesfield Garden Centre

Bullocks Lane, Sutton, Macclesfield,
Cheshire SK11 0HE
Tel: 01625 618933
Specialities: Giftware, Bamboos, Annuals/Biennials/Bedding

49 Marbury Hall Nurseries

The Old Walled Garden, Marbury, Northwich,
Cheshire CW9 6AT
Tel: 01606 74168
Specialities: Exotics, Shrubs, Climbers

50 Meadow View Nurseries

Upper Raby Rd, Neston, Cheshire CH64 7TY
Tel: 01513 364423

51 Minshulls Nursery

Eardswick Lane, Minshull Vernon, Crewe,
Cheshire CW1 4RQ
Tel: 01270 522482
Specialities: Trees, Shrubs

52 Morgan Nursery

Stonehouse Lane, Bulkeley, Malpas,
Cheshire SY14 8BQ
Tel: 01829 720514

53 The Morley Nursery Ltd

Altrincham Road, Wilmslow, Cheshire SK9 4LY
Tel: 01625 548822

54 Morreys

Forest Nursery, Chester Road, Kelsall, Tarporley,
Cheshire CW6 0SW
Tel: 01829 751342

55 Ned Yates Garden Centre

Moor Lane, Wilmslow, Cheshire SK9 6DN
Tel: 01625 522128
Specialities: Garden Products, Garden Machinery, Landscape Gardening

56 New Hall Home & Garden Centre

Chester High Road, Neston, Cheshire CH64 3TE
Tel: 01513 538492

57 The New Widnes Home & Garden Centre

Birchfield Rd, Widnes, Cheshire WA8 9TX
Tel: 01514 247221
Specialities: Greenhouses and Sheds, Fencing, Decking

58 Okell's Nurseries

Burton Lane, Duddon, Tarporley, Cheshire CW6 0EP
Tel: 01829 741512

59 Ollerton Nursery

Chelford Rd, Ollerton, Knutsford, Cheshire WA16 8RJ
Tel: 01565 652247
Specialities: Shrubs

60 Parkers Garden Centre

Lees Lane, Newton, Macclesfield, Cheshire SK10 4LJ
Tel: 01625 520002

61 Parkgate Nurseries

Boathouse Lane, Parkgate, Neston, Cheshire CH64 6RD
Tel: 01513 364178
Specialities: Annuals/Biennials/Bedding, Herbaceous Perennials, Shrubs

62 Parkside Rose Nursery

Sudlow Lane, Tabley, nr Knutsford, Cheshire WA16 0TP
Tel: 01565 652598 Mobile: 07754 210385 (on site)

Specialities:

Roses; Conifers; Shrubs

Description:

The Hart family established Parkside Rose Nursery more than 40 years ago and now grow more than 150 varieties of roses, including bush, climbers, patio, miniatures and standards. They also stock a good selection of shrubs and conifers.

Directions:

The nursery is located off the A556 west of Knutsford, off Northwich Road, 1½ miles down Sudlow Lane.

63 The Pottery Barn

Station Yard, Nantwich Road, Calveley, Tarporley, Cheshire CW6 9JT
Tel: 01829 262529

64 Primrose Hill Nurseries

Warrington Road, High Legh, Knutsford, Cheshire WA16 0SQ
Tel: 01925 752635

65 Raby Nurseries

Benty Heath Lane, Willaston, Neston, Cheshire CH64 1SB
Tel: 01513 274221
Specialities: Annuals/Biennials/Bedding, Herbaceous Perennials

66 Ramswood Nurseries

196 Manchester Rd, Rixton, Warrington, Cheshire WA3 6EA
Tel: 01925 821002

67 Reaseheath Garden Centre

at Reaseheath College, Worleston Road, Nantwich, Cheshire CW5 6DF
Tel: 01270 613283
e-mail: enquiries@reaseheath.ac.uk
website: www.reaseheath.ac.uk

Specialities:

Bedding Plants & Hanging Baskets; Rare and Unusual Herbaceous and Shrub Plants.

Products and Services:

- Trees and Shrubs
- Car Park
- Credit Cards Accepted
- Home Grown Cut Flowers
- Disabled Access
- Fruit Trees
- Garden Produce
- Giftware
- Home Produced Cheese, Butter & Ice Cream
- Houseplants
- Plant Guarantee
- Seeds and Bulbs
- Greeting Cards

Directions:

The Centre is on the B5074 (Worleston Road), 1 mile north of Nantwich and 15 minutes from Junction 16 of the M6

Description:

Much of the produce on sale in the Garden Centre comes from the college's own growing areas and is supplemented by products such as locally made preserves and chutneys. The Garden Centre also sells home-produced dairy produce, a range of giftware and greeting cards. Reaseheath College was

Roots 'N' Shoots

Whitchurch Road, Great Boughton,
Chester, Cheshire CH3 5QD
Tel/Fax: 01244 336922

Opening Times

9am-5.30pm, Monday to Saturday
10am-4.30pm, Sunday

With over 30 years experience, the family run **Roots 'N' Shoots** nursery offers a wealth of friendly advice for all your gardening needs. Their specialities range from a large and varied selection of specimen mature trees and plants to a stylish selection of handmade Italian pots and the largest range of Darwin pots in the northwest.

Roots 'N' Shoots also provide a garden and landscape design service. They stock The Vitalizer range - totally chemical free and organic, the perfect compost for your plants. Call in and see the fantastic array of beautiful flowers, trees and shrubs on offer, this is something very special. Roots 'N' Shoots are sponsers of the RHS Gold award winners and sell and accept National Garden Gift Vouchers.

- Aquatics
- Seasonal Displays
- Credit Cards Accepted
- Cut Flowers
- Fruit Trees
- Garden Design
- Garden Products
- Houseplants
- Large Car Park
- Plants
- Seeds and Bulbs
- Trees

rated "A truly outstanding college" by Ofsted in November 2006. The college offers a wide range of short, full-time and part-time courses. It enjoys an enviable setting amongst 500 acres of farms, parklands, lake, woodland and sports facilities, including its own 9-hole golf course. Within the site is an Animal Care Unit with a wide range of mammals, reptiles and birds from commonly kept pets to exotic tropical species. There is also an Equine Unit with stabling for over 30 horses, two outdoor riding arenas, a cross country course and a local veterinary group equine clinic on site. Other facilities include a demonstration fruit and vegetable garden, plant nursery and greenhouses.

68 Rivendell Garden Centre

Mill Lane, Widnes, Cheshire WA8 3UL
Tel: 01514 232638
Specialities: Garden Furniture, Garden Products, Giftware

69 Roots 'N' Shoots

Whitchurch Road, Great Boughton, Chester, Cheshire CH3 5QD
Tel/Fax: 01244 336922

SEE ADVERTISEMENT PAGE 26

Specialities:

Large Selection of Specimen Mature Trees & Plants; Highest Quality Prices

Description:

The nursery offers a large and varied selection of specimen mature trees and plants as well as a garden and landscape design service. It also stocks a stylish selection of hand-made Italian pots and the largest range of Darwin pots in the northwest.

Directions:

From the M53 north at Exit 13, take the 3rd exit off roundabout. At the traffic lights take the first exit off the roundabout and the nursery is first right.

70 Rowlinson Garden Products

Unit 2, Green Lane, Wardle, Nantwich, Cheshire CW5 6BN
Tel: 01829 262080
Specialities: Garden Furniture, Greenhouses and Sheds

71 Smithy Farm

Congleton Road North, Scholar Green, Stoke-on-Trent, Cheshire ST7 3SZ
Tel: 01782 783963
Specialities: Annuals/Biennials/Bedding, Shrubs, Climbers

72 Stapeley Water Gardens

London Road, Nantwich, Cheshire CW5 7LH
Tel: 01270 623868
Specialities: Water Products and Features, Garden Products, Aquatics

73 Station House Nurseries

Station Road, Burton, Neston, Cheshire CH64 5SD
Tel: 01513 530022
Specialities: Dahlia

74 Stoneyford Cottage Nursery

Stoneyford Lane, Stoneyford, Cuddington, Northwich, Cheshire CW8 2TF
Tel: 01606 888970
Specialities: Herbaceous Perennials

75 Strikes Astbury Meadow Garden Centre

Newcastle Road, Astbury, Congleton, Cheshire CW12 4RL
Tel: 01260 276466

76 Strikes High Legh Garden Centre

High Legh, Ditchfield Lane, High Legh, Knutsford, Cheshire WA16 0QW
Tel: 01925 756991

77 Strikes Sycamore Park Garden Centre

Chester Road, Great Sutton, Ellesmere Port, Cheshire CH66 2LX
Tel: 01513 391289

78 Strikes Weavervale Garden Centre

Winnington Lane, Northwich, Cheshire CW8 4EE
Tel: 01606 79965

79 Strikes Wilmslow Garden Centre

Manchester Road, Wilmslow, Cheshire SK9 2JN
Tel: 01625 525700

80 Taylors Choice Garden & Pet Superstore

Gresty Lane (Off B5071), Shavington, Crewe, Cheshire CW2 5DD
Tel: 01270 213359
Specialities: Garden Products, Aquatics, Restaurant/Coffee Shop

81 Two Oaks Ltd

Hanns Hall Road, Willaston, Neston, Cheshire CH64 2TQ
Tel: 01513 361385

82 Walkers Nurseries

Aldford Road, Huntington, Chester, Cheshire CH3 6EA
Tel: 01244 320731
Specialities: Chrysanthemum

83 Walton Nurseries

Cherry Lane, Lymm, Cheshire WA13 0SY
Tel: 01925 759026
Specialities: Fuchsia

84 Water World

Chester High Road, Burton, Neston, Cheshire CH64 8TF
Tel: 01513 363616
Specialities: Pet Products, Water Products and Features, Aquatics

85 Wayside Fuchsias Nursery

Wayside, Chester Road, Acton, Nantwich, Cheshire CW5 8LD
Tel: 01270 625795

86 Weaver Vale Garden Centre#

Winnington Avenue, Winnington, Northwich, Cheshire CW8 4EE
Tel: 01606 79965
Specialities: Aquatics, Garden Products, Restaurant/Coffee Shop

87 Webbs Garden Centre

5 Halton Road, Halton Road, Runcorn, Cheshire WA7 5QS
Tel: 01928 500623

88 White Moss Nursery & Garden Centre

South Lane, Widnes, Cheshire WA8 3TZ
Tel: 01925 721111

89 Willow Pool Garden Centre

Burford Lane, Lymm, Cheshire WA13 0SH
Tel: 01925 757827
Specialities: Garden Products, Garden Furniture, Restaurant/Coffee Shop

90 Wilmslow Garden Centre

Manchester Road, Wilmslow, Cheshire SK9 2JN
Tel: 01625 525700
Specialities: Herbaceous Perennials, Annuals/Biennials/Bedding, Cut Flowers

91 Woolston Garden Centre

Unit 1 New Cut Lane, Woolston, Warrington, Cheshire WA1 4AG
Tel: 01925 822300

92 Woolston Nursery

Moss Lane, Woolston, Warrington, Cheshire WA3 6EP
Tel: 01925 811274

93 Wyevale Garden Centre (Bridgemere Garden World)

Bridgemere Garden World, Bridgemere, Nantwich, Cheshire CW5 7QB
Tel: 01270 521100

94 Wyevale Garden Centre (Tarporley)

Forest Road, Cotebrook, Tarporley, Cheshire CW6 9EE
Tel: 01829 760433

CORNWALL

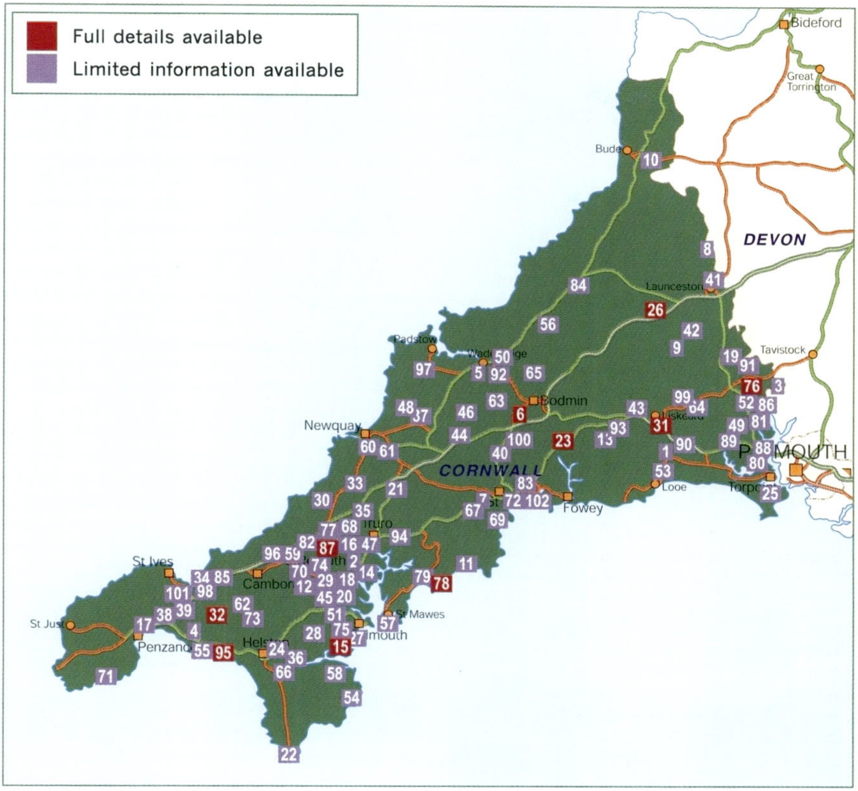

1 A J Brown

Chyvounder, Trewidland, Liskeard,
Cornwall PL14 4ST
Tel: 01503 240373

2 B.K Pengelly

Meadowside, Quenchwell, Carnon Downs, Truro,
Cornwall TR3 6LN
Tel: 01872 863340

3 Barracott Plants

Old Orchard, Calstock Road, Gunnislake,
Cornwall PL18 9AA
Tel: 01822 832234
Specialities: Homegrown, Annuals/Biennials/Bedding, Herbaceous Perennials

4 Bay Tree Nurseries

Relubbus Lane, St. Hilary, Penzance,
Cornwall TR20 9EF
Tel: 01736 763635

5 Beecroft

3 Clarence Terrace, Wadebridge,
Cornwall PL27 7NG
Tel: 01208 812892
Specialities: Bamboos

6 Bodmin Plant and Herb Nursery

Laveddon Mill, Bodmin, Cornwall PL30 5JU
Tel: 01208 72837 Fax: 01208 76491
e-mail: bodminnursery@aol
website: www.bodminnursery.co.uk

Specialities:
Herbs; Herbaceous Perennials; Hanging Baskets

Products and Services:
- Credit Cards Accepted
- Fruit Trees
- Plants
- Trees

Directions:
Open every day of the week, the nursery is located on the A389/A391 midway between Bodmin and Lanivet

Description:
Bodmin Plant and Herb Nursery is a 'proper' nursery, specialising in herbs and herbaceous perennials, and much, much more. It is the sort of Nursery that you thought no longer existed: traditional growing, fresh and healthy plants, a range that surprises and excites, as well as friendly advice on hand. Nestled in countryside on the site of an old water mill, the 4-acre site stocks more than 250 herb varieties, from culinary to medicinal, including the more unusual kinds such as Lemon Grass and Vietnamese Coriander. There's an even greater range of herbaceous perennials – in excess of 2000 varieties, including an excellent range of hardy geraniums, all raised in the nursery. A whole area is dedicated to planting in or around water, from Astilbes to Zantedeschias. The nursery also has a comprehensive Shrub range featuring hedging ideas for moorland and coast, plus specimen shrubs. A 'Bespoke and Beautiful' hanging basket service is available for summer and winter, or you can make up your own from the huge selection – or simply buy "off-the-peg". Opening times: 1st April - 31st October 9am -6pm; 1st Nov-31st March 9am-5pm; Sundays 10am-4pm.

7 Bodwen Nursery

Pothole, St. Austell, Cornwall PL26 7DW
Tel: 01726 883855
Specialities: Annuals/Biennials/Bedding, Shrubs, Homegrown

8 Braggs Wood Gardens & Nursery

Braggs Hill, Boyton, Launceston, Cornwall PL15 9RG
Tel: 01566 778668
Specialities: Annuals/Biennials/Bedding, Shrubs

9 Bregover Plants

Hillbrooke, Middlewood North Hill, Launceston, Cornwall PL15 7NN
Tel: 01566 782661
Specialities: Herbaceous Perennials, Shrubs

10 Brooks Garden Centre Limited

Stratton, Bude, Cornwall EX23 9NR
Tel: 01288 352897
Specialities: Floristry, Restaurant/Coffee Shop, Homegrown

11 Burncoose Nurseries

Caerhays Castle, Caerhays, St. Austell, Cornwall PL26 6LY
Tel: 01872 501310

12 Burncoose Nurseries

Burncoose, Gwennap, Redruth, Cornwall TR16 6BJ
Tel: 01209 860316
Specialities: Annuals/Biennials/Bedding, Shrubs

13 Calamazag Nursery

East Taphouse, Liskeard, Cornwall PL14 4NN
Tel: 01579 321799
Specialities: Shrubs

14 Carnon Downs Garden Centre

Quenchwell Road, Carnon Downs, Truro, Cornwall TR3 6LN
Tel: 01872 863058

15 Carwinion Garden

Mawnan Smith, nr Falmouth, Cornwall TR11 5JA
Tel: 01326 250250 Fax: 01326 250903
e-mail: jane@carwinion.co.uk
website: www.carwinion.co.uk

Specialities:
Bamboo; Cream Teas (summer only)

Description:
This peaceful 12-acre garden in the Helford Valley has outstanding collections of bamboo, camellias, hydrangeas and ferns together with wild flowers and exotics. The bamboo garden was created by the grandfather of the present owner, Antony Rogers, whose family have lived at Carwinion House for over 200 years. The bamboo garden now boasts more than 200 varieties, many of them rare in cultivation.

Directions:
Mawnan Smith is about 4 miles south of Falmouth. In the centre of the village is a thatched pub, the Red Lion. Carwinion Garden is 200 yards up on the left-hand side.

16 Chacewater Garden Centre

Chacewater, Truro, Cornwall TR4 8QG
Tel: 01872 560533
Specialities: Garden Products, Annuals/Biennials/Bedding

17 Churchtown Nurseries

Gulval, Penzance, Cornwall TR18 3BE
Tel: 01736 362626
Specialities: Herbaceous Perennials

18 Churchtown Nursery

Perranarworthal, Truro, Cornwall TR3 7QF
Tel: 01872 863033

19 Cockington Nursery

Stoke Climsland, Callington, Cornwall PL17 8NZ
Tel: 01579 370977
Specialities: Annuals/Biennials/Bedding, Seasonal Plants

20 Cornish Garden Nurseries

Barras Moor, Perranarworthal, Truro, Cornwall TR3 7PE
Tel: 01872 864380

21 The Cornish Rose Co

St. Newlyn, Mitchell, Newquay, Cornwall TR8 5AX
Tel: 01872 519146
Specialities: Rose

22 Cross Common Nursery

The Lizard, Helston, Cornwall TR12 7PD
Tel: 01326 290722

23 Duchy of Cornwall Nursery

Cott Road, Lostwithiel, Cornwall PL22 0HW
Tel: 01208 872668 e-mail: sales@duchyofcornwallnursery.co.uk
website: www.duchyofcornwallnursery.co.uk

SEE ADVERTISEMENT BELOW

Specialities:

Fuchsias; Camellias; Conifers

Description:

Set in a beautiful location in the heart of the Duchy of Cornwall, the nursery has the feel of a small specialist operation but is large enough to offer more than

4,500 varieties of plants – including 400 varieties of fuchsias alone.

Directions:

From the A390 just west of Lostwithiel follow the brown and white signs to the nursery which is 1½ miles off the A390.

24 Dutchy Flowers

West View, Trewennack, Helston,
Cornwall TR13 0PQ
Tel: 01326 561103
Specialities: Floristry

25 East of Eden Plants

St. Andrews Street, Millbrook, Torpoint,
Cornwall PL10 1BE
Tel: 01752 822782
Specialities: Seasonal Plants, Homegrown

26 Ellis Nurseries

Medrow Bungalow, Medrow, Polyphant, Launceston,
Cornwall PL15 7PS
Tel: 01566 86641
e-mail: timellis@ellisnurseries.wanadoo.co.uk
website: ellisnurseries.co.uk

Specialities:
Euphorbias; Hardy Perennials; Show Garden

Products and Services:

- Books and Stationery
- Display Gardens
- Garden Construction
- Garden Design
- Garden Products
- Plants
- Coffee Shop
- Seeds and Bulbs

Directions:

From the A30, 6 miles west of Launceston, turn right at Blackhill Quarry sign. Go up the hill, then left at the bottom of the green and follow the road to the left. The nurseries are 4th on the right.

Description:

Tim Ellis started gardening at the age of 14 and later trained at Merrist Wood Horticultural College. He has worked in horticulture all his life and moved to Cornwall in 1998 to realise his dream of building up his own nursery and garden. In 2002 he acquired a bare field and has gradually expanded to cover an acre. A tea room selling home-made cakes and cream teas and a white garden were added in 2007. One of Tim's specialities is his collection of hardy euphorbias which can be seen throughout the nursery. All the plants are organically grown as far as possible. The nursery makes its own compost and ducks roam the site on slug patrol. Tim is a well-known local speaker and is happy to visit garden clubs and talk on a range of subjects.

27 Falmouth Garden Centre

Swanpool Road, Goldenbank, Falmouth,
Cornwall TR11 5BH
Tel: 01326 315404
Specialities: Garden Products, Annuals/Biennials/Bedding

28 Fir Tree Farm Nursery

Tresahor, Constantine, Falmouth, Cornwall TR11 5PL
Tel: 01326 340593
Specialities: Herbaceous Perennials, Clematis

29 Fiveways Nursery

Pelean Cross, Ponsanooth, Truro, Cornwall TR3 7JQ
Tel: 01872 865185
Specialities: Annuals/Biennials/Bedding, Shrubs

30 Glen Carne Nursery

Glen Carne, Barkla Shop, St. Agnes, Cornwall TR5 0XN
Tel: 01872 552446
Specialities: Herbaceous Perennials, Shrubs

31 Goldenbank Nursery Garden Centre

Plymouth Road, Liskeard, Cornwall PL14 3PB
Tel: 01579 348622
website: www.goldenbanknursery.co.uk

SEE ADVERTISEMENT PAGE 33

Specialities:

Rare and unusual plants

Products and Services:

- Friendly expert advice
- Plant finding
- Planted baskets
- Cornish grown roses
- Choice stoneware and terracotta
- Well stocked shop
- Coffee Shop - local produce, freshly prepared

Description:

Growing a stunning selection of top quality plants with many rare and unusual species, all in a lovely garden setting.

Directions:

Just off A38 east exit to Liskeard. Between Morrisons and the Town Centre

32 Goldstephen Nursery

Pilgrim's Way, Fraddam, nr Hayle,
Cornwall TR27 6EQ
Tel: 01736 850363 Mob: 07784722374
e-mail: goldstephennursery@tiscali.co.uk

Specialities:

Roses; Herbaceous Perennials; Bulbs, Shrubs

Description:

The nursery offers a wide range of plants from seasonal bedding and tiny alpines to large specimen shrubs and ornamental trees. It also stocks a range of roses from the Cornish Rose Company chosen for scent and disease resistance. From March to September, opening times are 10am-5.30pm, Thursday to Monday; October to February, please telephone to check opening times.

Directions:

From the A30, take the Hayle exit at roundabout. Turn left at mini-roundabout into Guildford Road. Follow the lane as it merges with Wheal Alfred Road for approx. 2 miles to T-junction with the B3302. Turn left into Fraddam, then right at the crossroads into Pilgrim's Way (sign posted Townsend/Marazion) – the nursery is at the far end of this lane.

33 Goonhavern Garden Centre

Newquay Road, Goonhavern, Truro,
Cornwall TR4 9QQ
Tel: 01872 575088

34 Greenthings Hydroponics

51 Fore Street, Hayle, Cornwall TR27 4DX
Tel: 01736 757798
Specialities: Hydroponics

35 Gwarnick Mill Herbs

St. Allen, Truro, Cornwall TR4 9QU
Tel: 01872 540487
Specialities: Herbs

36 Gweek Nurseries

Parc Bottom, Gweek, Helston, Cornwall TR12 7AQ
Tel: 01326 221311
Specialities: Herbaceous Perennials, Seasonal Plants

37 H J Thomas

Lower Trenoweth Bungalow, St. Columb,
Cornwall TR9 6EW
Tel: 01637 881473

38 Hardy Exotics

Gilly Lane, Whitecross, Penzance,
Cornwall TR20 8BZ
Tel: 01736 740660
Specialities: Exotics, Hardy Garden Palms, Rare and Unusual Plants

39 Heather Lane Nursery

Back Lane West, Canonstown, Hayle,
Cornwall TR27 6NF
Tel: 01736 740198
Specialities: Seasonal Plants, Annuals/Biennials/Bedding

40 Hidden House Bulbs

Tanqueray, Bilberry Bugle, St. Austell,
Cornwall PL26 8QU
Tel: 01726 852709
Specialities: Bulbs

41 Homeleigh Garden Centre Limited

Dutson, Launceston, Cornwall PL15 9SP
Tel: 01566 773147

42 Illand Nurseries

Illand, Launceston, Cornwall PL15 7LS
Tel: 01566 782521
Specialities: Annuals/Biennials/Bedding, Shrubs

43 J B & A L Arnold

Venton Veor, Liskeard, Cornwall PL14 6SR
Tel: 01579 320409

44 J C Orchard

Castle-an-Dinas, St. Columb, Cornwall TR9 6JB
Tel: 01637 881300

45 J J Carnivorous Plants

27 St. Michaels Road, Ponsanooth, Truro,
Cornwall TR3 7ED
Tel: 01872 863391
Specialities: Carnivorous Plants

46 J R Nursery

Riverside Cottage, Rosenannon, Bodmin,
Cornwall PL30 5PJ
Tel: 01637 880108
Specialities: Annuals/Biennials/Bedding, Shrubs

47 Jane Adele Flowers & Nursery

Lemon Quay, Truro, Cornwall TR1 2LW
Tel: 01872 70101
Specialities: Annuals/Biennials/Bedding, Shrubs, Floristry

48 Japanese Garden & Bonsai Nursery

St. Mawgan Village, Newquay, Cornwall TR8 4ET
Tel: 01637 860116
Specialities: Bonsai, Bamboos, Japanese Maples

49 Kernock Park Plants

Kernock Park, Pillaton, Saltash, Cornwall PL12 6RY
Tel: 01579 350561
Specialities: Annuals/Biennials/Bedding

50 Killibury Nursery

Egloshayle, Wadebridge, Cornwall PL27 6EL
Tel: 01208 812722
Specialities: Shrubs, Annuals/Biennials/Bedding

51 L R Pangelly

Higher Kergilliack, Kergilliack, Falmouth,
Cornwall TR11 5PB
Tel: 01326 372271

52 Lanoyce Nurseries

St. Dominick, Saltash, Cornwall PL12 6TU
Tel: 01579 350297
Specialities: Herbaceous Perennials, Shrubs

53 Looe Garden Centre

St. Martin, Looe, Cornwall PL13 1NX
Tel: 01503 263866
Specialities: Garden Products, Shrubs

54 Losowek Herbs

Treventon, Well Lane, St. Keverne, Helston,
Cornwall TR12 6LZ
Tel: 01326 280514
Specialities: Herbs

55 Lower Kenneggy Nurseries

Lower Kenneggy Lane, Rosudgeon, Penzance,
Cornwall TR20 9AR
Tel: 01736 762959
Specialities: Herbaceous Perennials, Shrubs

56 Michaelstow Nurseries

Michaelstow, St. Tudy, Bodmin, Cornwall PL30 3PB
Tel: 01208 851262
Specialities: Annuals/Biennials/Bedding, Shrubs

57 Mr J de Maaker

Medlar Cottage, Polvarth Road, St. Mawes, Truro,
Cornwall TR2 5AY
Tel: 01326 270725
Specialities: Orchids

58 Myrtle Nursery

High Lane, Manaccan, Helston, Cornwall TR12 6HT
Tel: 01326 231604
Specialities: Annuals/Biennials/Bedding, Shrubs

59 New Season

Green Lane Nurseries, Green Lane, Redruth, Cornwall TR15 1LH
Tel: 01209 313245
Specialities: Annuals/Biennials/Bedding, Herbaceous Perennials, Shrubs

60 Newquay Bonsai

Unit 5E Treloggan Ind Est, Newquay, Cornwall TR7 2SX
Tel: 01637 879733
Specialities: Bonsai

61 Newquay Garden Centre

Little Trethiggey, Quintrell Downs, Newquay, Cornwall TR8 4LG
Tel: 01637 872199
Specialities: Garden Products, Annuals/Biennials/Bedding

62 The Nurseries

Praze, Camborne, Cornwall TR14 0PB
Tel: 01209 831402
Specialities: Seasonal Plants, Homegrown

63 Oak Lodge Woodland Garden & Nursery

Cotton Wood, Nanstallon, Bodmin, Cornwall PL30 5LQ
Tel: 01208 831292

64 Oak Park Nurseries

St. Ive, Liskeard, Cornwall PL14 3NA
Tel: 01579 383530
Specialities: Annuals/Biennials/Bedding, Homegrown

65 The Old Mill Herbary

Hellandbridge, Bodmin, Cornwall PL30 4QR
Tel: 01208 841206
Specialities: Herbs

66 The Old Withy Garden Nursery

Mawgan, Helston, Cornwall TR12 6BE
Tel: 01326 221171
Specialities: Herbaceous Perennials

67 Pengelly Plant Centre

Hewas Water, St. Austell, Cornwall PL26 7JG
Tel: 01726 883757

68 Penrose Water Gardens

Penrose, Tregavethan, Truro, Cornwall TR4 9ES
Tel: 01872 274852

69 Pentewan Valley Nurseries

Pentewan Road, Pentewan, St. Austell, Cornwall PL26 6DL
Tel: 01726 842360
Specialities: Shrubs, Herbaceous Perennials

70 Penventon Nursery

Trevsikey, Lanner, Redruth, Cornwall TR16 6AS
Tel: 01209 820049
Specialities: Seasonal Plants, Shrubs

71 Peter G Paris

St. Buryan, Penzance, Cornwall TR19 6DQ
Tel: 01736 810384
Specialities: Exotics, Hardy Garden Palms, Hardy Plants

72 Pine Lodge Gardens

St. Austell, Cornwall PL25 3RQ
Tel: 01726 73500

73 Polcrebo Nursery

Nancegollan, Helston, Cornwall TR13 0BG
Tel: 01209 831352
Specialities: Annuals/Biennials/Bedding, Shrubs

74 Poldice Plants

Little Poldice, Crofthandy, St. Day, Redruth, Cornwall TR16 5JP
Tel: 01209 822083
Specialities: Annuals/Biennials/Bedding, Herbaceous Perennials

75 Quality Daffodils

14 Roscarrack Close, Falmouth, Cornwall TR11 4PJ
Tel: 01326 317959
Specialities: Narcissus (Daffodil)

76 Rising Sun Nurseries

Harrowbarrow, Callington, Cornwall PL17 8JD
Tel/Fax: 01579 351231
website: www.risingsunnurseries.co.uk

Specialities:
House & Bedding Plants; Hanging Baskets; Ornamental Trees, Fruit Trees.

Directions:
Rising Sun Nurseries are ½ mile off the A390 Callington to Tavistock Road. Continue down the hill past Sleepy Hollow and the entrance is on the right.

Products and Services:

- Credit Cards Accepted
- Cut Flowers
- Bird Food
- Houseplants
- Plants
- Coffee Shop
- Seeds and Bulbs
- Trees

National Garden Gift Vouchers

Description:

The nurseries stock a huge selection of house plants, bedding plants, patio and plug plants. There's also a great choice of rockery and herbaceous plants as well as seeds, bulbs, fruit and ornamental trees, shrubs and a full range of garden tools and accessories. The nurseries are stockists of Tom Chambers hanging baskets, mangers and hayracks, including coir liners, and a wide selection of hanging basket accessories is available. The nurseries also offer a r*efill service for pots and baskets. Other items on site include an* excellent selection of bird feeders, bird tables and bird seed, and a very wide range of ceramic and terracotta pots and containers from around the world. The choice includes frost resistant pots, decorative pots and accessories, and ceramic ornaments. A popular amenity is the Coffee Shop which serves Cornish Cream Teas, a variety of cakes, Walls Ice Cream and Salcombe Dairy Ices.

77 Roseland House Nursery

Chacewater Hill, Chacewater, Truro,
Cornwall TR4 8QB
Tel: 01872 560451
Specialities: Clematis

78 The Roseland Nursery

Trewartha Chapel, Trewartha, Veryan,
Cornwall TR2 5QJ
Tel/Fax: 01872 501825 e-mail: 01872 501825
e-mail: info@theroselandnursery.co.uk
website: www.theroselandnursery.co.uk

Specialities:

Unusual Perennials; Hardy Exotics, Cacti & Succulents; Shrubs & Small Trees

Products and Services:

- Credit Cards Accepted
- Display Gardens
- Garden Products
- Houseplants
- Information Desk
- Large Car Park
- Plant Guarantee
- Seeds and Bulbs

Directions:

From the A3078 from Truro towards St Mawes turn left at Jet garage towards Veryan. Continue for about a 1 mile then turn left to Trewartha. The nursery is on this road on the right hand side.

Description:

The Roseland Nursery is located at one of the highest points on the Roseland and enjoys panoramic views over the peninsula across miles of open countryside to the Lizard. This small family-run traditional nursery specialises in unusual perennials and a host of hardy exotics. There is a very special collection of cacti and succulents, together with grasses, selected trees and shrubs, vegetable plants and produce. Seasonal hanging baskets and pots can be filled to order or bought off the shelf. Composts, terracotta pots, garden sundries (and ice cream!) are also available. Most of the nursery stock is grown here by traditional methods or otherwise sourced in Cornwall. Owners Nicola and Stuart Bush are properly qualified to give advice and are happy to deliver locally. And if they don't have what you are looking for, they will endeavour to find it for you.

79 Roseland Plant Centre

Ruan High Lanes, Truro, Cornwall TR2 5JR
Tel: 01872 501502
Specialities: Annuals/Biennials/Bedding, Shrubs

80 Shillingham Nurseries

Antony Passage, Shillingham, Saltash,
Cornwall PL12 4QU
Tel: 01752 844763
Specialities: Annuals/Biennials/Bedding, Shrubs

81 Smeaton Nurseries

St. Dominick, Saltash, Cornwall PL12 6SW
Tel: 01579 350349
Specialities: Annuals/Biennials/Bedding, Shrubs

82 St Day Nurseries

Greystones, Tolgullow, St. Day, Redruth,
Cornwall TR16 5PD
Tel: 01209 820271

83 St. Austell Garden Centre

Boscundle, St. Austell, Cornwall PL25 3RH
Tel: 01726 812197
Specialities: Shrubs, Paving

84 St. Kitts Herbery

Starapark, Camelford, Cornwall PL32 9XH
Tel: 01840 213442
Specialities: Herbs

85 Steamers Hill Nursery

40 Steamers Hill, Angarrack, Hayle,
Cornwall TR27 5JB
Tel: 01736 755411
Specialities: Annuals/Biennials/Bedding, Shrubs

86 Sunningdale Nurseries

Sunningdale, St. Dominick, Saltash,
Cornwall PL12 6TG
Tel: 01579 350309
Specialities: Seasonal Plants, Annuals/Biennials/Bedding

87 Sunny Corner Nurseries

The Square, Chacewater, Truro TR4 8PY
Tel: 01872 560084 Mobile: 07980 984421
e-mail: helen@sunny-corner.co.uk
website: www.sunny-corner.co.uk

Specialities:
Cornish Grown Plants; Floristry; Garden Services

Products and Services:

- Fresh Cut Flowers
- Floristry & Floral Tributes
- Fruit Trees
- Garden Construction
- Garden Design
- Garden Products
- Giftware
- Houseplants
- Large Car Park
- Bird Products
- Herbaceous Perennials
- Vegetable Plants
- Herbs
- Plants
- Seeds and Bulbs
- Trees

Directions:
Chacewater is located between Truro and Redruth on the Threemilestone to Scorrier road. The nursery is in the centre of the village behind the King's Head pub.

Description:
Approached through a stone archway and wrought iron gates, Sunny Corner Nurseries are situated within a listed walled garden which during the late 1800s served as the village market. The original walls are hosts to house sparrows and wall bees. The solid granite pillars are still in place although the lean-to roof was renovated in 1973 following the collapse of the mining industry. With plenty of undercover space in the greenhouses, polytunnels and shop area, which looks out onto raised plant beds, this quaint little nursery is a must in any weather. The wide range of plants on offer include evergreen and deciduous shrubs, perennials, herbaceous and alpine plants together with trees, climbers and succulents for every location. The nursery also stocks a comprehensive range of gardening products and has a full floristry service. Sunny Corner is open from 9am-5pm, Monday to Saturday, and from 10.30am-4pm on Sunday.

88 Tamar View Nurseries Ltd

Callington Road, Carkeel, Saltash,
Cornwall PL12 6PH
Tel: 01752 847366

89 Tartan Down Nurseries

Tartan Down, Landrake, Saltash,
Cornwall PL12 5AF
Tel: 01752 851431
Specialities: Homegrown, Annuals/Biennials/Bedding, Herbaceous Perennials

90 Trebrown Palm Tree Nurseries

Horningtops, Liskeard, Cornwall PL14 3PU
Tel: 01503 240170
Specialities: Palms, Annuals/Biennials/Bedding

91 Trecanna Nursery

Latchley, Gunnislake, Cornwall PL18 9AY
Tel: 01822 834680
Specialities: Annuals/Biennials/Bedding, Shrubs

92 Trelawney Garden Centre

Sladesbridge, Wadebridge, Cornwall PL27 6JA
Tel: 01208 893030

93 Trelose Waters

Pennellick Farm, East Taphouse, Liskeard,
Cornwall PL14 4NL
Tel: 01579 320848

94 Tresillian Garden Centre

1 Riverside Cottage, Tresillian, Truro,
Cornwall TR2 4BA
Tel: 01872 520544

Specialities: Landscape Gardening, Floristry, Garden Ornaments

95 Trevena Cross Nurseries & Garden Centre

Breage, Helston, Cornwall TR13 9PS
Tel: 01736 763880 Fax: 01736 762828
e-mail: sales@trevenacross.co.uk
website: www.trevenacross.co.uk

SEE ADVERTISEMENT BELOW

Specialities:

Exotic Plants & Seaside Gardens; Tree Ferns; Plants from southern Hemisphere

Products and Services:

- Aquatics
- Children's Play Area
- Seasonal Displays
- Credit Cards Accepted
- Disabled Facilities
- Dried/Artificial Flowers
- Garden Furniture
- Garden Products
- Giftware
- Houseplants
- Large Car Park
- Pet Products
- Plant Guarantee
- Plants
- Restaurant/Coffee Shop
- Seeds and Bulbs
- Trees

Directions:

The nurseries are on the A394 between Penzance and Helston in the village of Breage.

96 Trevenson Moor Garden Centre

Cross Roads, Tehidy, Camborne,
Cornwall TR14 0EP
Tel: 01209 610909

Specialities: Seasonal Plants, Seeds

97 Trevisker Nursery

Higher Trevisker, Padstow, Cornwall PL28 8LD
Tel: 01841 541361

Specialities: Seasonal Plants, Annuals/Biennials/ Bedding

98 Trevassack Nurseries & Garden Centre

4 Trevassack Hill, Hayle, Cornwall TR27 5HP
Tel: 01736 752119

99 Troutbeck Plants

Troutbeck Farm, St. Ive, Liskeard,
Cornwall PL14 3ND
Tel: 01579 343500

Specialities: Homegrown, Seasonal Plants

100 Wall Cottage Nurseries

Lockengate Bugle, St. Austell, Cornwall PL26 8RU
Tel: 01208 832234
Specialities: Exotics, Fuchsia

101 Wyevale Garden Centre (Nut Lane)

Nut Lane, Hayle, Cornwall TR27 6LG
Tel: 01736 753731

102 Wyevale Garden Centre (St Austell)

Par Moor Road, Par, Cornwall PL24 2SQ
Tel: 01726 814854

CUMBRIA

1 Aero Nurseries & Garden Centre

Harker Road End, Carlisle, Cumbria CA6 4HQ
Tel: 01228 674612

2 Alan Clarke Rhododendrons

Longthwaite House, Wigton, Cumbria CA7 8NR
Tel: 01697 342592
Specialities: Rhododendron (Inc Azalea)

3 Allerby Community Gardens

Allerby, Wigton, Cumbria CA7 2NL
Tel: 01900 815003

4 Baines Paddock Nursery

Haverthwaite, Ulverston, Cumbria LA12 8PF
Tel: 01539 558236
Specialities: Annuals/Biennials/Bedding, Shrubs

5 Bank Mill Nurseries

Bank House, Beckfoot Silloth, Wigton,
Cumbria CA7 4LF
Tel: 01900 881340
Specialities: Annuals/Biennials/Bedding, Shrubs

6 Beechcroft Just Trees

Bongate, Appleby-in-Westmorland,
Cumbria CA16 6UE
Tel: 01768 351201
Specialities: Trees

7 Beetham Nurseries

Beetham, Milnthorpe, Cumbria LA7 7AP
Tel: 01539 563630
Specialities: Restaurant/Coffee Shop, Garden Furniture, Annuals/Biennials/Bedding

8 Boonwood Garden Centre

Gosforth, Seascale, Cumbria CA20 1BP
Tel: 01946 725330
Specialities: Seaside Speciality, Climbers, Trees

9 Broom Horticultural Supplies

Broom La Cottage, Long Marton,
Appleby-in-Westmorland, Cumbria CA16 6JP
Tel: 01768 361650

10 Brownthwaite Hardy Plants

Fell Yeat, Casterton, Carnforth, Cumbria LA6 2JW
Tel: 01524 271340
Specialities: Herbaceous Perennials

11 Bush Green Cottage Garden & Nursery

Foxfield Rd, Broughton-in-Furness,
Cumbria LA20 6BY
Tel: 01229 716724
Specialities: Cottage Garden Plants, Annuals/Biennials/Bedding, Herbaceous Perennials

12 Carr Bank Garden Centre

Sanside Road, Carr Bank, Nr Milnthorpe,
Cumbria LA7 7JY
Tel: 01524 762313
Specialities: Pet Products, Greenhouses and Sheds

13 Caths Garden Plants

The Walled Garden, Heaves Hotel, Heaves,
Nr Levens, Kendal, Cumbria LA8 8EF
Tel/Fax: 015395 61126
e-mail: cath@cathsgardenplants.co.uk
website: www.cathsgardenplants.co.uk

Specialities:

Herbaceous Perennials, Shrubs, Ferns, Grasses, Alpines, Herbs, Climbers

Description:

A traditional nursery within the walls of an old walled garden in a peaceful setting, Caths Garden Plants grows an eclectic range of quality plants naturally and has been awarded Gold medals at National Flower Shows. Help and advice always available and courses and events held regularly.

Directions:

Leave M6 at junction 36. Follow A590/A591 Barrow/South Lakes along duel carriageway. Take first left onto A590 Barrow at roundabout, then 2nd right turn across duel carriageway into **Heaves Hotel** (approx. 500yds).

14 Chiltern Seeds

Bortree Stile, Ulverston, Cumbria LA12 7PB
Tel: 01229 581137
Specialities: Seeds

15 Cochrane's Nurseries

Longtown, Carlisle, Cumbria CA6 5LZ
Tel: 01228 791220
Specialities: Annuals/Biennials/Bedding, Herbaceous Perennials

16 Cottage Garden Plants

1 Sycamore Close, Whitehaven, Cumbria CA28 6LE
Tel: 01946 695831
Specialities: Cottage Garden Plants

17 The Courtyard Garden Nurseries

Underley Home Farm, Kearstwick, Carnforth,
Cumbria LA6 2DY
Tel: 01524 272444

18 Crooklands Garden Centre

Crooklands Brow, Dalton-in-Furness,
Cumbria LA15 8JH
Tel: 01229 464225
Specialities: Landscape Gardening, Restaurant/Coffee Shop

19 Dixon Landscapes

Cocklakes Yard, Cocklakes Cumwhinton, Carlisle,
Cumbria CA4 0BQ
Tel: 01228 562777
Specialities: Landscape Gardening

20 Douglas Nursery

Newby Cross, Carlisle, Cumbria CA5 6JP
Tel: 01228 710789
Specialities: Annuals/Biennials/Bedding, Herbaceous Perennials

21 E.G Bachem & Son

Eden Vale Nursery, Long Marton, Appleby-in-Westmorland, Cumbria CA16 6BU
Tel: 01768 361605
Specialities: Annuals/Biennials/Bedding, Trees, Shrubs

22 Eden Cottage Nursery

Eden Cottage Nursery, Kirkby Stephen, Cumbria CA17 4AP
Tel: 01768 371630
Specialities: Trees & Shrubs, Acers, Dwarf and Rare Conifers

23 Elizabeth Gardens

Church Rigg Farm, Wigton, Cumbria CA7 0AB
Tel: 01697 345470

24 Foxfield Garden Centre

The Laurels, Foxfield, Broughton-in-Furness, Cumbria LA20 6BT
Tel: 01229 716318

25 G Fowler

Stoney Bridge, Cark In Cartmel, Grange-over-Sands, Cumbria LA11 7PE
Tel: 01539 558110

26 Garden House Nurseries

D.W. Hickson & Son, Garden House Nurseries, Dalston, Carlisle, Cumbria CA5 7LL
Tel: 01228 710298
e-mail: stephicoso@hotmail.com
website: www.gardenhousenursery.com

Specialities:

Hardy Geraniums; Hostas; Herbaceous Perennials

Description:

A small family business which specialises in herbaceous perennials with an extensive selection of hardy geraniums, grasses, hostas, hemerocallis and irises. It also stocks a variety of basket, tub and patio plants as well as conifers and annual bedding plants.

Directions:

Located in the centre of Dalston opposite the church. Dalston is 4 miles south of Carlisle on the B5299 and is approx. 4 miles from junction 42 of the M6.

27 Garden of Eden

Thiefside Garden Centre, Thiefside, Calthwaite, Penrith, Cumbria CA11 9RN
Tel: 01768 885183
Specialities: Garden Furniture

28 Gilithwaite Rigg Nursery

Lickbarrow Road, Windermere, Cumbria LA23 2EB
Tel: 01539 443472
Specialities: Landscape Gardening

29 Graceland Nurseries

Kirklinton, Carlisle, Cumbria CA6 6BZ
Tel: 01228 675515
Specialities: Annuals/Biennials/Bedding, Shrubs

30 Grange Plant Centre

off Main Street, Grange-over-Sands, Cumbria LA11 6DY
Tel: 01539 533510
Specialities: Shrubs, Herbaceous Perennials

31 Grasmere Garden Centre Ltd

Churchsile, Grasmere, Ambleside, Cumbria LA22 9SW
Tel: 01539 435255
Specialities: Garden Furniture, Annuals/Biennials/Bedding, Trees

32 Greystone Garden Centre

Greystone Lane, Dalton-in-Furness, Cumbria LA15 8QQ
Tel: 01229 463005
Specialities: Garden Furniture, Garden Machinery

33 Halecat Garden Nurseries

North Lodge, Witherslack, Grange-over-Sands, Cumbria LA11 6SD
Tel: 01539 552536
Specialities: Herbaceous Perennials, Grasses, Annuals/Biennials/Bedding

34 Hartside Nursery Garden

Nr Alston, Alston, Cumbria CA9 3BL
Tel: 01434 381372
Specialities: Annuals/Biennials/Bedding, Shrubs

35 Hayes Garden World

Lake Road, Ambleside, Cumbria LA22 0DW
Tel: 01539 433434

36 Hazel Dene Garden Centre

Culgaith, Culgaith, Penrith, Cumbria CA10 1QF
Tel: 01768 882520

37 Highfield Nurseries

High Harrington, Workington, Cumbria CA14 5RT
Tel: 01946 831322
Specialities: Annuals/Biennials/Bedding, Shrubs

38 Hill Foot Garden Centre

County Road, Ulverston, Cumbria LA12 7SE
Tel: 01229 587282

39 Hilltop Nursery

Sedbergh Rd, New Hutton, Kendal, Cumbria LA8 0AE
Tel: 01539 731079

40 Holker Hall & Gardens

Cark In Cartmel, Grange-over-Sands, Cumbria LA11 7PL
Tel: 01539 558328

41 Holmes of Natland

The Nurseries, Natland, Kendal, Cumbria LA9 7QL
Tel: 01539 560224
Specialities: Homegrown

42 Inglefield Plants Ltd

Windermere Road, Staveley, Nr Kendal, Cumbria LA8 9LY
Tel: 01539 821142

Specialities:

Specimen Trees; Mediterranean Trees; Architectural Trees; Shrubs & Perennials

Description:

This family-run company is devoted to bringing the more exotic trees, shrubs, plants and containers from around the world. A particular passion is for architectural plants whose colourful foliage and stems combine to create dramatic contrasts whatever the season.

Directions:

Leave the M6 at Junction 36 and follow the A591 towards Windermere. About 5 miles past Kendal, turn right into Staveley village. Stay on this road through the village and you will see the nursery signposted to the left.

43 JRG Dahlias

22 Summerville Road, Milnthorpe, Cumbria LA7 7DF
Tel: 01539 562691
Specialities: Dahlia

44 Keith Singleton

Sea View Nurseries, Nethertown, Egremont, Cumbria CA22 2UQ
Tel: 01946 820412
Specialities: Annuals/Biennials/Bedding, Shrubs

45 Lakeland Hanging Baskets

5 Castle Oval, Kendal, Cumbria LA9 7BD
Tel: 01539 738177
Specialities: Hanging Baskets

46 The Lakes Home Centre

Low Road, Cockermouth, Cumbria CA13 0HH
Tel: 01900 827282

47 Lamberlea Pond and Garden Centre

Nethertown, Egremont, Cumbria CA22 2UH
Tel: 01946 820536

48 Larch Cottage Nurseries

Melkinthorpe, Penrith, Cumbria CA10 2DR
Tel: 01931 712404

49 Little Dodd Garden Centre Ltd

Underskiddaw, Keswick, Cumbria CA12 4PY
Tel: 01768 773491

50 Longwood Garden Centre

Wigton Road, Silloth, Wigton, Cumbria CA7 4PQ
Tel: 01697 332465
Specialities: Garden Furniture, Garden Products

51 Mike Capstick Nursery

41 Brow Nelson, Dalston, Carlisle, Cumbria CA5 7LE
Tel: 01228 710926

52 Moota Garden Centre

Moota, Cockermouth, Cumbria CA13 0QE
Tel: 01697 321550
Specialities: Water Products and Features, Garden Products

53 Moss House Nurseries

Justicetown, Westlinton, Carlisle, Cumbria CA6 6AN
Tel: 01228 674469
Specialities: Annuals/Biennials/Bedding, Shrubs

54 Nextness Nurseries

Glenvilla, Next Ness, Ulverston, Cumbria LA12 7QW
Tel: 01229 583622
Specialities: Annuals/Biennials/Bedding, Grasses, Shrubs

55 Norcroft Nurseries

Roadends, Intack, Southwaite, Carlisle, Cumbria CA4 0LH
Tel: 01697 473933 Fax: 01697 473969
e-mail: stellaandkeithbell@sbell44.fsnet.co.uk
Specialities: Hardy herbaceous, ornamental grasses, hostas

56 Oakbank Nurseries

Longtown Road, Brampton, Cumbria CA8 2AB
Tel: 01697 73908

57 Oakhurst Garden Centre & Squirrels Pantry Tearoom

Lamplugh Road, Cockermouth, Cumbria CA13 0QN
Tel: 01900 822180 Fax: 01900 825806
e-mail: oakhurstgardencentre@btinternet.com
website: www.oakhurstgardencentre.com
Specialities: Patio Clematis, Herbaceous Perennials

58 Penton Mill Rhododendrons

Penton, Carlisle, Cumbria CA6 5QU
Tel: 01228 577336
Specialities: Rhododendron (Inc Azalea)

59 The Pot Place

Station Yard, Plumpton, Penrith, Cumbria CA11 9PA
Tel: 01768 885500
Specialities: Pots

60 Proud Plants

East of Eden Nurseries, Ainstable, Carlisle, Cumbria CA4 9QN
Tel: 01768 896604
Specialities: Herbaceous Perennials, Shrubs, Alpines/Rock Plants

61 Robert Lesley & Sons

Holly House Nurseries, Durdar Road, Carlisle, Cumbria CA2 4TR
Tel: 01228 521476
Specialities: Trees

62 Scented Roses

Stewart Hill Cottage, Hesket Newmarket, Wigton, Cumbria CA7 8HX
Tel: 01768 484172
Specialities: Rose

63 Scott's Wildflowers

Swallow Hill Barn, 31 Common Side, Distington, Workington, Cumbria CA14 4PU
Tel: 01946 830486
e-mail: scotts.wildflowers@virgin.net
website: www.scottswildflowers.co.uk
Specialities: Wild flowers

64 Southern Lakeland Nurseries

Cark in Cartmel, Station Road, Flookburgh, Grange-over-Sands, Cumbria LA11 7JZ
Tel: 01539 558237
Specialities: Annuals/Biennials/Bedding, Grasses, Shrubs

65 Strikes Houghton Hall Garden Centre

Houghton, Houghton, Carlisle, Cumbria CA6 4JB
Tel: 01228 400610

66 Summerdale Garden Nursery

Summerdale House, Cow Brow Lupton, Carnforth, Cumbria LA6 1PE
Tel: 01539 567210
Specialities: Annuals/Biennials/Bedding, Herbaceous Perennials

67 Tarn Road Nurseries

Tarn Road, Brampton, Cumbria CA8 1QY
Tel: 01697 73781

68 Tithe Barn Nurseries

Long Lane, Stainton With Adgarley, Barrow-in-Furness, Cumbria LA13 0NG
Tel: 01229 464338

69 Westwood Nurseries & Garden Centre

Orton Grange, Carlisle, Cumbria CA5 6LB
Tel: 01228 712123
Specialities: Shrubs, Conifers, Annuals/Biennials/Bedding

70 Whitesyke Garden Centre

Longtown, Carlisle, Cumbria CA6 5TR
Tel: 01228 792718
Specialities: Garden Furniture, Garden Products

71 Woodlands Nurseries & Garden Centre

Crooklands, Milnthorpe, Cumbria LA7 7NJ
Tel: 01539 567273

DERBYSHIRE

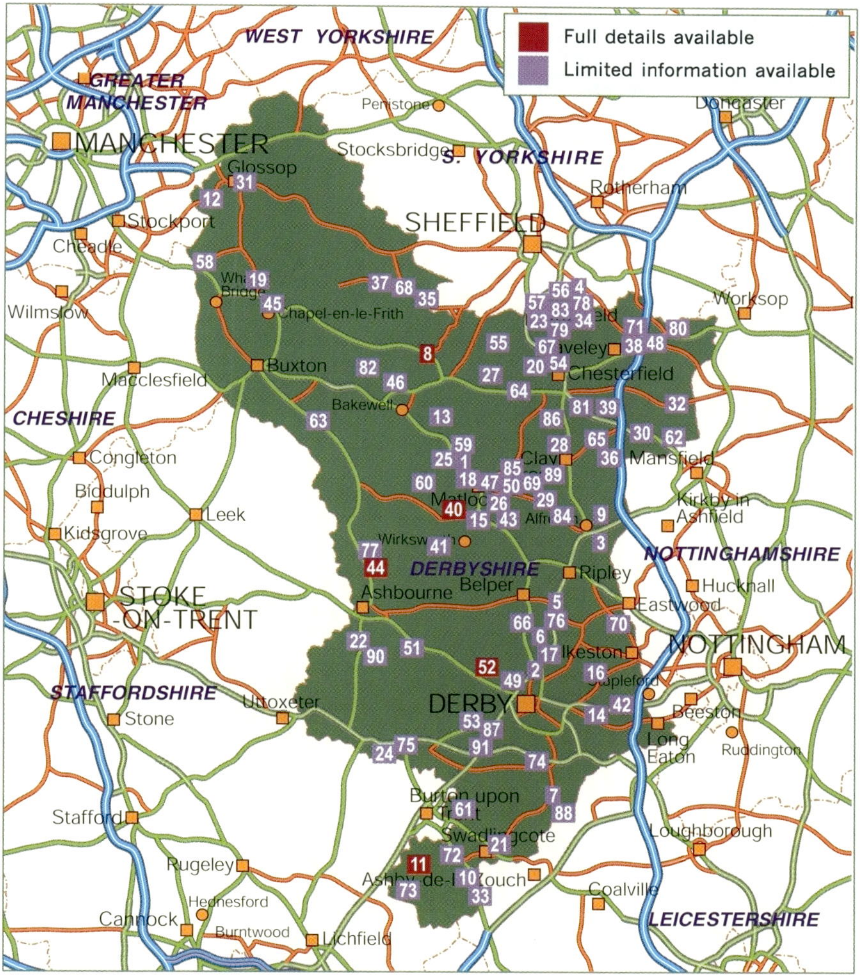

1 Abbey Brook Cactus Nursery

Old Hackney Lane, Hackney, Matlock, Derbyshire DE4 2QJ
Tel: 01629 580306
Specialities: Shrubs, Cacti And Succulents

2 Abbey Hill Nurseries

Home Nook Farm, North Avenue, Darley Abbey, Derby, Derbyshire DE22 1EZ
Tel: 01332 557652

3 Agave Nursery

15 Sleetmoor Lane, Somercotes, Alfreton, Derbyshire DE55 1RB
Tel: 01773 605843
Specialities: Annuals/Biennials/Bedding, Herbaceous Perennials

4 Allards Plant Centre

21-24 Cattle Market, Chequers Road, Derby, Derbyshire DE21 6EP
Tel: 01332 205135

5 Bensley Nurseries

7 Bramble Way, Kilburn, Belper,
Derbyshire DE56 0LH
Tel: 01332 781961
Specialities: Hardy Plants, Annuals/Biennials/Bedding

6 Birchwood Farm Nurseries

Port Way, Coxbench, Derby, Derbyshire DE21 5BE
Tel: 01332 880685
Specialities: Seasonal Plants, Shrubs

7 Bradgate Nurseries

The Stables, Cockshut Lane, Melbourne, Derby,
Derbyshire DE73 8DG
Tel: 01332 865650
Specialities: Herbaceous Perennials, Seasonal Plants

8 Calver Sough Garden Centre & Aquatics

Hassop Road, Calver, Hope Valley,
Derbyshire S32 3XH
Tel/Fax: 01433 630692
e-mail: calvernurseries@onetel.net

Specialities:

Plants & Trees; Terracotta Pots; Ornamental water and aquatics features.

Description:

A family business. This attractively laid out site offers a fine range of healthy plants as well as an extensive range of terracotta pots in all sizes, stylish garden furniture and a host of garden accessories.

Directions:

The Centre is just off the A619 west of Chesterfield in the village of Calver.

9 Carnfield Hall Garden & Craft Centre

Carnfield Hill, South Normanton, Alfreton,
Derbyshire DE55 2BE
Tel: 01773 834577
Specialities: Giftware, Specimen Trees

10 Castle Nurseries

71 Mount Pleasant Road, Castle Gresley,
Swadlincote, Derbyshire DE11 9JE
Tel: 01283 221733
Specialities: Shrubs, Herbaceous Perennials

11 Chapman's Nurseries

Burton Road, Rosliston, Swadlincote,
Derbyshire DE12 8JX
Tel: 01283 543546 Fax: 01283 565215
e-mail: kevinjenkinson@hotmail.co.uk

Specialities:

Plants

Products and Services:

- Seasonal Displays
- Credit Cards Accepted
- Disabled Facilities
- Fruit Trees
- Garden Furniture
- Garden Products
- Greenhouses and Sheds
- Houseplants

Directions:

Leave Burton-on-Trent southwards on the A444. At Castle Gresley (approx 7 miles), turn right and follow the signs for the National Forestry Centre. The nursery is opposite the Centre.

Description:

The Chapman family have owned and run the Garden Centre since it began more than 70 years ago. It started by growing tomatoes and then specialised in chrysanthemums and bedding plants before developing into the fully-fledged nursery and garden centre it is today. It is generally known as *THE* plant centre of south Derbyshire and offers a huge selection of plants ranging from bedding plants to large specimen trees. The Chapmans back up their products with a wealth of horticultural knowledge and are happy to offer information and advice. Not only do they offer traditional plants but also a bespoke plant sourcing service for that very special variety. A popular amenity at the Centre is the log cabin tea room offering quality tea and coffee, and an array of fine food. The shop area stocks all the usual garden sundries, house plants, wild bird products, bulbs and garden furniture.

12 Charlesworth Nursery and Garden Centre

Glossop Road, Charlesworth, Glossop, Derbyshire SK13 5HB
Tel: 01457 863998

13 Chatsworth Garden Centre Ltd

Calton Lees, Beeley, Matlock, Derbyshire DE4 2NX
Tel: 01629 734004

14 Collyers Nurseries

260 Nottingham Road, Borrowash, Derby, Derbyshire DE72 3FR
Tel: 01332 662035
Specialities: Rare and Unusual Plants

15 Cromford Garden Centre

Derby Road, Cromford, Matlock, Derbyshire DE4 3RN
Tel: 01629 824990
Specialities: Garden Products, Garden Furniture

16 Dale Abbey Plants

Hagg Lane, Dale Abbey, Ilkeston, Derbyshire DE7 4PR
Tel: 01159 322728
Specialities: Annuals/Biennials/Bedding, Shrubs

17 Derby Garden Centre

Alfreton Road, Little Eaton, Derby, Derbyshire DE21 5DB
Tel: 01332 831666

18 Derwent Treescapes Ltd

Walker Croft, Deep Carr Lane, Matlock, Derbyshire DE4 3NQ
Tel: 01629 582437
Specialities: Trees

19 Dewsnaps

Chinley, High Peak, Derbyshire SK23 6AW
Tel: 01663 751175

20 Dunston Hall Garden Centre

Dunston Road, Chesterfield, Derbyshire S41 9RL
Tel: 01246 268468

21 Edwards Nurseries

133 Burton Road, Woodville, Swadlincote, Derbyshire DE11 7JW
Tel: 01283 215565
Specialities: Annuals/Biennials/Bedding, Shrubs

22 Fairways Garden Centre

Hollies Farm, Clifton, Ashbourne, Derbyshire DE6 2GN
Tel: 01335 347900

23 Ferndale Nursery & Garden Centre

Dyche Lane, Coal Aston, Dronfield, Derbyshire S18 3BJ
Tel: 01246 412763

24 Firs Farm Nursery

Scropton, Derby, Derbyshire DE65 5PN
Tel: 01283 520331

25 Forest Nurseries

Oddford Lane, Two Dales, Matlock, Derbyshire DE4 2EX
Tel: 01629 733225

26 Fox Cottage Plants

Yew Tree Farm, Thatchers Lane, Tansley, Matlock, Derbyshire DE4 5FD
Tel: 01629 57493
Specialities: Seasonal Plants, Shrubs

27 Freebirch Nurseries

Quarry Farm, Freebirch, Eastmoor, Chesterfield, Derbyshire S42 7DQ
Tel: 01246 582412
Specialities: Herbaceous Perennials, Garden Products

28 G B S Nurseries

West Street, Clay Cross, Chesterfield, Derbyshire S45 9QX
Tel: 01246 250010

29 Garthwaite Nurseries

Cornerways, Matlock Road, Brackenfield, Alfreton, Derbyshire DE55 6DD
Tel: 01629 534354
Specialities: Specialist Bedding, Herbaceous Perennials

30 Glapwell Nurseries

Glapwell Lane, Glapwell, Chesterfield, Derbyshire S44 5PY
Tel: 01623 812191
Specialities: Annuals/Biennials/Bedding, Herbaceous Perennials, Shrubs

31 Glossop Garden Centre

Norfolk Street, Glossop, Derbyshire SK13 8BS
Tel: 01457 869140
Specialities: Garden Products

32 Golden Cottage Plants

Scarcliffe Lanes, Upper Langwith, Mansfield, Derbyshire NG20 9RQ
Tel: 07952 804077
Specialities: Annuals/Biennials/Bedding, Cottage Garden Plants

33 Grotto Farm Nursery

Main Street, Overseal, Swadlincote, Derbyshire DE12 6LG
Tel: 01283 760277
Specialities: Seasonal Plants, Herbaceous Perennials

34 Handley Rose Nurseries

Lightwood Road, Marsh Lane, Sheffield, Derbyshire S21 5RG
Tel: 01246 432921
Specialities: Rose

35 Hathersage Nurseries

Brook Cottage, Station Road, Hathersage, Hope Valley, Derbyshire S32 1DD
Tel: 01433 650886
Specialities: Herbaceous Perennials, Annuals/Biennials/Bedding, Shrubs

36 The Herb Garden

Hall View, Hardstoft, Pilsley, Chesterfield, Derbyshire S45 8AH
Tel: 01246 854268
Specialities: Herbs

37 High Peak Garden Centre

Hope Road, Bamford, Hope Valley, Derbyshire S33 0AL
Tel: 01433 651484
Specialities: Trees, Garden Products, Climbers

38 Highfield Nursery

Chesterfield Road, Barlborough, Chesterfield, Derbyshire S43 4TX
Tel: 01246 810636
Specialities: Seasonal Plants, Annuals/Biennials/Bedding

39 Hobson D.T & J.A

Palterton Lane, Sutton Scarsdale, Chesterfield, Derbyshire S44 5UT
Tel: 01246 240111

40 Hollies Farm Plant Centre

Uppertown, Bonsall, Matlock DE4 2AW
Tel: 01629 822734

Specialities:

Award-winning hanging baskets; Herbaceous perennials; Planted Containers

Description:

Set in glorious countryside 900ft above the village of Cromford, Hollies Farm Plant Centre specialises in herbaceous perennials, basket container and bedding plants, alpines, shrubs and compost – all at very reasonable prices. The Centre is also well-known for its colourful award-winning hanging baskets.

Directions:

From the A5012, take the turning towards Bonsall. Turn off the High Street into Uppertown Road and Hollies Farm Plant Centre can be found after about 350 yards.

41 J.C Poyser

Bancroft Nursery, Hopton, Wirksworth, Matlock, Derbyshire DE4 4DF
Tel: 01629 540554
Specialities: Annuals/Biennials/Bedding, Shrubs

42 Lanes Garden Centre

Near Meadow Farm, Breaston Lane, Risley, Derby, Derbyshire DE72 3TT
Tel: 01159 397128
Specialities: Garden Products, Annuals/Biennials/Bedding

43 Lea Rhododendron Gardens Ltd

Lea, Matlock, Derbyshire DE4 5GH
Tel: 01629 534380
Specialities: Rhododendron (Inc Azalea)

44 Leatherbritches Nursery

Fenny Bentley, Ashbourne, Derbyshire DE6 1LF
Tel: 07713 743295 or 017738 527090
e-mail: leatherbritchesnursery@yahoo.com
Specialities: Cottage garden plants, Shrubs, Garden design

45 Longdendale Nursery

Blackbrook, Chapel-en-Le-Frith, High Peak, Derbyshire SK23 0PU
Tel: 01298 813940
Specialities: Herbaceous Perennials, Shrubs

46 Longstone Hardy Plants Nursery

Stancil House, Barn Furlong, Great Longstone, Bakewell, Derbyshire DE45 1TR
Tel: 01629 640136
Specialities: Herbaceous Perennials, Hardy Plants

47 The Lorna Cross Landscapes Nursery

Nottingham Rd, Tansley, Matlock, Derbyshire DE4 5FY
Tel: 01629 582722
Specialities: Landscape Gardening, Shrubs, Trees

48 Low Road Nurseries

61 Low Road, Clowne, Chesterfield, Derbyshire S43 4LQ
Tel: 01246 810488
Specialities: Annuals/Biennials/Bedding, Herbaceous Perennials

49 Markeaton Garden Centre & Nursery

Markeaton Lane, Derby, Derbyshire DE22 4NH
Tel: 01332 292554

50 Matlock Garden Waterlife & Pet Centre

Nottingham Road, Tansley, Matlock, Derbyshire DE4 5FR
Tel: 01629 580500
Specialities: Aquatics, Water Products and Features

51 Mercaston Tree Company

Hollington Lane, Ednaston, Ashbourne, Derbyshire DE6 3AE
Tel: 01335 360947
Specialities: Trees

52 Meynell Langley Gardens

Lodge Lane (off Flagshaw Lane), Kirk Langley, nr Ashbourne, Derby DE6 4NT
Tel/Fax: 01332 824358
e-mail: enquiries@meynell-langley-gardens.co.uk
website: www.meynell-langley-gardens.co.uk

SEE ADVERTISEMENT PAGE 50

Specialities:

Bedding & Pot Plants; Herbaceous Perennials; Trees & Shrubs

Description:

A family-run nursery since 1941, Meynell Langley Gardens offers a superb range of quality plants coupled with sound advice from the people who grow them. Originally supplying the markets and local garden-centres, it now sells all of its plants from the nursery.

Directions:

From the A52 in Kirk Langley turn onto Flagshaw Lane and follow signs for Kedleston.

53 Mickleover Garden Centre

Hospital Lane, Mickleover, Derby, Derbyshire DE3 0DR
Tel: 01332 514350
Specialities: Restaurant/Coffee Shop, Annuals/Biennials/Bedding, Garden Products

54 The Mill Garden Centre

Whittington Way, Whittington Moor, Chesterfield, Derbyshire S41 9AG
Tel: 01246 260718
Specialities: Greenhouses and Sheds, Water Products and Features

55 Millthorpe Nursery

Millthorpe Lane, Holmesfield, Dronfield, Derbyshire S18 7SA
Tel: 01142 584007
Specialities: Annuals/Biennials/Bedding, Shrubs

56 Moss Valley Nurseries

88 Main Road, Ridgeway, Sheffield, Derbyshire S12 3XR
Tel: 01142 483343

57 New Leaf Plant Centre

Dyche Lane, Coal Aston, Dronfield, Derbyshire S18 3AA
Tel: 01246 413311

58 New Mills Garden Centre

The Bridge, Albion Road, New Mills, High Peak, Derbyshire SK22 3EX
Tel: 01663 744007

59 The Nurseries

Holt Road, Hackney, Matlock, Derbyshire DE4 2QD
Tel: 01629 733224
Specialities: Trees, Shrubs

60 Peak Cottage Plants

Wensley Road, Winster, Matlock, Derbyshire DE4 2DH
Tel: 01629 650428
Specialities: Annuals/Biennials/Bedding, Shrubs

61 Planters at Bretby

Bretby Lane, Bretby, Burton-on-Trent,
Derbyshire DE15 0QS
Tel: 01283 703355

62 Pleasley Vale Plant Nursery

Common Lane, Pleasley Vale, Mansfield,
Derbyshire NG19 8RZ
Tel: 01623 810188
Specialities: Annuals/Biennials/Bedding, Shrubs

63 Pomeroy Nursery

Street Farm, Flagg, Buxton, Derbyshire SK17 9QG
Tel: 01298 83245
Specialities: Annuals/Biennials/Bedding, Shrubs

64 Proctors Nurseries

Chatsworts Road, Chesterfield, Derbyshire S40 3PW
Tel: 01246 566554
Specialities: Annuals/Biennials/Bedding, Shrubs

65 Richards Nurseries

Windy Ridge, Tibshelf Road, Holmewood,
Chesterfield, Derbyshire S42 5TA
Tel: 01246 852596
Specialities: Annuals/Biennials/Bedding, Shrubs

66 Riverside Garden Centre

Makeney, Milford, Belper, Derbyshire DE56 0RU
Tel: 01332 841191

67 Riverside Garden Centre

Sheffield Road, Chesterfield, Derbyshire S41 9ED
Tel: 01246 261631
Specialities: Restaurant/Coffee Shop, Landscape Gardening

68 Riverside Herb Centre

Castleton Road, Hathersage, Hope Valley,
Derbyshire S32 1EG
Tel: 01433 651435
Specialities: Herbs

69 Scotland Nurseries

Stretton Road, Tansley, Matlock,
Derbyshire DE4 5GF
Tel: 01629 583036
Specialities: Shrubs, Herbaceous Perennials

70 Shipley Garden Centre

Hassock Lane North, Shipley, Heanor,
Derbyshire DE75 7JB
Tel: 01773 713596
Specialities: Garden Products, Garden Furniture, Herbaceous Perennials

71 Springbank Nurseries

Westfield Lane, Barlborough, Chesterfield,
Derbyshire S43 4TP
Tel: 01246 810393
Specialities: Shrubs, Annuals/Biennials/Bedding

72 Stanton Nurseries

Woodland Road, Stanton, Burton-on-Trent,
Derbyshire DE15 9TH
Tel: 01283 214144
Specialities: Herbaceous Perennials, Annuals/Biennials/Bedding, Trees

73 Summerfields Nurseries

Catton, Swadlincote, Derbyshire DE12 8LW
Tel: 01827 383214
Specialities: Annuals/Biennials/Bedding, Herbaceous Perennials

74 Swarkestone Nursery

Lowes Lane, Swarkestone, Derby,
Derbyshire DE73 7GQ
Tel: 01332 700800
Specialities: Annuals/Biennials/Bedding, Herbaceous Perennials, Shrubs

75 Sycamore Farm

Brook Lane, Foston, Derby, Derbyshire DE65 5PW
Tel: 01283 815635

76 T Cutler

41 Derby Road, Lower Kilburn, Belper,
Derbyshire DE56 0NG
Tel: 01332 881080

77 Tissington Nursery

Tissington, Ashbourne, Derbyshire DE6 1RA
Tel: 01335 390650
Specialities: Annuals/Biennials/Bedding, Herbaceous Perennials

78 Troway Halls Honey Trail & Nursery

Troway Hall, Troway Marsh Lane, Sheffield,
Derbyshire S21 5RU
Tel: 01246 292425
Specialities: Xmas Trees

79 Unstone Plant Centre

Main Road, Unstone, Dronfield, Derbyshire S18 4AB
Tel: 01246 414490

80 Van Dyke's Garden Centre

Worksop Road, Clowne, Chesterfield,
Derbyshire S43 4TD
Tel: 01246 813388

81 Van Gemeren Nureseries and Garden Centre

Mansfield Road, Hasland, Chesterfield,
Derbyshire S41 0JL
Tel: 01246 234710

82 The Victorian Gardens Nursery

Cressbrook, Buxton, Derbyshire SK17 8SY
Tel: 01298 871552
Specialities: Annuals/Biennials/Bedding, Shrubs

83 Ward's Nurseries & Garden Centre Ltd

Eckington Road, Coal Aston, Dronfield,
Derbyshire S18 3BA
Tel: 01246 412622
Specialities: Annuals/Biennials/Bedding, Garden Furniture

84 Wessington Garden Centre

Matlock Road, Wessington, Alfreton,
Derbyshire DE55 6DS
Tel: 01773 832517
Specialities: Garden Furniture, Garden Products

85 Whitelea Nursery

Whitelea Lane, Tansley, Matlock,
Derbyshire DE4 5FL
Tel: 01629 55010
Specialities: Shrubs, Herbaceous Perennials

86 Wingerworth Garden Centre

Birkin Lane, Wingerworth, Chesterfield,
Derbyshire S42 6RD
Tel: 01246 204214

87 Woodgrow Horticulture Ltd

Landscapes Supply Depot, Burton Road, Findern,
Derby, Derbyshire DE65 6BE
Tel: 01332 517600
Specialities: Trees, Shrubs

88 Woodhouse Farm Nurseries

Woodhouses, Melbourne, Derby,
Derbyshire DE73 8DN
Tel: 01332 862645
Specialities: Seasonal Plants, Herbaceous Perennials

89 Woolley Moore Nurseries Ltd

Ashover Rd, Woolley Moor, Alfreton,
Derbyshire DE55 6FF
Tel: 01246 590320
Specialities: Shrubs, Seasonal Plants

90 Wyaston Nursery

Nursery Cottage, Wyaston, Ashbourne,
Derbyshire DE6 2DR
Tel: 01335 346164

91 Wyevale Garden Centre (Derby)

Burton Road, Findern, Derby, Derbyshire DE65 6BE
Tel: 01332 514268

DEVON

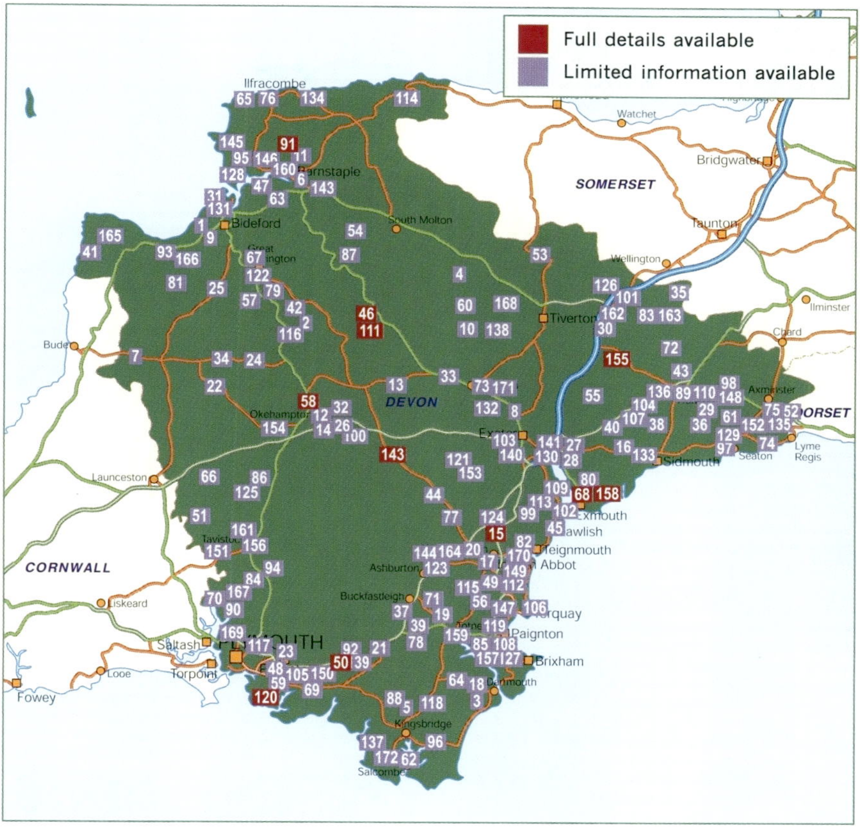

1 Abbeycombe Gardens & Nursery

Abbotsham, Bideford, Devon EX39 5AP
Tel: 01237 424264
Specialities: Seasonal Plants, Annuals/Biennials/Bedding, Shrubs

2 Alba Trees Plc

Barton Cottage Dowland, Dowland, Winkleigh, Devon EX19 8PB
Tel: 01805 804218
Specialities: Trees

3 Ash Tree Farm Nursery

Ash Tree Farm, Ashburton, Dartmouth, Devon TQ6 0LR
Tel: 01803 712437
Specialities: Annuals/Biennials/Bedding, Shrubs

4 Ashmoor Nursery

Lower Ash Moor, Rose Ash, South Molton, Devon EX36 4RF
Tel: 01884 860355
Specialities: Annuals/Biennials/Bedding, Shrubs, Herbaceous Perennials

5 Avon Mill Garden Centre

Woodleigh Road, Loddiswell, Kingsbridge,
Devon TQ7 4DD
Tel: 01548 550338

6 B J 's Garden Centre

Unit 6-7, Seven Brethren Bank, Sticklepath,
Barnstaple, Devon EX31 2AS
Tel: 01271 327966
Specialities: Garden Products

7 Barton Gate Nursery

Barton Gate, Pancrasweek, Holsworthy,
Devon EX22 7JT
Tel: 01288 381252
Specialities: Annuals/Biennials/Bedding, Shrubs

8 Bernaville Nurseries

Three Horseshoes, Cowley, Exeter, Devon EX5 5EU
Tel: 01392 851326

9 Bideford Garden Centres

Clovelly Rd Industrial Estate, Bideford,
Devon EX39 3HN
Tel: 01237 424348
Specialities: Garden Furniture, Garden Products, Water Products and Features

10 The Big Grass Co.

Hookhill Plantation, Woolfardisworthy, Crediton,
Devon EX17 4RX
Tel: 01363 866146
Specialities: Turf

11 Blakewell Water Garden Centre

Muddiford, Barnstaple, Devon EX31 4ET
Tel: 01271 344533
Specialities: Aquatics, Water Products and Features, Garden Furniture

12 Bostock Garden Centre

North Road, North Tawton, Okehampton,
Devon EX20 1BQ
Tel: 01837 53248

13 Bow Aquatic Centres

Bow, Crediton, Devon EX17 6LA
Tel: 01363 82438
Specialities: Aquatics, Water Products and Features

14 Bowden Hostas

Sticklepath, Okehampton, Devon EX20 2NL
Tel: 01837 840989
Specialities: Hosta

15 Briartor Nurseries

Ideford Combe, Newton Abbot,
Devon TQ12 3GR
Tel: 01626 354567
e-mail: linda.blight1@virgin.net

Specialities:
Fuchsias & Geraniums; Hanging Baskets & Patio Containers; Pot Plants

Products and Services:
- Plants

Directions:
Briartor Nurseries are located on the B3195 at Ideford Combe near the village of Kingsteignton.

Description:
Owners Geoff and Linda Blight started their business more than 40 years ago. They began by growing tomatoes and flowers and then extended their range to specialise in bedding and pot plants to sell to the trade. They have recently opened their retail outlet. The site covers about an acre and includes modern aluminium glasshouses, the retail area and a parking area. All the plants for sale are healthy and vigorous, and grown on site. They include a full range of seasonal bedding and patio plants. The nursery also sells hanging baskets and patio containers.

16 Brooklands Cross Nursery

Newton Poppleford, Sidmouth, Devon EX10 0BY
Tel: 01395 567030
Specialities: Annuals/Biennials/Bedding, Herbaceous Perennials

17 Buckfast Nursery

Unit 29 Market Hall, Market Street, Newton Abbot, Devon TQ12 2RJ
Tel: 01626 201009
Specialities: Annuals/Biennials/Bedding, Shrubs

18 Bugford Nursery

Bugford, Dartmouth, Devon TQ6 0LT
Tel: 01803 712490
Specialities: Seasonal Plants, Annuals/Biennials/Bedding

19 Bunch of Herbs

Staverton Bridge, Staverton, Totnes, Devon TQ9 6NU
Tel: 01803 762678
Specialities: Herbs

20 Burnham Nurseries

Forches Cross, Newton Abbot, Devon TQ12 6PZ
Tel: 01626 352233
Specialities: Orchids

21 C J S Rosegrowers

Station Cottage, Wrangaton, South Brent, Devon TQ10 9HG
Tel: 01364 72297
Specialities: Rose

22 Chapel Farm House Nursery

Halwill Junction, Halwill, Beaworthy, Devon EX21 5UF
Tel: 01409 221594
Specialities: Specialist Bedding

23 Chaplins Garden Centre

Galileo Close, Newnham Industrial Estate Plympton, Plymouth, Devon PL7 4JW
Tel: 01752 202700
Specialities: Xmas Trees, Restaurant/Coffee Shop

24 Charles Dumpleton

Highampton, Beaworthy, Devon EX21 5LE
Tel: 01409 231302

25 Cherrywood Plant Centre

Cherrywood Park, Langtree, Torrington, Devon EX38 8LQ
Tel: 01805 601504
Specialities: Seasonal Plants, Specimen Plants

26 Cleave House

Sticklepath, Okehampton, Devon EX20 2NL
Tel: 01837 840481
Specialities: Hosta

27 Clyst St Mary Garden Centre

Sidmouth Road, Clyst St. Mary, Exeter, Devon EX5 1AE
Tel: 01392 876281

28 Clyst Valley Seeds

104 Clyst Valley Road, Clyst St. Mary, Exeter, Devon EX5 1DE
Tel: 01392 874543
Specialities: Seeds, Flower, Vegetable

29 Colour Wheel Plants

Tower Road, Tower Cross, Honiton, Devon EX14 9TN
Tel: 01404 42975
Specialities: Herbaceous Perennials, Seasonal Plants

30 Cottage Garden Nursery

Uffculme Road, Willand, Cullompton, Devon EX15 2SA
Tel: 01884 820971
Specialities: Cottage Garden Plants, Annuals/Biennials/Bedding

31 Cottage Garden Plants and Herbs

4 Lundy View, Northam, Bideford, Devon EX39 1BE
Tel: 01237 470370
Specialities: Cottage Garden Plants, Herbs

32 Country Lanes Garden Centre

Exeter Road, Stockley, Okehampton, Devon EX20 1QH
Tel: 01837 52489
Specialities: Aquatics, Water Products and Features

33 Crediton Garden Centre

Barnstable Cross, Crediton, Devon EX17 2ER
Tel: 01363 773110
Specialities: Rose, Herbaceous Perennials

34 Crooked Oak Nursery

Brandis Corner, Holsworthy, Devon EX22 7YA
Tel: 01409 221638
Specialities: Herbaceous Perennials

35 Culm View Nursery

Waterloo Farm, Clayhidon, Cullompton, Devon EX15 3TN
Tel: 01823 680698
Specialities: Herbaceous Perennials

36 Dale Horticraft

1 Dale View, Northleigh, Colyton, Devon EX24 6BS
Tel: 01404 871394

37 Dean Prior Garden Centre

Whites Farm, Lower Dean, Buckfastleigh, Devon TQ11 0LS
Tel: 01364 644451
Specialities: Pots, Restaurant/Coffee Shop, Garden Furniture

38 Devon Violet Nursery

10 Sunny Hill, Ottery St. Mary, Devon EX11 1DZ
Tel: 01404 813701

39 Devon Violet Nursey

Rattery, South Brent, Devon TQ10 9LG
Tel: 01364 643033
Specialities: Santipaulia (African Violet)

40 Devonshire Lavenders & Herbs

Ottery St. Mary, Devon EX11 1JZ
Tel: 01404 822223
Specialities: Herbs, Lavender

41 Docton Mill Gardens

Docton Mill, Lymebridge, Hartland, Bideford, Devon EX39 6EA
Tel: 01237 441369

42 Dolton Nursery

Rear of Buckland Farm, Dolton, Winkleigh, Devon EX19 8RE
Tel: 01805 603618

43 Don Hatch Conifers

Combe Raleigh, Honiton, Devon EX14 4TQ
Tel: 01404 42981
Specialities: Conifers

44 East Wrey Nursery

Moretonhampstead Road, Lustleigh, Newton Abbot, Devon TQ13 9SN
Tel: 01647 277437
Specialities: Pot Plants, Annuals/Biennials/Bedding

45 Easter Hill Nursery

77 Exeter Road, Dawlish, Devon EX7 0AG
Tel: 01626 864399
Specialities: Shrubs, Herbaceous Perennials

46 Eggesford Gardens

Eggesford, Chulmleigh, Devon EX18 7QU
Tel: 01769 580250 Fax: 01769 581041
e-mail: enquiries@eggesfordgardens.com
website: www.eggesfordgardens.com

SEE ADVERTISEMENT PAGE 57

Specialities:

Herbaceous Perennials; Wildlife Gardening; Fuchsias/Pelargoniums

Products and Services:

- Barbeques
- Books and Stationery
- Camping Equipment
- Seasonal Displays
- Clothing
- Credit Cards Accepted
- Disabled Facilities
- Display Gardens
- Dried & Artificial Flowers
- Fruit Trees
- Garden Construction
- Garden Design
- Garden Furniture
- Garden Products
- Giftware
- Houseplants
- Information Desk
- Large Car Park
- Paving
- Pet Products
- Plant Guarantee
- Plants
- Restaurant/Coffee Shop
- Seeds and Bulbs
- Trees

Directions:

Eggesford Gardens is located just of the A377 Exeter to Barnstaple Road at Eggesford Railway Station.

47 Egmont Water Gardens

1 Griggs Garden, Fremington, Barnstaple, Devon EX31 3BQ
Tel: 01271 321424

48 Elburton Flowers Ltd

Vinery Lane, Plymouth, Devon PL9 8DD
Tel: 01752 492939
Specialities: Floristry, Cut Flowers

49 Elmcroft Gardens

Old Newton Road, Kingskerswell, Newton Abbot, Devon TQ12 5LB
Tel: 01803 872526

50 Endsleigh Garden & Leisure

Ivybridge, Devon PL21 9JL
Tel: 01752 898989
website: www.g-l.co.uk

Specialities:

Plants for Dry or Shady areas, Fragrant Flowering Plants, Container Plants

Products and Services:

- Aquatics
- Barbeques
- Books & Stationery
- Clothing
- Credit Cards Accepted
- Disabled Facilities
- Display Gardens
- Garden Furniture
- Garden Products
- Giftware
- Houseplants
- Information Desk
- Large Car Park
- Pet Products
- Plants
- Plant Guarantee
- Restaurant/Coffee Shop
- Seasonal Displays
- Seeds and Bulbs

SEE ADVERTISEMENT PAGE 58

Directions:

Take the Ivybridge exit off the A38. Not far from Plymouth or Exeter.

51 Endsleigh Gardens

Milton Abbot, Tavistock, Devon PL19 0PG
Tel: 01822 870235
Specialities: Trees, Shrubs

52 Entwood Farm Plants

Uplyme, Harcombe, Lyme Regis, Devon DT7 3RN
Tel: 01297 444034
Specialities: Grasses, Herbaceous Perennials

53 Exbridge Nurseries

Exbridge, Dulverton, Devon TA22 9AZ
Tel: 01398 323503

54 Fairhaven Nursery

Chittlehampton, Umberleigh, Devon EX37 9QT
Tel: 01769 540528
Specialities: Annuals/Biennials/Bedding, Seasonal Plants

55 Feebers Hardy Plants

1 Feeber Cottage, Westwood, Broadclyst, Exeter, Devon EX5 3DQ
Tel: 01404 822118
Specialities: Herbaceous Perennials, Hardy Plants

56 Fermoys Garden Centre

Totnes Road, Ipplepen, Newton Abbot, Devon TQ12 5TN
Tel: 01803 813504

57 Fernwood Nursery

Peters Marland, Torrington, Devon EX38 8QG
Tel: 01805 601446
Specialities: Annuals/Biennials/Bedding, Herbaceous Perennials

58 Folly Gate Plants

The Old Post Office, Folly Gate,
nr Okehampton, Devon EX20 3AF
Tel: 01837 659164
e-mail: perrylamb@hotmail.com

Specialities:

Hardy Herbaceous Perennials; Grasses; Shrubs

Directions:

The hamlet of Folly Gate is situated 2 miles north of Okehampton on the A386 to Hatherleigh. A small lane runs beside the Crossways Inn and Folly Gate Plants can be found 100 yards down this lane on the right. Customers are welcome to use the car park located through the black gates beside the house.

Description:

The owner of Folly Gate Plants, Sara Lamb, is a trained horticulturalist and her small friendly home business offers a more unusual selection of hardy plants including geraniums, kniphofias, euphorbias, grasses and herbs.

59 Fordbrook Nursery

Brixton, Plymouth, Devon PL8 2BQ
Tel: 01752 880315
Specialities: Annuals/Biennials/Bedding, Shrubs

60 G J Webber

New House, Witheridge, Tiverton, Devon EX16 8NU
Tel: 01884 860229

61 The Garden Shop

Tannery Yard, King Street, Colyton, Devon EX24 6PD
Tel: 01297 551113

62 The Garden Store

67 Fore Street, Salcombe, Devon TQ8 8BU
Tel: 01548 844449

63 Gardeners Delight Nursery

Rooksbridge, Old Bideford Road, Bickington, Barnstaple, Devon EX31 2PA
Tel: 01271 861461
Specialities: Annuals/Biennials/Bedding, Shrubs

64 Gardentime Limited

Hemborough Post, Blackawton, Totnes, Devon TQ9 7DE
Tel: 01803 712345
Specialities: Annuals/Biennials/Bedding

65 The Gate House

Lee, Ilfracombe, Devon EX34 8LR
Tel: 01271 862409

66 Glenvale Nursery

North Road, Lifton, Devon PL16 0DS
Tel: 01566 784719
Specialities: Seasonal Plants, Annuals/Biennials/Bedding

67 Greenfields Nursery

Calvesford Road, Torrington, Devon EX38 7DQ
Tel: 01805 622685
Specialities: Shrubs, Annuals/Biennials/Bedding

68 Greenfingers Garden Centre

Pound Lane, Exmouth, Devon EX8 3JT
Tel: 01395 274632
e-mail: info@greenfingers-gardencentre.co.uk
website: www.greenfingers-gardencentre.co.uk

SEE ADVERTISEMENT BELOW

Specialities:

Shrubs; Trees; Roses

Products and Services:

- Aquatics
- Barbecues
- Seasonal Displays
- Conservatories
- Credit Cards Accepted
- Disabled Facilities
- Display Gardens
- Fencing
- Fruit Trees
- Garden Construction
- Garden Furniture
- Garden Products
- Greenhouses and Sheds
- Houseplants
- Large Car Park
- Paving
- Plant Guarantee
- Plants
- Seeds and Bulbs
- Trees

Directions:

Approaching Exmouth on the A376 turn left into Hulham Road which is just before the traffic lights. Pound Lane is the right turn off the first mini roundabout.

70 Hewton Nursery

Bere Alston, Bere Alston, Yelverton,
Devon PL20 7BW
Tel: 01822 840460
Specialities: Hardy Plants, Annuals/Biennials/Bedding

71 Hill House Nursery & Tearooms

Landscove, Ashburton, Newton Abbot,
Devon TQ13 7LY
Tel: 01803 762273
Specialities: Annuals/Biennials/Bedding, Seasonal Plants

72 Hill View Nursery

Dunkeswell, Honiton, Devon EX14 4SZ
Tel: 01404 891376
Specialities: Annuals/Biennials/Bedding, Herbaceous Perennials

73 Home Farm Plants

Devonshire Traditional Breed Centre, Downes,
Crediton, Devon EX17 3PL
Tel: 01363 772430
Specialities: Herbaceous Perennials

74 Hortus Nursery

Shrubbery Bungalow, Rousdon, Lyme Regis,
Devon DT7 3XW
Tel: 01297 444019
Specialities: Landscape Gardening, Annuals/Biennials/Bedding, Shrubs

75 Hunters Croft Nursery

Charmouth Road, Axminster, Devon EX13 5ST
Tel: 01297 33366

76 Ilfracombe Garden Shop

11 Portland St, Ilfracombe, Devon EX34 9NL
Tel: 01271 867002
Specialities: Garden Products

77 Jeffery's Country Store

Station Yard, Station Road, Bovey Tracey,
Newton Abbot, Devon TQ13 9AL
Tel: 01626 832250

78 Joyners Plants Ltd

Cockwells Nursery, Harberton, Totnes,
Devon TQ9 7LP
Tel: 01364 73958

79 Kenwith Nursery

Blinsham Beaford, Beaford, Winkleigh,
Devon EX19 8NT
Tel: 01805 603274
Specialities: Annuals/Biennials/Bedding, Shrubs

80 Kings Garden & Leisure Ltd

Higher Hulham Road, Exmouth, Devon EX8 5DZ
Tel: 01395 271911
Specialities: Annuals/Biennials/Bedding

81 Kismeldon Nursery

West Putford, Holsworthy, Devon EX22 7XH
Tel: 01409 241311
Specialities: Shrubs, Seasonal Plants

82 Lea Park Nurseries

Ash Hill, Bishopsteignton, Teignmouth,
Devon TQ14 9PY
Tel: 01626 775237
Specialities: Seasonal Plants, Annuals/Biennials/Bedding

83 Little Southey

Northcott Ashill, Cullompton, Devon EX15 3LT
Tel: 01884 840545

84 Long Ash Garden Centre

Coppice Town Road, Yelverton, Devon PL20 7LL
Tel: 01822 852840
Specialities: Annuals/Biennials/Bedding, Seasonal Displays

85 Longcombe Nursery & Garden Centre

Longcombe Cross, Longcombe, Totnes,
Devon TQ9 6PP
Tel: 01803 863098
Specialities: Annuals/Biennials/Bedding, Floristry

86 Lydford Alpine Nursery

2 Southern Cottages, Lydford, Okehampton,
Devon EX20 4BL
Tel: 01822 820398
Specialities: Alpines/Rock Plants, Herbaceous Perennials

87 M & M Plants

Lloret, Chittlehamholt, Umberleigh,
Devon EX37 9PD
Tel: 01769 540448
Specialities: Seasonal Plants, Shrubs

88 M G M Nurseries

Brent Road, Loddiswell, Kingsbridge,
Devon TQ7 4DU
Tel: 01548 550754
Specialities: Hardy Plants

89 M G Webber

30 Hale Lane, Honiton, Devon EX14 1HN
Tel: 01404 42711

90 Maristow Nursery Gardens

Maristow, Roborough, Plymouth, Devon PL6 7BZ
Tel: 01752 736779

91 Marwood Hill Gardens

Marwood, Barnstaple, North Devon EX31 4EB
Tel: 01271 342528
e-mail: info@marwoodhillgarden.co.uk
website: www.marwoodhillgarden.co.uk

SEE ADVERTISEMENT ABOVE

Specialities:

Herbaceous plants, trees and shrubs.

Description:

Over the course of some 30 years, 18 acres of pastureland have been transformed into a spectacular water garden with 3 lakes that are surrounded by herbaceous plants, trees and shrubs. The Gardens also contain the largest bog garden in the West Country.

Directions:

From the A361, 3 miles northwest of Barnstaple, follow the brown and white tourist signs for the gardens which are near the village of Guineaford.

92 Meadow Cottage Plants

Pitt Hill, Ivybridge, Devon PL21 0JJ
Tel: 01752 894532

93 Merry Harriers Nurseries

Woolsery, Bideford, Devon EX39 5QH
Tel: 01237 431611

94 Monkswell Farm Garden Centre

Monkswell Farm, Horrabridge, Yelverton, Devon PL20 7QY
Tel: 01822 853603
Specialities: Garden Furniture, Garden Products

95 Moor Lane Nursery

Saunton Rd, Braunton, Devon EX33 1HG
Tel: 01271 814171
Specialities: Annuals/Biennials/Bedding, Shrubs

96 Mr. Mervyn Thomas

Aintree, Green Park Way, Chillington, Kingsbridge, Devon TQ7 2HY
Tel: 01548 580381
Specialities: Aster (Michaelmas Daisy)

97 Mrs S Reynolds

Blue Firs, Wessiters Drive, Seaton, Devon EX12 2PJ
Tel: 01297 625342
Specialities: Penstemon

98 Nursery of Miniatures

Hilltop, Hutgate Road, Honiton, Devon EX14 9UX
Tel: 01404 42617
Specialities: Bonsai

99 Obelisk Gardens Flowers & Plants

Obelisk Hill, Mamhead, Exeter, Devon EX6 8HG
Tel: 01626 866236
Specialities: Annuals/Biennials/Bedding, Shrubs, Floristry

100 Okehayes Nursery

Okehayes, Sticklepath, Okehampton, Devon EX20 2NT
Tel: 01837 840687

101 Old Well Garden Centre

Waterloo Cross, Uffculme, Cullompton, Devon EX15 3ES
Tel: 01884 840873
Specialities: Garden Products

102 Olde Barn Nurseries

Exeter Rd, Dawlish, Devon EX7 0NZ
Tel: 01626 890445

103 Orchard Nurseries

Tedburn Road, Whitestone, Exeter, Devon EX4 2HD
Tel: 01392 811508
Specialities: Annuals/Biennials/Bedding, Herbaceous Perennials

104 Otter Nurseries

Gosford Road, Ottery St. Mary, Devon EX11 1LZ
Tel: 01404 815815
Specialities: Shrubs, Annuals/Biennials/Bedding

105 Otter Nurseries

Chittleburn Hill, Brixton, Plymouth, Devon PL8 2BH
Tel: 01752 405422

106 Otter Nurseries

250 Babbacombe Road, Torquay, Devon TQ1 3TA
Tel: 01803 214294

107 The Palm House

8 North Street, Ottery St. Mary, Devon EX11 1DR
Tel: 01404 815450
Specialities: Hardy Garden Palms, Exotics

108 Park Bay Centre

Brixham Road, Paignton, Devon TQ4 7BA
Tel: 01803 666682
Specialities: Specimen Plants, Garden Furniture, Palms

109 The Perfume Garden Plant Centre

Powderham Park, Powderham, Exeter, Devon EX6 8JQ
Tel: 01626 891456
Specialities: Annuals/Biennials/Bedding, Herbaceous Perennials

110 Perrie Hale Forest Nursery

Northcote Hill, Honiton, Devon EX14 9TH
Tel: 01404 43344

111 Pine Cottage Plants

Pine Cottage, Fourways, Eggesford, nr Chulmleigh, Devon EX18 7QZ
Tel: 01769 580076
e-mail: pcplants@supanet.com
website: www.pcplants.co.uk

Specialities:

Agapanthus and other South African Bulbs; Trees; Shrubs

Description:

This small nursery overlooking the Taw Valley grows a limited range of trees and shrubs, some hardy and tender perennials, but its main speciality is its National Collection of Agapanthus of which it stocks a large range.

Directions:

The nursery is located almost midway between Exeter and Barnstaple, one mile up a hill off the A377 at Eggesford station.

112 Plant World

St Marychurch Road, Coffinswell, Newton Abbot, Devon TQ12 4SE
Tel: 01803 872800
Specialities: Seeds, Herbaceous Perennials

113 Plants Direct South West Ltd

Mowlish Manor, Kenton, Exeter, Devon EX6 8HS
Tel: 01626 891240

114 Plantwise Nursery

Kibsworthy Farm, Lynton, Devon EX35 6JT
Tel: 01598 753766
Specialities: Seasonal Plants, Annuals/Biennials/Bedding

115 Pleasant View Nursery

Two Mile Oak, Ogwell, Newton Abbot, Devon TQ12 6DG
Tel: 01803 813388
Specialities: Annuals/Biennials/Bedding, Shrubs

116 Plovers Plants

Plovers Barn, Meeth, Okehampton, Devon EX20 3EP
Tel: 01837 810577
Specialities: Shrubs, Herbaceous Perennials

117 Plymouth Garden Centre

Fort Austin Avenue, Plymouth, Devon PL6 5NU
Tel: 01752 771820
Specialities: Annuals/Biennials/Bedding, Greenhouses and Sheds, Fencing

118 Potters Nurseries

East Allington, Totnes, Devon TQ9 7PZ
Tel: 01548 521274
Specialities: Hardy Plants, Herbaceous Perennials

119 The Potting Shed Nursery

Moles Lane, Marldon, Paignton, Devon TQ3 1SY
Tel: 01803 872809
Specialities: Shrubs, Herbaceous Perennials

120 Pounsley Plants

Pounsley Combe, Spriddlestone, Brixton, Plymouth, Devon PL9 0DW
Tel: 01752 402873
e-mail: pou599@aol.com
website: www.pounsleyplants.com

Specialities:

Old Roses; Hardy South African Plants; Unusual Perennials

Description:

Jane Hollow trained at Bicton College as a mature student and started her business in 1991. A year later she decided to specialise in roses of which she has a lifetime's knowledge. She is always happy to share her love and experience of roses with her customers.

Directions:

West of Brixton on A379 turn South signposted Brixton Torr and Otter Garden Centre. Continue for 1 mile. At brow of hill turn left, follow nursery signs. Grid ref: SX521538

121 Reedy Meadow Nursery

Dunsford, Exeter, Devon EX6 7AB
Tel: 01647 253380
Specialities: Annuals/Biennials/Bedding, Shrubs

122 RHS Rosemoor Plant Centre

RHS Gardens, Rosemoor, Great Torrington, Devon EX38 8PH
Tel: 01805 624067

123 Riseley Nursery

Goodstone House, Bickington, Newton Abbot, Devon TQ12 6LJ
Tel: 01626 821494
Specialities: Annuals/Biennials/Bedding, Pot Plants

124 Rock Nursery

Station Hill, Chudleigh, Newton Abbot, Devon TQ13 0EE
Tel: 01626 852134
Specialities: Annuals/Biennials/Bedding, Conifers

125 Rowden Gardens

Brentor, Tavistock, Devon PL19 0NG
Tel: 01822 810275
Specialities: Aquatics, Rare and Unusual Plants

126 Sampford Shrubs

Westleigh, Tiverton, Devon EX16 7EW
Tel: 01884 821164
Specialities: Shrubs

127 Samuel Dobie & Son

Long Road, Paignton, Devon TQ4 7SX
Tel: 08701 123623

128 Sandy Lane Farm Nursery

Velator, Braunton, Devon EX33 2NU
Tel: 01271 812288
Specialities: Annuals/Biennials/Bedding, Shrubs

129 Scalwell Lane Nursery

54 Scalwell Lane, Seaton, Devon EX12 2DL
Tel: 01297 22738
Specialities: Seasonal Plants, Specimen Plants

130 Seabrook Nursery

Topsham Road, Exeter, Devon EX2 7DT
Tel: 01392 875070

131 The Secret Garden

Cross Street, Northam, Bideford, Devon EX39 1BS
Tel: 01237 474403
Specialities: Floristry

132 Sherwood Nursery

Newton St. Cyres, Exeter, Devon EX5 5BT
Tel: 01392 851216
Specialities: Buddleja (Buddleia), Magnolia, Seasonal Plants

133 Sidmouth Garden Centre

Stowford, Sidmouth, Devon EX10 0NA
Tel: 01395 516142

134 Silverdale Nurseries

Shute Lane, Combe Martin, Ilfracombe, Devon EX34 0EU
Tel: 01271 882539
Specialities: Shrubs, Herbaceous Perennials

135 Sinclair Garden Nursery

Yawl Hill Lane, Uplyme, Lyme Regis, Devon DT7 3RP
Tel: 01297 443079
Specialities: Annuals/Biennials/Bedding, Shrubs

136 Slades Countrywise Ltd

Turks Head Corner, Exeter Road, Honiton, Devon EX14 1AZ
Tel: 01404 45966

137 South Milton Nurseries

South Milton, Kingsbridge, Devon TQ7 3JP
Tel: 01548 560303
Specialities: Annuals/Biennials/Bedding, Herbaceous Perennials

138 South Yeo Nursery

South Yeo Farmhouse, Poughill, Crediton, Devon EX17 4LF
Tel: 01363 866740
Specialities: Annuals/Biennials/Bedding, Shrubs

139 Springfield Nursery

Filham, Ivybridge, Devon PL21 0DN
Tel: 01752 894034
Specialities: Annuals/Biennials/Bedding, Seasonal Plants

140 Springwell Nursery

Old Ide Lane, Ide, Exeter, Devon EX2 9RY
Tel: 01392 410545

141 St Bridget Nurseries

Old Rydon Lane, Exeter, Devon EX2 7JY
Tel: 01392 873672

142 St Johns Garden Centre

St. Johns Lane, Barnstaple, Devon EX32 9DD
Tel: 01271 343884

143 Stone Lane Gardens, Tree Nursery & Arboretum

Stone Farm, Chagford, Devon TQ13 8JU
Tel: 01647 231311
e-mail: june@mythicgarden.eclipse.co.uk
website: www.mythicgarden.com

Specialities:

Birch & Alder; Wild Origin – Bare Root and in Pots; The Mythic Garden; Sculpture Exhibition

Products and Services:

- Large Car Park
- Mail Order

Directions:

From the north, east or west, leave the A30 at Whiddon Down junction and take the Moretonhampstead road (A382). Immediately upon leaving Whiddon Down, take the left turn for Drewsteignton. Follow the signs for Stone Lane Gardens. From the south, take the A382 from Moretonhampstead heading towards Whiddon Down. After passing Sandy Park, take the next right turn and follow the signs for Stone Lane Gardens. OS Map 191. Grid 708908.

Description:

Stone Lane Gardens is a family-owned wholesale and retail specialist nursery growing Birch and Alder trees from around the world. It has an acclaimed 5-acre arboretum with National Collections of wild origin Birch and Alder. Visitors can view mature trees in the attractive garden before buying. There's a wide selection of many rare and unusual species and clones. Helpful, friendly advice on selection and planting is

available from one of the country's leading birch and alder experts. The nursery also has a choice selection of specially selected grafted cultivars available. The birches in particular are attractive ornamental trees with a stunning variety of coloured barks and good autumn leaf colour, whilst the alders produce an array of abundant catkins. Another major attraction here is The Mythic Garden, an imaginative sculpture exhibition which is open from May to September.

144 Stormsdown Nursery

Bickington, Newton Abbot, Devon TQ12 6LQ
Tel: 01626 821507

145 Streamways

Georgeham, Braunton, Devon EX33 1JQ
Tel: 01271 890215
Specialities: Hanging Baskets

146 Stuart Holder

Inanda, Lower Park Road, Braunton,
Devon EX33 2LQ
Tel: 01271 815269

147 Styles Garden Centre

Moles Lane, Marldon, Paignton, Devon TQ3 1SY
Tel: 01803 873056

148 Sundial Garden

A35 Offwell, Honiton, Devon EX14 9RR
Tel: 01404 831549
Specialities: Restaurant/Coffee Shop, Fuchsia

149 Sunray Plants

Haccombe, Newton Abbot, Devon TQ12 4SJ
Tel: 01626 873776
Specialities: Seasonal Plants, Hardy Plants

150 Sunridge Nursery

Worston, Yealmpton, Plymouth, Devon PL8 2LN
Tel: 01752 880438
Specialities: Annuals/Biennials/Bedding, Herbaceous Perennials

151 Tamar Organics

Gulworthy, Tavistock, Devon PL19 8JE
Tel: 01822 834887
Specialities: Organic, Organic Trees

152 Tarhan Nursery

Shiphay Farm, Colyton, Devon EX24 6EU
Tel: 01297 553570

153 Teign Valley Nursery

Bridford Mill, Bridford, Exeter, Devon EX6 7LB
Tel: 01647 252654
Specialities: Trees, Herbaceous Perennials, Conifers

154 Thorndon Cross Nursery

Thorndon Cross, Okehampton, Devon EX20 4NF
Tel: 01837 861347
Specialities: Annuals/Biennials/Bedding, Specimen Plants

155 Thornhayes Nursery

St Andrews Wood, Dulford, Cullompton,
Devon EX15 2DF
Tel: 01884 266746 Fax: 01884 266739
e-mail: trees@thornhayes-nursery.co.uk
website: www.thornhayes-nursery.co.uk

Specialities:

Fruit Trees, Ornamental Trees, Hedging

Description:

Thornhayes offers one of the largest ranges of trees available in the UK. Free descriptive catalogue (catalogue also online). Delivery throughout UK and Ireland. Expert advice from experienced staff. Consultancy and design service available. Open Mon-Fri 8am-4pm, Sat (Sept-Apr) 9.30-2pm.

Directions:

Three miles southeast of Cullompton, off the A373, signposted from Dulford Cross. Location map on website and in catalogue.

156 Thyme For Herbs

9 Hawthorn Road, Tavistock, Devon PL19 9DL
Tel: 01822 613624
Specialities: Herbs

157 Torbay and South Hams Garden Centre

Stoke Road, Stoke Gabriel, Totnes, Devon TQ9 6QB
Tel: 01803 782882

158 Total Eden

Knowle Orchards, Knowle Hill, Budleigh Salterton, Devon EX9 7AL
Tel: 01395 444455
website: www.hardyexoticplants.co.uk

SEE ADVERTISEMENT BELOW

Specialities:

Palms, Bamboos, Hardy Exotics, Landscaping.

Description:

Their aim is to open eyes to the vast possibilities for our climate and inspire a new generation of gardens. An exiting, exotic looking garden is possible even in the depths of winter. Let them help you realise your gardens potential and let your imagination run.

Directions:

From Exmouth take Salterton Road (B3178) for 2.48 miles. At the roundabout, take the second exit onto Knowle Hill (B3178) toward Budleigh Salterton. Total Eden is signed on the right.

159 Totnes Pet & Garden Ltd

41 Fore Street, Totnes, Devon TQ9 5HN
Tel: 01803 863370

160 Trelawney at Ashford

Strand Lane, Ashford, Barnstaple, Devon EX31 4BW
Tel: 01271 342880

161 Triffid Nurseries

22 Glanville Rd, Tavistock, Devon PL19 0EB
Tel: 01822 616871
Specialities: Annuals/Biennials/Bedding, Shrubs

162 Turnpike Nurseries

Leonard Moor Cross, Uffculme, Cullompton, Devon EX15 3EX
Tel: 01884 840980
Specialities: Annuals/Biennials/Bedding, Herbaceous Perennials

163 Ward Alpines Nursery

Newton Farm Hemyock, Hemyock, Cullompton, Devon EX15 3QS
Tel: 01823 680410
Specialities: Alpines/Rock Plants, Shrubs

164 Welcome Nursery

Liverton, Newton Abbot, Devon TQ12 6JA
Tel: 01626 821098
Specialities: Seasonal Plants, Herbaceous Perennials

165 Westcountry Lupins

Ford Hill Forge, Hartland, Bideford,
Devon EX39 6EE
Tel: 01237 441208
Specialities: Specialist Bedding

166 Westcountry Nurseries

Donkey Meadow, Woolsery, Bideford,
Devon EX39 5QH
Tel: 01237 431111
Specialities: Annuals/Biennials/Bedding, Shrubs

167 Wildside Nursery

Green Lane, Buckland Monachorum, Yelverton,
Devon PL20 7NP
Tel: 01822 855755
Specialities: Herbaceous Perennials

168 Withleigh Nurseries

Sunrise Hill, Withleigh, Tiverton, Devon EX16 8JG
Tel: 01884 253351
Specialities: Annuals/Biennials/Bedding, Shrubs

169 Woodvale Nursery

Truro Drive, Plymouth, Devon PL5 4LA
Tel: 01752 701645
Specialities: Annuals/Biennials/Bedding, Seasonal Plants, Shrubs

170 Wyevale Garden Centre (Teignmouth - Jack's Patch)

Newton Road, Bishopsteignton, Teignmouth,
Devon TQ14 9PN
Tel: 01626 776996

171 Wykehill Gardens

Shobrooke, Crediton, Devon EX17 1AN
Tel: 01392 851652
Specialities: Shrubs, Herbaceous Perennials, Xmas Trees

172 Yarde Gate Nursery

Salcombe Road, Malborough, Kingsbridge,
Devon TQ7 3BX
Tel: 01548 842741

DORSET

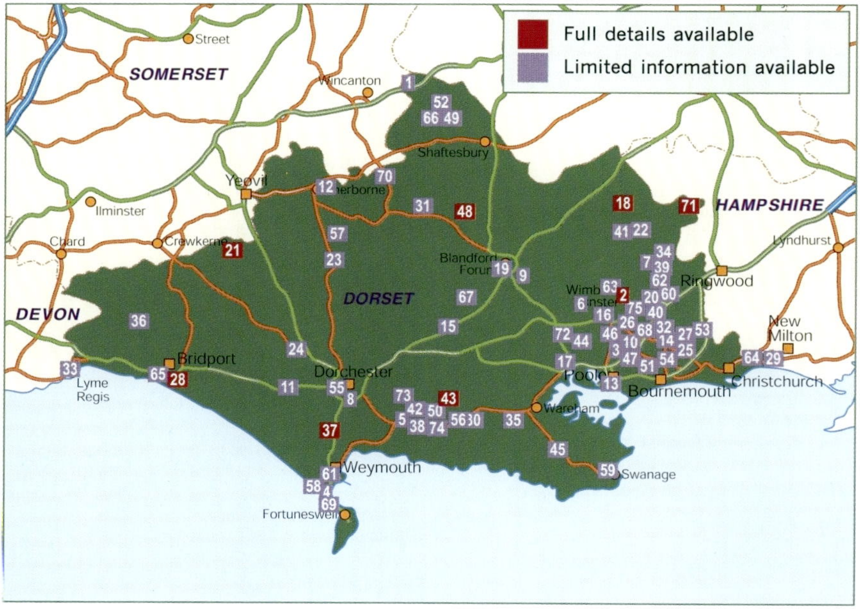

1 Abbey Plants

Chaffeymoor, Bourton, Gillingham, Dorset SP8 5BY
Tel: 01747 840841

2 Barthelemy & Co.

262 Wimborne Road West, Stapehill,
Wimborne, Dorset BH21 2DZ
Tel: 01202 874283 Fax: 01202 897482
e-mail: kenere@tiscali.co.uk

Specialities:

Japanese Maples; Cornus; Conifers

Description:

Established in 1920, this family-run business stocks a huge selection of ornamental Japanese Maples as well as a good choice of the Cornus Florida group – one of the most spectacular of flowering trees. Also available are conifers, beeches, wisterias and hibiscus.

Directions:

From the A31 Wimborne Minster bypass, about 3 miles east of Wimborne at the roundabout take the B3073. Barthelemy & Co. is a short distance along this road, on the right hand side.

3 Beecroft Garden Centre

Queen Anne Drive, Wimborne, Dorset BH21 3BA
Tel: 01202 693705
Specialities: Garden Products, Herbaceous Perennials, Garden Furniture

4 Bennetts Water Gardens

Putton Lane, Chickerell, Weymouth, Dorset DT3 4AF
Tel: 01305 785150
Specialities: Aquatics, Nymphaea (Water Lilies), Water Products and Features

5 Benville Nursery

25C Moreton Road, Owermoigne, Dorchester, Dorset DT2 8HY
Tel: 01305 852386
Specialities: Shrubs, Annuals/Biennials/Bedding

6 Bonsai Garden Centre

549 Abbott Street, Pamphill, Wimborne, Dorset BH21 4EF
Tel: 01202 887797
Specialities: Herbaceous Perennials, Bonsai

7 Brackendale Nurseries

Horton Road, Three Legged Cross, Wimborne, Dorset BH21 6SD
Tel: 01202 822349
Specialities: Conifers, Hedging, Shrubs

8 Brooklands Plants

25 Treves Road, Dorchester, Dorset DT1 2HE
Tel: 01305 265846
Specialities: Shrubs, Herbaceous Perennials

9 C W Abbott & Son

Langton Long, Blandford Forum, Dorset DT11 9HR
Tel: 01258 452513

10 Canford Magna Garden Centre

170 Magna Road, Wimborne, Dorset BH21 3AP
Tel: 01202 579571

11 Carol Budd Conifers

Winterbourne Abbas, Dorchester, Dorset DT2 9LJ
Tel: 01305 889280
Specialities: Conifers

12 Castle Gardens Centre

New Road, Sherborne, Dorset DT9 5NR
Tel: 01935 814633

13 Chestnut Nursery

75 Kingland Road, Poole, Dorset BH15 1TN
Tel: 01202 685999
Specialities: Annuals/Biennials/Bedding, Herbaceous Perennials

14 Coppins Plant Centre

Christchurch Road, Ferndown, Dorset BH22 8TB
Tel: 01202 574665
Specialities: Annuals/Biennials/Bedding, Shrubs

15 Cottage Garden Nurseries

3 The Rings, Milborne St. Andrew, Blandford Forum, Dorset DT11 0HY
Tel: 01258 837639
Specialities: Cottage Garden Plants

16 Country Affairs

14 Poole Road, Wimborne, Dorset BH21 1QG
Tel: 01202 887878

17 The Courtyard Garden Centre

Huntick Road, Lytchett Minster, Poole, Dorset BH16 6BA
Tel: 01202 622709

18 Cranborne Manor Garden Centre

Cranborne Manor, Cranborne, Dorset BH21 5PP
Tel: 01725 517248 Fax: 01725 517862
e-mail: gardencentre@cranborne.co.uk
website: www.cranborne.co.uk

SEE ADVERTISEMENT BELOW

Specialities:

Roses; Clematis; Pots & Containers

Description:

This is a small independent country garden centre located in the old walled garden of Cranborne

Manor. It specialises in old-fashioned roses, clematis, herbaceous plants and shrubs but also stocks a range of garden sundries and ornaments. There's also a gift shop, clothing department and tea room.

Directions:

From the A338 at Fordingbridge, take the B3078 to Cranborne. In the village, follow the signs to the centre.

19 Digwell Nursery

The Villa, Market Place, Blandford Forum,
Dorset DT11 7EB
Tel: 01258 454714

20 Dorset Garden Supplies

Unit 46-47 Uddens Trading Estate, Wimborne,
Dorset BH21 7NL
Tel: 01202 871432
Specialities: Garden Products

21 Dorset Water Lily Company

Yeovil Road, Halstock, nr Yeovil,
Somerset BA22 9RR
Tel: 01935 891668 Fax: 01935 891946
e-mail: dorsetwaterlily@uwclub.net
website: www.dorsetwaterlily.co.uk

SEE ADVERTISEMENT BELOW

Specialities:

Hardy Water Lilies; Tropical Water Lilies; Hardy Water Marginals

Description:

A Hampton Court Gold and Tudor Rose Winner in 2006, this nursery offers a wide variety of hardy water lilies to suit all water features. The water lily stock is well supported by a range of marginal, herbaceous, bogside and oxygenating plants to suit all sites.

Directions:

We are approx 3 miles from the A37 Dorchester to Yeovil road 5 miles south of Yeovil.

Standard Hardy Lily
Connoisseur Hardy Lily
Tropical Lily
Lotus (Nelumbo)
Marginals
Bogside Plants
Oxygenating Plants
Herbaceous Plants
Award winning gardens

Dorset Water Lily Company was established in 1987 and is set in 3 acres of lovely Dorset countryside. It is a Hampton Court Gold and Tudor Rose Winner in 2006, and offers a wide variety of hardy water lilies to suit all water features. The water lily stock is well-supported by a range of marginal, herbaceous, bogside and oxygenating plants to suit all sites.

Since 2001 they have successfully created award winning show gardens at Hampton Court Flower Show, and have exhibited at Tatton Park for the first time in 2006. The Tatton Park garden illustrated that you do not need a large garden or pond to be able to have a lily, there are varieties that can be put into tubs.

FOR ALL YOUR WATER LILIES AND MARGINAL PLANTS.....

DORSET WATER LILY COMPANY

YEOVIL ROAD, HALSTOCK, NR YEOVIL, SOMERSET BA22 9RR
TEL: 01935 891668 FAX: 01935 891946
E-MAIL: DORSETWATERLILY@UWCLUB.NET
WEBSITE: WWW.DORSETWATERLILY.CO.UK

OPENING TIMES:
9.00 TO 4.00 MONDAY, FRIDAY AND SATURDAY

22 Forest Edge Nurseries

Verwood Road, Woodlands, Wimborne,
Dorset BH21 8LJ
Tel: 01202 824387
Specialities: Annuals/Biennials/Bedding, Shrubs

23 Forest Nursery

Middlemarsh, Sherborne, Dorset DT9 5QN
Tel: 01963 210257
Specialities: Annuals/Biennials/Bedding, Shrubs

24 Frampton Roses

34 Dorchester Road, Frampton, Dorchester,
Dorset DT2 9NF
Tel: 01300 320453
Specialities: Rose

25 Gardens Nursery

Muscliffe Lane, Bournemouth, Dorset BH9 3NN
Tel: 01202 510429
Specialities: Shrubs, Annuals/Biennials/Bedding

26 Glenacres Nursery

232 Wimborne Rd West, Wimborne,
Dorset BH21 2DY
Tel: 01202 872069
Specialities: Annuals/Biennials/Bedding, Shrubs

27 Golden Acres Nursery

359 Christchurch Road, West Parley, Ferndown,
Dorset BH22 8SJ
Tel: 01202 570033

28 C. W. Groves & Son Ltd

West Bay Road, Bridport, Dorset DT6 4BA
Tel: 01308 422654 Fax: 01308 420888
e-mail: garden@grovesnurseries.co.uk
website: www.grovesnurseries.co.uk

Specialities:

National Collection of Sweet Garden Violets & Parma Violets; More than 130 Varieties of Roses; Home-made Cakes in Citrus Grove Café.

Products and Services:

- Aquatics
- Barbecues
- Books and Stationery
- Seasonal Displays
- Conservatories
- Credit Cards Accepted
- Cut Flowers
- Disabled Facilities
- Display Gardens
- Dried and Artificial Flowers
- Fencing
- Fruit Trees
- Garden Furniture
- Garden Machinery
- Garden Products
- Giftware
- Greenhouses and Sheds
- Houseplants
- Large Car Park
- Paving
- Pet Products
- Plants
- Restaurant/Coffee Shop
- Seeds and Bulbs
- Trees

Directions:

The Centre is adjacent to the Crown Pub roundabout on the A35 Bridport by-pass

Description:

The nursery is owned and run by the Groves family – Clive and Diana, and son Charles – who are the 5th and 6th generation of the family since the business was established in 1866. Clive qualified at Cannington College, Charles at the University of Leeds. The 6-acre site is beautifully laid out with eye-catching features such as the koi ponds, rose gardens and pergolas. An experimental Potager Garden culminates in a Solar Dome which houses trials of tomato varieties and chilli peppers during the summer. In winter, it contains the National Collection of *viola odorata cultivans* (sweet violets) and double Parma violets. The nursery has won Gold at the International Violet Competition on two occasions. The Garden Centre shop is well-stocked with all the usual products, with particularly extensive selections of tools, pet products and garden furniture. From October to December, the nursery has a special Christmas shop.

29 Grundens Nursery

316 Lymington Road, Highcliffe, Christchurch,
Dorset BH23 5EY
Tel: 01425 272452
Specialities: Garden Furniture

30 Hanging Gardens Nursery

Winfrith Newburgh, Dorchester, Dorset DT2 8DD
Tel: 01305 851632
Specialities: Shrubs, Annuals/Biennials/Bedding

31 Harts Garden Supplies Ltd

Station Road, Sturminster Newton,
Dorset DT10 1BD
Tel: 01258 472788
Specialities: Garden Products

32 Haskins Garden Centre

Longham, Ferndown, Dorset BH22 9DG
Tel: 01202 591919

33 Haye Lane Nurseries

Haye Lane, Lyme Regis, Dorset DT7 3NQ
Tel: 01297 442604
Specialities: Herbaceous Perennials, Shrubs

34 Holly Hedge Nurseries

Verwood Road, Three Legged Cross, Wimborne, Dorset BH21 6RJ
Tel: 01202 822447
Specialities: Shrubs, Trees, Seasonal Plants

35 Holme Nurseries

West Holme Farm, West Holme, Wareham, Dorset BH20 6AQ
Tel: 01929 554716
Specialities: Farm Shop

36 The Iris Garden

Yard House, Pilsdon, Bridport, Dorset DT6 5PA
Tel: 01308 868797
Specialities: Iris

37 Island Garden Nursery

Church Street, Upwey, Dorset DT3 5QB
Tel: 01305 814345

Specialities:
Fuchsias, Geraniums, Hanging Baskets & Tubs.

Products and Services:
- Credit Cards Accepted
- Display Gardens
- Plants

Directions:
From the A354, about 4 miles south of Dorchester, turn right on the B3159 into Upwey village, Follow this road as it bends right into Church Street and you will shortly see the nursery on the left. The entrance is via Westbrook Farm.

Description:
Once a watercress farm and converted to a nursery in 1982, Island Garden enjoys a picturesque setting beside a river. Owner Janet Willment offers her customers a very varied selection with hundreds of varieties available from February onwards. The choice includes fuchsias, geraniums, perennials, shrubs, heathers, alpines, conifers and patio plants. The nursery also supplies and refills hanging baskets and tubs. A plant lover's paradise, the nursery is open every day from 9am to 6pm.

38 J G Mills

19 Moreton Road, Owermoigne, Dorchester, Dorset DT2 8HT
Tel: 01305 853404

39 John Brown Garden Centre

Ringwood Road, Three Legged Cross, Wimborne, Dorset BH21 6RD
Tel: 01202 822203

40 Knoll Gardens

Stapehill Road, Wimborne, Dorset BH21 7ND
Tel: 01202 873931
Specialities: Grasses, Herbaceous Perennials

41 Koirin Azalea Centre

Woodlands, Wimborne, Dorset BH21 8LN
Tel: 01202 824629
Specialities: Rhododendron (Inc Azalea)

42 Mill House Nurseries

Mill House, 33 Moreton Road, Owermoigne, Dorchester, Dorset DT2 8HZ
Tel: 01305 852220

43 Moreton Gardens & Plant Centre

Moreton, Dorchester, Dorset DT2 8RH
Tel/Fax: 01929 405084
e-mail: thomas.penny@btconnect.com
website: www.moretondorset.co.uk

Specialities:
Unique and Tranquil 3.5 acres of Landscaped Gardens; Camellias, Azaleas, Rhododendrons; Knowledgeable Advice & Garden Design.
Landscaping Division -www.tompennyandsons.co.uk

Products and Services:

- Credit Cards Accepted
- Disabled Facilities
- Fruit Trees
- Garden Construction
- Garden Design
- Garden Products
- Giftware
- Houseplants
- Information Desk
- Large Car Park
- Plants
- Restaurant/Coffee Shop
- Seeds and Bulbs
- Trees

Directions:

Moreton Gardens are situated in the old village of Moreton just off the B3390 which runs between the A35 from Affpuddle (approx. 3 miles) towards Weymouth and the A352

Description:

The E-shaped walled garden covers approximately 7 acres. One half of the 'E' encloses the landscaped gardens open to the public; the other half contains the nurseries and is not open to the public. In the centre is the intact Bothy – an octagonal building where the gardeners for Moreton House lived when it was the kitchen garden. Moreton Brook runs through the centre with sluice gates, weirs and a large pond. The plant shop is of traditional overlap timber construction and houses the refreshment area and shop selling accessories, gifts and souvenirs. Within the gardens, which are rich in butterflies and insects, there are formal areas, a fountain and rose garden, bog garden, woodland garden and large herbacious borders. The plant centre is staffed by helpful and knowledgeable gardeners, stocks a wide range of plants with camellias, azaleas and rhododendrons as some of the centre's specialities. Others include specimen trees and shrubs, along with all sorts of unique, unusual and interesting plants.

Opening hours:
Mar - Oct Everyday 10am-5pm
Nov - Dec Wednesday - Sunday 10am-4pm
Jan - Feb Weekends only 10am-4pm

44 Naked Cross Nurseries

Waterloo Road, Corfe Mullen, Wimborne, Dorset BH21 3SP
Tel: 01202 693256
Specialities: Annuals/Biennials/Bedding, Herbaceous Perennials

45 Norden Farm Shop & Mini Garden Centre

Norden Farm, Norden Corfe Castle, Wareham, Dorset BH20 5DS
Tel: 01929 480098
Specialities: Farm Shop, Annuals/Biennials/Bedding, Garden Products

46 The Oaks Garden Centre & Nursery

Queen Anne Drive, Wimborne, Dorset BH21 3BA
Tel: 01202 603322
Specialities: Garden Furniture, Garden Products, Water Products and Features

47 Oakview Nursery

69 Mandale Road, Bournemouth, Dorset BH11 8HZ
Tel: 01202 580522
Specialities: Annuals/Biennials/Bedding, Herbaceous Perennials

48 Oasis Plant Centre

Shaftesbury Road, Child Okeford, Dorset DT11 8EQ
Tel: 01258 861325
e-mail: john@oasisplantcentre.co.uk

Specialities:

Delphiniums – including cutting-raised named cultivars; Hellebores; Home Brew & Wine-making

Products and Services:

- Credit Cards Accepted
- Fruit Trees
- Large Car Park
- Plants
- Trees

Directions:

The village of Child Okeford is half-way between Shaftesbury and Blandford, just off the A350. The nursery is ½ mile from the village centre on the road joining with the A350 at Iwerne Minster

Description:

Oasis Plant Centre is owned and run by John Harbour, a retired farmer, and his wife Valerie who trained as a PE teacher and now runs the Oasis Health & Beauty Centre on the same site. They started the nursery as a retirement project but it has steadily grown. It is situated under Hambledon Hill in a walled enclosure with about 1 acre open to the public and a further acre

for future expansion if required. John and Valerie try to keep available any plant that is requested with the exception of lime-hating plants which will not thrive in the soil here. They stock a good range of shrubs and perennials and quite a few trees, including fruit trees. Also available are hanging baskets, planters and composts. The Centre is open from 9am to 5pm, Monday to Friday, and from 10am to 4pm on Saturdays, Sundays and bank holidays.

49 Orchard Park Ltd

Shaftesbury Road, Gillingham, Dorset SP8 5JG
Tel: 01747 835544

50 Owermoigne Nurseries

20 Wareham Road, Owermoigne, Dorchester, Dorset DT2 8HN
Tel: 01305 852483
Specialities: Herbaceous Perennials

51 Parks Perennials

242 Wallisdown Road, Bournemouth, Dorset BH10 4HZ
Tel: 01202 524464
Specialities: Herbaceous Perennials

52 Plantworld Nursery

Milton-on-Stour, Gillingham, Dorset SP8 5QA
Tel: 01747 824015
Specialities: Climbers, Shrubs, Herbaceous Perennials

53 Plowmans Garden Nursery & Plant Centre Ltd

392 Christchurch Road, West Parley, Ferndown, Dorset BH22 8SW
Tel: 01202 582169

54 The Potting Shed

911 Wimborne Road, Bournemouth, Dorset BH9 2BJ
Tel: 01202 514870

55 Poundbury Gardens

Bridport Rd, Poundbury, Dorchester, Dorset DT1 3RT
Tel: 01305 257250

56 R.J Hansford

West Burton Farm, Winfrith Newburgh, Dorchester, Dorset DT2 8DD
Tel: 01305 852956

57 Ryewater Nursery

Bishops Down, Sherborne, Dorset DT9 5PL
Tel: 01963 210645

58 School Hill Garden Nursery

School Hill, Chickerell, Weymouth, Dorset DT3 4BA
Tel: 01305 771760
Specialities: Annuals/Biennials/Bedding, Shrubs

59 Seagreen Nursery

39 Station Road, Swanage, Dorset BH19 1AD
Tel: 01929 422990
Specialities: Floristry, Garden Products

60 Shelley Nurseries

77 Pinehurst Road, West Moors, Ferndown, Dorset BH22 0AJ
Tel: 01202 873283
Specialities: Annuals/Biennials/Bedding, Shrubs

61 St Annes Garden Centre Ltd

20 Coldharbour, Chickerell, Weymouth, Dorset DT3 4BG
Tel: 01305 775566
Specialities: Garden Products, Garden Furniture

62 Stanford House Nurseries

Three Cross Road, West Moors, Wimborne, Dorset BH21 6QW
Tel: 01202 892266
Specialities: Annuals/Biennials/Bedding

63 Stewarts Country Garden Centre

God's Blessing Lane, Holt, Wimborne, Dorset BH21 7DF
Tel: 01202 882462

64 Stewarts Gardenlands

Lyndhurst Road, Somerford, Christchurch, Dorset BH23 4SA
Tel: 01425 272244

65 Swallowdale Nurseries

West Bay Fields, Bridport, Dorset DT6 5LB
Tel: 01308 426450
Specialities: Annuals/Biennials/Bedding, Climbers

66 Thorngrove Garden Centre

Common Mead Lane, Gillingham, Dorset SP8 4RE
Tel: 01747 822242

67 Three Bears Nursery

Winterborne Clenston, Blandford Forum, Dorset DT11 0NY
Tel: 01258 880779
Specialities: Annuals/Biennials/Bedding, Herbaceous Perennials

68 Trehane Camellia Nursery

Stapehill Road, Wimborne, Dorset BH21 7ND
Tel: 01202 873490
Specialities: Camellia, Rhododendron (Inc Azalea)

69 Weymouth Garden Centre

Egdon Hall, Lynch Lane, Weymouth, Dorset DT4 9DN
Tel: 01305 766336
Specialities: Garden Furniture, Garden Products, Trees

70 Williams Nursery

Station Road, Stalbridge, Sturminster Newton, Dorset DT10 2RQ
Tel: 01963 362355

71 Wolvercroft World of Plants

Fordingbridge Road, Alderholt, Dorset SP6 3BE
Tel: 01425 652437 Fax: 01425 655525
e-mail: gardens@wolvercroft.co.uk
website: www.wolvercroft.co.uk

Specialities:

Trees; Clematis; Tearoom

Products and Services:

- Seasonal Displays
- Credit Cards Accepted
- Disabled Facilities
- Fencing
- Fruit Trees
- Garden Construction
- Garden Furniture
- Garden Products
- Giftware
- Houseplants
- Information Desk
- Large Car Park
- Paving
- Plant Guarantee
- Plants
- Restaurant/Coffee Shop
- Seeds and Bulbs
- Trees

Directions:

From Cranborne, take the B3078 south towards Fordingbridge. Drive through the village of Alderholt, then follow the main road to Fordingbridge for about ½ a mile. Wolvercroft World Of Plants is on the right hand side at the top of the hill.

Description:

Located on the edge of the New Forest National Park, this 2-acre nursery produces high quality bedding, herbaceous, shrubs and climbing plants. The owners also source high quality plants from nurseries throughout the UK and Europe giving you the best choice anywhere in the country.

But, it's not just plants. There are many other items to make your garden complete – furniture; composts; trellis; fencing; containers in wood,

terracotta, stone, resin and plastic; and all manner of garden sundries. The excellent staff are helpful and friendly, many with a vast local knowledge of the plants which would be suitable for your garden.

Make sure you leave enough time to visit the "Sticky Bun Tea Rooms", where a warm welcome awaits you. Enjoy a revitalising cup of coffee, cream tea or a tempting cake or teacake, whilst browsing around the renowned selection of greetings cards and gifts for any occasion.

72 Woodlands Nurseries

Blandford Road, Lytchett Matravers, Poole, Dorset BH16 6AG
Tel: 01258 857163
Specialities: Annuals/Biennials/Bedding, Herbaceous Perennials

73 Woodsford Nursery

Crossways, Dorchester, Dorset DT2 8BT
Tel: 01305 852764
Specialities: Annuals/Biennials/Bedding, Herbaceous Perennials, Shrubs

74 Wyevale Garden Centre (Galton)

24 Wareham Road, Galton, Dorchester, Dorset DT2 8BY
Tel: 01305 852324

75 Wyevale Garden Centre (Wimborne)

Wimborne Road West, Wimborne, Dorset BH21 2DN
Tel: 01202 874208

DURHAM

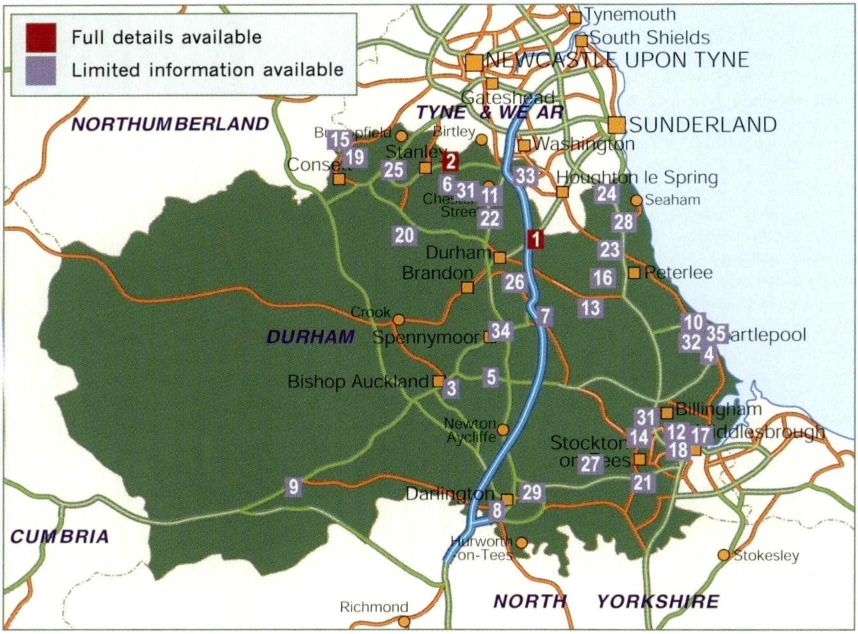

1 A N Sanderson Ltd

Chestnut Cottage Ramside, Belmont, County Durham DH1 1SU
Tel/Fax: 01913 842086
e-mail: daemonans@aol.com
website: www.sandersonsnurseries.co.uk
Specialities: Conifers, Shrubs, Trees

2 Beamish Clematis Nursery

Burntwood Cottage, Stoney Lane, Beamish, County Durham DH9 0SJ
Tel/Fax: 01913 700202
website: www.beamishclematisnursery.co.uk
Specialities: Clematis, Climbers, Acers

3 Beech Tree Nurseries

Adelaide Bank, Shildon, County Durham DL4 1DF
Tel: 01388 772076
Specialities: Shrubs, Trees, Alpines/Rock Plants

4 Blackford Nurseries

Meryl Gardens, Stockton Road, Hartlepool, County Durham TS25 2PL
Tel: 01429 268509
Specialities: Annuals/Biennials/Bedding, Shrubs

5 Blossoms Nursery

The Gardens, Windlestone, Ferryhill, County Durham DL17 0NA
Tel: 01388 720219
Specialities: Floristry, Flower

6 Congburn Nurseries

Edmondsley, Durham, County Durham DH7 6DY
Tel: 01913 719070
Specialities: Annuals/Biennials/Bedding, Shrubs

7 Coxhoe Direct Paving

1 Wesley Place, Coxhoe, Durham, County Durham DH6 4LG
Tel: 01913 773272
Specialities: Paving

8 Elmridge Gardens

Coniscliffe Road, Darlington, County Durham DL3 8DH
Tel: 01325 462710
Specialities: Annuals/Biennials/Bedding, Shrubs

9 Equatorial Plant Co.

7 Gray Lane, Demesnes, Barnard Castle,
County Durham DL12 8PD
Tel: 01833 690519
Specialities: Exotics

10 Foster & Armstrong

12 Murray Street, Hartlepool,
County Durham TS26 8PE
Tel: 01429 422002

11 Foxwood Garden Centre Ltd

Holm Hill Lane, Chester Moor, Chester Le Street,
County Durham DH2 3RQ
Tel: 01913 890044
Specialities: Restaurant/Coffee Shop, Conservatories, Garden Products

12 GCH Gardens

Cowpen Lane, Billingham,
County Durham TS23 4AT
Tel: 01642 564627
Specialities: Trees, Shrubs

13 Green Lane Nurseries

Green Lane, Trimdon Station,
County Durham TS29 6NU
Tel: 01429 882727
Specialities: Annuals/Biennials/Bedding, Shrubs

14 Harpers Garden Centre

Junction Road, Stockton-on-Tees,
County Durham TS20 1SU
Tel: 01642 674636

15 Herb Patch

Ebchester, Consett, County Durham DH8 9JA
Tel: 01207 562099
Specialities: Herbs

16 Horns Garden Centre

Dixon Estate, Shotton Colliery, Durham,
County Durham DH6 2PX
Tel: 01915 262987

17 Howden House Farm Nurseries

Howden House Farm, Limetrees Close,
Middlesbrough, County Durham TS2 1SL
Tel: 01642 673598

18 Hugh & I Garden Shop

1-7 Roscoe Road, Billingham,
County Durham TS23 1HP
Tel: 01642 365454

19 J Beveridge Nursery

Ebchester Hill, Shotley Bridge, Consett,
County Durham DH8 0SX
Tel: 01207 560228
Specialities: Annuals/Biennials/Bedding, Shrubs

20 Lanchester Nurseries

Ford Raod, Lanchester, Durham,
County Durham DH7 0SS
Tel: 01207 521133
Specialities: Pots, Homegrown, Garden Furniture

21 Peter Barratts Garden Centre

Yarm Road, Stockton-on-Tees,
County Durham TS18 3SQ
Tel: 01642 613433
Specialities: Annuals/Biennials/Bedding, Trees, Garden Furniture

22 Pity Me Nursery

Front St, Pity Me, Durham,
County Durham DH1 5GZ
Tel: 01913 832925
Specialities: Annuals/Biennials/Bedding, Shrubs

23 Plants 'R' Ross Nursery

South Hetton Road, Peterlee,
County Durham SR8 3BZ
Tel: 01915 270855
Specialities: Annuals/Biennials/Bedding, Seasonal Plants, Herbaceous Perennials

24 Pococks Roses

Seaton Lane, Seaton, Seaham,
County Durham SR7 0LT
Tel: 01914 151548
Specialities: Rose

25 Pontop Nursery

Pontop Pike, Dipton, Stanley,
County Durham DH9 9AS
Tel: 01207 571141
Specialities: Homegrown, Annuals/Biennials/Bedding

26 Poplar Tree Garden Centre

Hall Lane, Shincliffe, Durham,
County Durham DH1 2NG
Tel: 01913 847553

27 R Taylor & Son

Nine Acres, Whinney Hill, Stockton-on-Tees,
County Durham TS21 1BL
Tel: 01642 582718

28 Richardsons Garden Centre

Stockton Road, Cold Hesledon, Seaham,
County Durham SR7 8RW
Tel: 01915 818181
Specialities: Garden Products

29 Robinsons Nurseries

Mill Lane, Darlington, County Durham DL1 2XQ
Tel: 01325 332966
Specialities: Seasonal Plants, Herbaceous Perennials

30 Sir Plants-Alot

Sandy Lane West, Wynyard Raod, Billingham,
County Durham TS22 5NB
Tel: 01740 644977
Specialities: Herbaceous Perennials, Annuals/Biennials/Bedding, Xmas Trees

31 St Barnabas Nursery

Waldridge, Chester Le Street,
County Durham DH2 3RY
Tel: 01913 871233
Specialities: Annuals/Biennials/Bedding, Shrubs

32 Stranton Nursery

Tanfield Rd, Hartlepool, County Durham TS25 5DD
Tel: 01429 233950
Specialities: Annuals/Biennials/Bedding, Herbaceous Perennials

33 Strikes Garden Centre

Lambton Park, Chester Le Street,
County Durham DH3 4PZ
Tel: 01913 855154

34 Thinford Farm Nurseries

Thinford, Durham, County Durham DH6 5JX
Tel: 01740 655704
Specialities: Annuals/Biennials/Bedding, Shrubs

35 Whitegates Garden Centre

Ladysmith Street, Hartlepool,
County Durham TS25 1PX
Tel: 01429 868656
Specialities: Annuals/Biennials/Bedding, Garden Furniture

ESSEX

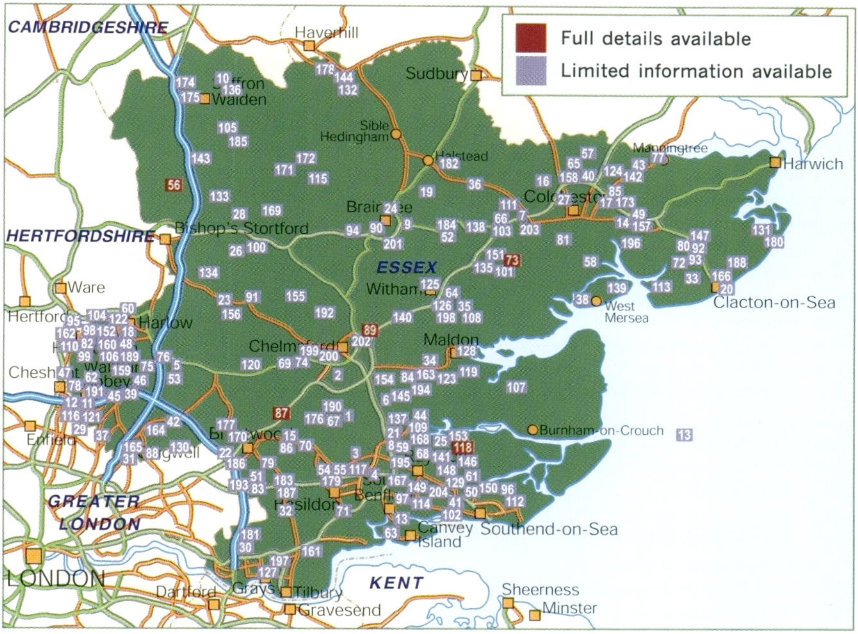

1 A.E Harnett & Sons

Smallgains Lane, Stock, Ingatestone, Essex CM4 9PR
Tel: 01277 840302
Specialities: Annuals/Biennials/Bedding, Herbaceous Perennials

2 Abercorn Plant & Garden Centre

Beehive Lane, Chelmsford, Essex CM2 8LX
Tel: 01245 257398
Specialities: Annuals/Biennials/Bedding, Garden Products

3 Alpha Garden Centre

238 London Road, Wickford, Essex SS12 0JX
Tel: 01268 766093

4 Alton Garden Centre

Arterial Road, North Benfleet, Wickford, Essex SS12 9JG
Tel: 01268 726421

5 Art Nursery

Vicarage Lane West, North Weald, Epping, Essex CM16 6AL
Tel: 01992 523346

6 B H N Pot Plants

East Hanningfield Road, Rettendon Common, Chelmsford, Essex CM3 8EW
Tel: 01245 400360
Specialities: Pot Plants

7 Barnplants

Turkeycock Lane, Stanway, Colchester, Essex CO3 8ND
Tel: 01206 210486
Specialities: Annuals/Biennials/Bedding, Herbaceous Perennials, Restaurant/Coffee Shop

8 Battlesbridge Mills Garden Centre

Old Chelmsford Road, Wickford, Essex SS11 8TR
Tel: 01268 764001
Specialities: Garden Furniture, Garden Products

9 Bayree Nurseries

Bayree Farm, Coggeshall Road, Stisted, Braintree, Essex CM77 8AE
Tel: 01376 344301
Specialities: Annuals/Biennials/Bedding, Herbaceous Perennials

10 Beeches Nursery

Walden Road, Ashdon, Saffron Walden, Essex CB10 2NB
Tel: 01799 584362
Specialities: Annuals/Biennials/Bedding, Herbaceous Perennials

11 Beechside Nursery

Mott Street, Essex E4 7RW
Tel: 01992 715521
Specialities: Annuals/Biennials/Bedding, Shrubs

12 Beechview Nursery

Avey Lane, Waltham Abbey, Essex EN9 3QH
Tel: 01992 714065
Specialities: Annuals/Biennials/Bedding, Herbaceous Perennials

13 Benfleet Garden Centre

74 High Street, Benfleet, Essex SS7 1ND
Tel: 01268 795559
Specialities: Garden Products

14 The Beth Chatto Gardens Ltd

Elmstead Market, Colchester, Essex CO7 7DB
Tel: 01206 822007
Specialities: Herbaceous Perennials

15 Billericay Nurseries

London Rd, Billericay, Essex CM12 9HT
Tel: 01277 622083

16 Blackheath Bulb Co

Nayland Road, West Bergholt, Colchester, Essex CO6 3DE
Tel: 01206 271261
Specialities: Bulbs

17 Blenheim Nurseries

Bromley Road, Ardleigh, Colchester, Essex CO7 7SF
Tel: 01206 870605

18 Bonnies Oak Garden Centre

Water Lane, Roydon, Harlow, Essex CM19 5DR
Tel: 01279 792231
Specialities: Garden Furniture, Greenhouses and Sheds, Annuals/Biennials/Bedding

19 Bourne Brook Nurseries

Stisted Road, Greenstead Green, Halstead, Essex CO9 1RB
Tel: 01787 476369
Specialities: Annuals/Biennials/Bedding, Fuchsia

20 Bow-Kay Plants

14B High Street, Clacton-on-Sea, Essex CO15 1NS
Tel: 01255 430003

21 The Boxtree Nursery

Beeches Road, Rawreth, Wickford, Essex SS11 8TG
Tel: 01268 560056
Specialities: Trees

22 Brentwood Garden Centre

Vicarage Close, Brentwood, Essex CM14 5NF
Tel: 01277 262303
Specialities: Greenhouses and Sheds

23 Bretts Garden Centre

Chelmsford Road, White Roding, Dunmow, Essex CM6 1RF
Tel: 01279 876876

24 Broadfield Nursery

Broad Road, Braintree, Essex CM7 5NJ
Tel: 01376 343672
Specialities: Shrubs

25 Bushukan Bonsai

Lower Road, Hockley, Essex SS5 5NL
Tel: 01702 201029
Specialities: Bonsai

26 Canfield Nursery Shop

Bullocks Lane, Takeley, Bishop's Stortford, Essex CM22 6TA
Tel: 01279 870352
Specialities: Bulbs, Herbaceous Perennials, Shrubs

27 Cants of Colchester

Nayland Road, Mile End, Colchester, Essex CO4 5EB
Tel: 01206 844008
Specialities: Rose

28 Chapel End Nursery

Chapel End, Broxted, Dunmow, Essex CM6 2BW
Tel: 01279 850771
Specialities: Trees, Shrubs, Herbaceous Perennials

29 Chapelfield Nursery

Sewardstone Road, Essex E4 7RE
Tel: 020 8529 1840

30 Chestnut Farm Nursery

South Road, South Ockendon, Essex RM15 6DT
Tel: 01708 670023
Specialities: Annuals/Biennials/Bedding, Shrubs

31 Chigwell Nursery

245 High Road, Chigwell, Essex IG7 5BL
Tel: 020 8500 2690
Specialities: Specimen Plants, Shrubs, Annuals/Biennials/Bedding

32 The Christabella Charitable Trust

Barnards Farm, Brentwood Road, West Horndon, Brentwood, Essex CM13 3LX
Tel: 01277 811262
Specialities: Ornamentals

33 Clacton Garden Centre

St. Johns Road, Clacton-on-Sea, Essex CO16 8DY
Tel: 01255 425711
Specialities: Garden Furniture

34 Claremont Nurseries

Bryants Lane, Woodham Mortimer, Maldon, Essex CM9 6TB
Tel: 01245 222643
Specialities: Annuals/Biennials/Bedding, Herbaceous Perennials

35 Colburn Lys

Gt Totham Road, Wickham Bishops, Witham, Essex CM8 3NP
Tel: 01621 891587

36 Colne Valley Nurseries

Boley Rd, White Colne, Colchester, Essex CO6 2PT
Tel: 01787 223344

37 Connaught Farm Nurseries

Epping New Road, Loughton, Essex IG10 4AA
Tel: 020 8504 5168
Specialities: Shrubs, Herbaceous Perennials

38 Cooks Garden Centre

14 Church Road, West Mersea, Colchester, Essex CO5 8QH
Tel: 01206 382154
Specialities: Garden Products

39 Copped Hall Nursery

High Road, Epping, Essex CM16 4DH
Tel: 01992 571533
Specialities: Annuals/Biennials/Bedding, Shrubs

40 The Cottage Garden

Langham Road, Boxted, Colchester, Essex CO4 5HU
Tel: 01206 272269
Specialities: Cottage Garden Plants

41 Cottage Garden Nursery

573 Prince Avenue, Westcliff-on-Sea, Essex SS0 0JL
Tel: 01702 352191
Specialities: Cottage Garden Plants

42 Coveney Nurseries

Ongar Road, Abridge, Romford, Essex RM4 1AA
Tel: 01708 688479

43 Craft Nurseries Garden Plant Centre

34 Harwich Road, Lawford, Manningtree, Essex CO11 2LS
Tel: 01206 230699
Specialities: Restaurant/Coffee Shop, Annuals/Biennials/Bedding

44 Crouch Vale Nurseries Ltd

Burnham Road, South Woodham Ferrers, Chelmsford, Essex CM3 5QP
Tel: 01245 323422
Specialities: Annuals/Biennials/Bedding, Shrubs

45 Crown Hill Nursery

Crown Hill, Waltham Abbey, Essex EN9 3TF
Tel: 01992 700877
Specialities: Annuals/Biennials/Bedding, Shrubs

46 Crowther Landscapes

Unit 6, Lindsey Street, Epping, Essex CM16 6RE
Tel: 08458 800165
Specialities: Landscape Gardening

47 D.G.N Stubbins Nursery

Stubbins Hall Lane, Holyfield, Waltham Abbey, Essex EN9 2EE
Tel: 01992 762543

48 De Rosa Garden Centre

Tylers Rd, Roydon, Harlow, Essex CM19 5LJ
Tel: 01992 893117
Specialities: Garden Products, Trees

49 Deans Bark & Mulch

Harwich Road, Great Bromley, Colchester,
Essex CO7 7JQ
Tel: 01206 250485

50 Dickens Plants

45 Eastwoodbury Crescent, Southend-on-Sea,
Essex SS2 6XG
Tel: 01702 544878
Specialities: Annuals/Biennials/Bedding, Shrubs

51 Dobson & Co

Little Warley Hall Lane, Little Warley, Brentwood,
Essex CM13 3HA
Tel: 01277 263232

52 Dutch Nursery & Garden Centre

West Street, Coggeshall, Colchester, Essex CO6 1NT
Tel: 01376 561287
Specialities: Restaurant/Coffee Shop, Garden Products

53 Elite Garden Nursery

Cold Hall Lodge, Kiln Road, North Weald, Epping,
Essex CM16 6AD
Tel: 01992 523649
Specialities: Annuals/Biennials/Bedding

54 The Elms Nursery and Garden Centre Ltd

Hardings Elms Road, Crays Hill, Billericay,
Essex CM11 2UH
Tel: 01268 521721
Specialities: Water Products and Features, Annuals/Biennials/Bedding

55 The English Country Garden

Daniels Farm, Wash Road, Basildon,
Essex SS15 4AZ
Tel: 01268 271849

56 F. W. Whyman

Hillcrest, Hill Green, Clavering, Essex CB11 4QS
Tel: 01799 550568
e-mail: gerald@whymansnurseries.fsnet.co.uk
website: www.whymansnurseries.co.uk

Specialities:

Herbaceous Plants; Hanging Baskets; Seasonal Bedding

Description:

Whyman's Nurseries is a family run business established in 1967. It specialises in herbaceous plants, shrubs, climbers and seasonal items such as seed potatoes. In addition, the nursery also stocks a good range of pots and tubs, garden statuary, Thompson and Morgan seeds

Directions:

The nursery is on the B1038 in the picturesque village of Clavering, 7 miles southwest Saffron Walden.

57 Fillpots Nursery

52 Straight Road, Boxted, Colchester,
Essex CO4 5RB
Tel: 01206 273834

58 Flora Exotica

Pasadena, South Green Road, Fingringhoe,
Colchester, Essex CO5 7DR
Tel: 01206 729414
Specialities: Exotics

59 Forge Nurseries

Church Road, Rawreth, Wickford, Essex SS11 8SH
Tel: 01268 734970
Specialities: Alpines/Rock Plants, Fuchsia, Patio Plants

60 Four Acres Nursery

Old House Lane, Roydon, Harlow, Essex CM19 5DL
Tel: 01279 793383

61 Four Seasons Nursery & Farm Shop

Southend Rd, Rochford, Essex SS4 1HX
Tel: 01702 530765
Specialities: Farm Shop

62 Fourways Nursery

Breach Barns Lane, Waltham Abbey,
Essex EN9 2AD
Tel: 01992 713232
Specialities: Annuals/Biennials/Bedding, Shrubs

63 Garden World Garden Centre

Canvey Road, Canvey Island, Essex SS8 0QD
Tel: 01268 692595

64 Glen Chantry Garden and Perennial Nursery

Ishams Chase, Wickham Bishops, Witham,
Essex CM8 3LG
Tel: 01621 891342
Specialities: Herbaceous Perennials, Rare and Unusual Plants

65 Go Tropical

30a Straight Road, Boxted, Colchester,
Essex CO4 5HN
Tel: 01206 272783
Specialities: Exotic Fruit Trees, Palms, Tropical

66 Godbolts Nurseries

Coggeshall Rd, Marks Tey, Colchester,
Essex CO6 1LJ
Tel: 01206 212344
Specialities: Annuals/Biennials/Bedding, Shrubs

67 Greenacre Nurseries

Smallgains Lane, Stock, Ingatestone,
Essex CM4 9PR
Tel: 01277 840867
Specialities: Annuals/Biennials/Bedding, Pot Plants, Shrubs

68 Greenacres Nursery

Folly Chase, Hockley, Essex SS5 4SF
Tel: 01702 202422

69 Greenbrook Nurseries

Wyses Road, Highwood, Chelmsford,
Essex CM1 3SN
Tel: 01245 248871

70 Greenleas Nursery

Greenleas Farm, London Road, Billericay,
Essex CM12 9HP
Tel: 01277 636377
Specialities: Annuals/Biennials/Bedding, Shrubs

71 Greenwise Nurseries

Vange Park Road, Vange, Basildon, Essex SS16 5LA
Tel: 01268 555040
Specialities: Trees

72 Greenwood Nursery

Clay Lane, St. Osyth, Clacton-on-Sea,
Essex CO16 8HH
Tel: 01255 830600
Specialities: Annuals/Biennials/Bedding, Shrubs

73 Growing Together

Layer Marney Nursery, Smythes Green,
Layer Marney, Colchester CO5 9XW
Tel: 01206 330695 Fax: 01206 330053
e-mail: enquiries@growing-together.net
website: www.growing-together.net

SEE ADVERTISEMENT BELOW

Specialities:

Unusual Herbaceous Perennials; Display Gardens; Garden School with Workshops

Description:

This small independent nursery specialises in providing an eclectic range of plants that provide form, texture, colour and impact. In 2006, the nursery opened its garden school offering courses that range from pottery workshops to "how to create a natural pool".

Directions:

Smythes Green is on the B1022, about 7 miles southwest of Colchester.

74 Hanging Gardens Nurseries Ltd

Ongar Road West, Writtle, Chelmsford, Essex CM1 3NT
Tel: 01245 422245
Specialities: Hanging Baskets, Shrubs, Herbaceous Perennials

75 Happy Grow Nursery and Garden Centre

High Road, Thornwood, Epping, Essex CM16 6LX
Tel: 01992 575387

76 Harlow Garden Centre

Canes Lane, Hastingwood, Harlow, Essex CM17 9LD
Tel: 01279 419039

77 Hearts Delight Garden Centre

Long Road, Lawford, Manningtree, Essex CO11 2EF
Tel: 01206 392539

78 Heathfield Nursery

Galley Hill, Waltham Abbey, Essex EN9 2AG
Tel: 08700 600420

79 Hill Crest Nurseries

Brentwood Road, Ingrave, Brentwood, Essex CM13 3NZ
Tel: 01277 810385

80 Hilltop Nurseries

Clacton Road, Weeley Heath, Clacton-on-Sea, Essex CO16 9DN
Tel: 01255 831083
Specialities: Annuals/Biennials/Bedding, Herbaceous Perennials

81 Hillwood Nurseries

Hillwood, Bounstead Road, Blackheath, Colchester, Essex CO2 0DE
Tel: 01206 564882

82 Hoe Lane Nursery

Hoe Lane, Nazeing, Waltham Abbey, Essex EN9 2RJ
Tel: 01992 892181
Specialities: Annuals/Biennials/Bedding

83 Homefield Nursery

Arterial Road, Little Warley, Brentwood, Essex CM13 3ES
Tel: 01277 812240

84 Horseshoe Nurseries

White Elm Road, Danbury, Chelmsford, Essex CM3 4LR
Tel: 01245 223789

85 Hull Farm Nursery

Spring Valley Lane, Ardleigh, Colchester, Essex CO7 7SA
Tel: 01206 230045
Specialities: Conifers

86 Hunters Chase Garden Centre

Rayleigh Road, Hutton, Brentwood, Essex CM13 1SN
Tel: 01277 623793
Specialities: Garden Furniture, Herbaceous Perennials, Pots

87 Ingatestone Garden Centre

Ingatestone Garden Centre, Roman Road, Ingatestone, Essex CM4 9AU
Tel: 01277 353268 Fax: 01277 352113
e-mail: igc123@tiscali.co.uk
website: www.ingatestone-gardencentre.co.uk

SEE ADVERTISEMENT PAGE 85

Specialities:

Ornamental Trees, Shrubs, Perennials.

Description:

This 8-acre site is a veritable Aladdin's Cave for gardeners. In addition to the huge range of plants and trees, the centre stocks just about anything you could find in a garden – from aquatics to zinnias – as well as home accessories, giftware, clothing and collectibles. There's also the Café du Jardin and a children's play area.

Directions:

From Chelmsford, leave the A12 at Junction 13. From Brentwood, leave the A12 at Junction 12. Then follow brown tourism direction signs to Ingatestone Garden Centre. Located on the Southwest/ Brentwood side of Ingatestone Village

88 Jennikings Garden Centre

212 Manor Road, Chigwell, Essex IG7 4JX
Tel: 020 8501 2328
Specialities: Garden Furniture, Annuals/Biennials/Bedding

89 Jewels Garden Centre

Hodges Cottage, Colchester Road, Chelmsford, Essex CM2 5PR
Tel: 01245 460878

Specialities:
Landscaping; Turfing & Fencing; Rustic Garden Furniture

Products and Services:
- Credit Cards Accepted
- Fencing
- Garden Construction
- Garden Design
- Greenhouses and Sheds
- Paving
- Plants
- Trees & Speciality Trees

Directions:
First service station on the A12 from London (Boreham interchange), opposite BMW and next to Sainsbury.

Description:
This family-owned and operated business, in addition to its wide range of quality plants, also sells an extensive selection of hand-made rustic garden ware such as trellises, benches and decking. Other items on sale include pots, planters, top soil and cultivated turf, gas canisters, coal, logs and charcoal, aggregates, garden ornaments and hutches. At Christmas time, it has a full range of Christmas trees and wreaths. The nursery also stocks water features and ponds for which it will give advice and install in your garden. A full garden landscaping service is also available. An expert consultant will visit you at your garden within a 25-mile radius. The nursery is open year round from 8.30am-6.30pm, Monday to Saturday; 10am-4pm on Sunday. It is closed on Christmas Eve, Christmas Day and New Year's Day.

90 John Ray Nursery

36 Station Road, Braintree, Essex CM7 3QJ
Tel: 07866 296531
Specialities: Annuals/Biennials/Bedding, Herbaceous Perennials

91 Keers Green Nursery

Keers Green, Dunmow, Essex CM6 1PQ
Tel: 01279 876050
Specialities: Annuals/Biennials/Bedding, Shrubs

92 Ken Muir Nursery

Honeypot Farm, Rectory Road, Weeley Heath, Clacton-on-Sea, Essex CO16 9BJ
Tel: 01255 830181

93 Kidby Nurseries Ltd

Hilltop Crescent, Weeley Heath, Clacton-on-Sea, Essex CO16 9HF
Tel: 01255 830249
Specialities: Annuals/Biennials/Bedding, Herbaceous Perennials, Shrubs

94 King & Co

Woodstock, Pods Lane, Rayne, Braintree, Essex CM77 6WF
Tel: 01376 340469
Specialities: Topiary, Shrubs, Trees

95 Kinglea Plants

Pecks Hill, Nazeing, Waltham Abbey, Essex EN9 2NY
Tel: 01992 899410

96 L S Mummeries Nurseries

Barrow Hall Road, Little Wakering, Southend-on-Sea, Essex SS3 0NG
Tel: 01702 219278

97 L S Mummery (Nurseries) Ltd

195 Kiln Road, Benfleet, Essex SS7 1SJ
Tel: 01702 552152

98 Lakeview Nursery

Dobbs Weir Road, Roydon, Harlow, Essex CM19 5JX
Tel: 01992 445502
Specialities: Annuals/Biennials/Bedding, Shrubs, Herbaceous Perennials

99 Langley Nursery

9 Nazeing Road, Nazeing, Waltham Abbey, Essex EN9 2JN
Tel: 01992 892439

100 Langthorns Plantery

High Cross Lane, Little Canfield, Dunmow, Essex CM6 1TD
Tel: 01371 872611

101 Layer Marney Nurseries

Smythes Green, Layer Marney, Colchester, Essex CO5 9XW
Tel: 01206 330695

102 Leigh Garden Supplies

1 Leigham Drive, Leigh-on-Sea, Essex SS9 1PS
Tel: 01702 474376
Specialities: Garden Products

103 Livelands Nursery

121 London Road, Marks Tey, Colchester, Essex CO6 1EB
Tel: 01206 210504
Specialities: Annuals/Biennials/Bedding, Shrubs

104 Lowerhill Nursery

Sedge Green, Low Hill Road, Roydon, Harlow, Essex CM19 5JN
Tel: 01992 451477

105 Manor Nursery

Thaxted Road, Wimbish, Saffron Walden, Essex CB10 2UT
Tel: 01799 513481
Specialities: Annuals/Biennials/Bedding, Shrubs

106 Matthew's Plants

Tylers Rd, Roydon, Harlow, Essex CM19 5LJ
Tel: 01279 793539
Specialities: Annuals/Biennials/Bedding, Hanging Baskets, Shrubs

107 Mayflower Garden Centre

Mill Road, Mayland, Chelmsford, Essex CM3 6EQ
Tel: 01621 740269
Specialities: Specialist Bedding, Rose

108 Maypole Pet & Garden Centre

Maypole Road, Wickham Bishops, Witham, Essex CM8 3LX
Tel: 01621 892411

109 Meadow Croft Garden Centre

Woodham Road, Battlesbridge, Wickford, Essex SS11 7QU
Tel: 01245 320314

110 Messrs D Rhodes & J Rockliffe

2 Nursery Road, Nazeing, Waltham Abbey, Essex EN9 2JE
Tel: 01992 451598
Specialities: Begonia

111 Mill Race Garden Centre

New Road, Aldham, Colchester, Essex CO6 3QT
Tel: 01206 242521
Specialities: Aquatics, Restaurant/Coffee Shop

112 Morleys Nurseries

Southend Road, Great Wakering, Southend-on-Sea, Essex SS3 0PU
Tel: 01702 585668
Specialities: Aquatics, Pet Products, Greenhouses and Sheds

113 Mr B Wickenden

164 Point Clear Road, St. Osyth, Clacton-on-Sea, Essex CO16 8JB
Tel: 01255 821744

114 Mr M Haslett

12 Strangman Avenue, Thundersley, Benfleet, Essex SS7 1RB
Tel: 01702 558775
Specialities: Carnivorous Plants

115 N F & M C Kempster

Cross Farm Cottage, Walthams Cross, Great Bardfield, Braintree, Essex CM7 4QH
Tel: 01371 810125

116 Netherhouse Nurseries

Sewardstone Road, Essex E4 7RJ
Tel: 020 8529 1051
Specialities: Annuals/Biennials/Bedding, Herbaceous Perennials

117 Nevendon Nursery & Garden Centre

Stevensons Farm, Arterial Road, Basildon, Essex SS14 3JH
Tel: 01268 285330

118 Newhall Garden Centre

Lower Road, Hockley, Essex SS5 5JU
Tel: 01702 20007 Fax: 01702 203277
e-mail: jane-newhall@btconnect.com
website: www.newhallgardencentre.co.uk

SEE ADVERTISEMENT PAGE 87

Specialities:

Customer Service; Free Local Delivery; Good Selection of Healthy Plants

Description:

In addition to an extensive range of plants, shrubs and trees, the Centre also stocks hanging baskets, pots and aggregates. There's also the Blumeringue Gift Shop with a wide selection of hand-made crafts, and the Artful Dodger Art Gallery which also has a framing service.

Directions:

The Centre is located between the A130 and A127.

119 Newhall Vineyards

Chelmsford Road, Purleigh, Chelmsford, Essex CM3 6PN
Tel: 01621 828343
Specialities: Vines

120 The New Maltings

Chelmsford Road, Norton Mandeville, Ingatestone, Essex CM4 0LN
Tel: 01277 824747

121 Northfield Nursery

Sewardstone Road, Essex E4 7RG
Tel: 020 8529 0367

122 The Nursery

Old House Lane, Roydon, Harlow, Essex CM19 5DN
Tel: 01279 793331
Specialities: Annuals/Biennials/Bedding, Shrubs

123 Oak Nursery

Hackmans Lane, Purleigh, Chelmsford, Essex CM3 6RJ
Tel: 01621 828651
Specialities: Shrubs, Herbaceous Perennials

124 Oakview Nurseries

Dead Lane, Ardleigh, Colchester, Essex CO7 7RH
Tel: 01206 231134
Specialities: Shrubs

125 Old Mill House Garden Nursery

Old Millhouse, Guithavon Valley, Witham, Essex CM8 1HF
Tel: 01376 512396
Specialities: Shade Loving Plants, Herbaceous Perennials

126 Olivers Nurseries Ltd

Maldon Road, Witham, Essex CM8 3HY
Tel: 01376 513239
Specialities: Annuals/Biennials/Bedding, Shrubs

127 Orchard Garden Centre

Chadwell Road, Grays, Essex RM17 5SY
Tel: 01375 372195
Specialities: Garden Products

128 Orchids

Rivendell, White Horse Lane, Maldon, Essex CM9 5QP
Tel: 01621 874647
Specialities: Exotics, Orchids

129 P E Woods

Cherry Orchard Lane, Rochford, Essex SS4 1PN
Tel: 01702 545841

130 P R Spruzen

Tysea Hill, Stapleford Abbotts, Romford, Essex RM4 1JP
Tel: 01708 342717

131 Parker's Garden Company

Frinton Road, Kirby Cross, Frinton-on-Sea, Essex CO13 0PD
Tel: 01255 674838

132 Parker's Garden Company

Birdbrook, Halstead, Essex CO9 4BQ
Tel: 01440 788540

133 Parkins Plants

Mill Road, Henham, Bishop's Stortford, Essex CM22 6AA
Tel: 01279 851937
Specialities: Shrubs, Herbaceous Perennials

134 Pathview Nursery

Feathers Hill, Hatfield Broad Oak,
Bishop's Stortford, Essex CM22 7HB
Tel: 01279 718431
Specialities: Flower, Winter Pansies

135 Perrywood Nurseries

Kelvedon Road, Inworth, Tiptree, Essex CO5 9SX
Tel: 01376 570777
Specialities: Annuals/Biennials/Bedding, Herbaceous Perennials, Pot Plants

136 Pinewood Nursery

Walden Road, Ashdon, Saffron Walden,
Essex CB10 2NB
Tel: 01799 584165

137 Planthome Garden Centre

Woodham Road, Battlesbridge, Wickford,
Essex SS11 7QW
Tel: 01245 320263

138 Poplar Nurseries

Coggeshall Road, Marks Tey, Colchester,
Essex CO6 1HR
Tel: 01206 210374

139 The Pot & Grass Co

2 Shop Lane, East Mersea, Colchester,
Essex CO5 8TR
Tel: 01206 382861
Specialities: Grasses, Pots

140 Pot Shop

Damases Lane, Boreham, Chelmsford,
Essex CM3 3AL
Tel: 01245 465895
Specialities: Pots

141 Pottash Garden Centre

Main Road, Hawkwell, Hockley, Essex SS5 4JN
Tel: 01702 201120

142 Prestige Plant Propagators

Home Farm Lane, 20 Harwich Road, Ardleigh,
Colchester, Essex CO7 7LT
Tel: 01206 231032

143 R & R Saggers Plants

Waterloo House, High Street, Newport,
Saffron Walden, Essex CB11 3PG
Tel: 01799 540858
Specialities: Floristry

144 R Bolton & Son

Birdbrook, Halstead, Essex CO9 4BQ
Tel: 01440 785246

145 R H S E Plant Centre & Shop

Hyde Hall Road, Rettendon Common, Chelmsford,
Essex CM3 8ET
Tel: 01245 402113

146 R.G Copeman

Glebeland Nursery, Ironwell Lane, Hockley,
Essex SS5 4JY
Tel: 01702 202320
Specialities: Annuals/Biennials/Bedding, Herbaceous Perennials, Shrubs

147 Rainbow Nurseries

23 Colchester Road, Weeley Heath, Clacton-on-Sea,
Essex CO16 9JT
Tel: 01255 830505

148 Rawal Pindi Nursery

66 Main Road, Hawkwell, Hockley, Essex SS5 4JH
Tel: 01702 206244
Specialities: Annuals/Biennials/Bedding, Herbaceous Perennials

149 Rayleigh View Nursery

63 Kingsley Lane, Benfleet, Essex SS7 3TZ
Tel: 01268 747885
Specialities: Annuals/Biennials/Bedding, Shrubs

150 Reads (Sutton) Ltd

Sutton Road, Rochford, Essex SS4 1LQ
Tel: 01702 611211

151 Red House

School Road, Messing, Colchester, Essex CO5 9TH
Tel: 01621 815219
Specialities: Annuals/Biennials/Bedding, Shrubs

152 Reeves Lane Nursery

Reeves Lane, Roydon, Harlow, Essex CM19 5LE
Tel: 01279 793417
Specialities: Annuals/Biennials/Bedding

153 Riverside Nursery

Hayfield Lodge, Lower Road, Hockley,
Essex SS5 5LE
Tel: 01702 201100

154 Robin Savill Clematis Specialist

The Old Nursery, Butts Green Road, Sandon, Chelmsford, Essex CM2 7RN
Tel: 01245 224576
Specialities: Clematis

155 Robin Savill Clematis Specialist

2 Bury Cottages, Bury Road, Pleshey, Chelmsford, Essex CM3 1HB
Tel: 01245 237380
Specialities: Clematis

156 The Rodings Plantery

Anchor Lane, Abbess Roding, Ongar, Essex CM5 0JW
Tel: 01279 876421
Specialities: Bamboos, Rare and Unusual Plants, Trees

157 Rolts Nursery & Garden Centre

Clacton Road, Elmstead, Colchester, Essex CO7 7DD
Tel: 01206 822427
Specialities: Greenhouses and Sheds, Annuals/Biennials/Bedding, Shrubs

158 Rooster Nursery

7 Accommodation Road, Boxted, Colchester, Essex CO4 5HR
Tel: 01206 272232

159 Rose Cottage Plants

Bay Tree Farm, Epping Green, Epping, Essex CM16 6PU
Tel: 01992 573775
Specialities: Annuals/Biennials/Bedding, Shrubs

160 Roydon Hamlet Water Garden Centre

Tylers Rd, Roydon, Harlow, Essex CM19 5LJ
Tel: 01279 792235
Specialities: Aquatics, Water Products and Features

161 Sandown Nurseries

Sandown Road, Orsett, Grays, Essex RM16 3DD
Tel: 01375 891343
Specialities: Annuals/Biennials/Bedding, Herbaceous Perennials

162 Sedge Green Nursery

Sedge Green, Nazeing, Waltham Abbey, Essex EN9 2PA
Tel: 01992 441175
Specialities: Annuals/Biennials/Bedding, Shrubs

163 Sheil Nurseries

Hyde Farm Nursery, Hyde Lane, Danbury, Chelmsford, Essex CM3 4LN
Tel: 01245 226470

164 Sheila Chapman Clematis

c/o Coveney Nurseries Ltd, Ongar Road, Abridge, Romford, Essex RM4 1AA
Tel: 01708 688090
Specialities: Clematis

165 Shrubs & Tubs

114A High Road, Chigwell, Essex IG7 6NT
Tel: 020 8559 8595
Specialities: Shrubs

166 Silverton Garden Centre

118 Oxford Road, Clacton-on-Sea, Essex CO15 3TH
Tel: 01255 470324
Specialities: Landscape Gardening, Annuals/Biennials/Bedding

167 Slatfords Nuseries

Arterial Road, Rayleigh, Essex SS6 7TU
Tel: 01268 771569
Specialities: Annuals/Biennials/Bedding, Herbaceous Perennials

168 Smedleys Nurseries

Woodham Rd, Battlesbridge, Wickford, Essex SS11 7QU
Tel: 01268 733518
Specialities: Shrubs, Annuals/Biennials/Bedding

169 Snowhill Nursery

Snow Hill, Great Easton, Dunmow, Essex CM6 2DR
Tel: 01371 870030
Specialities: Annuals/Biennials/Bedding, Shrubs

170 Sow & Grow

Ongar Road, Pilgrims Hatch, Brentwood, Essex CM15 9JH
Tel: 01277 375252
Specialities: Seeds

171 Spains Hall Forest Nursery

Clockhouse Farm, Finchingfield Road,
Little Sampford, Saffron Walden, Essex CB10 2QN
Tel: 01799 586506

172 Spains Hall Forest Nursery

Spains Hall Road, Finchingfield, Braintree,
Essex CM7 4NJ
Tel: 01371 810156
Specialities: Annuals/Biennials/Bedding

173 Spring Valley Nursery

Crockleford Hall, Spring Valley Lane, Ardleigh,
Colchester, Essex CO7 7SA
Tel: 01206 230238
Specialities: Annuals/Biennials/Bedding, Shrubs

174 Springwell Nursery Limited

Little Chesterford, Saffron Walden, Essex CB10 1UE
Tel: 01799 530959
Specialities: Hanging Baskets, Annuals/Biennials/Bedding, Hardy Plants

175 Stalley's Barn

Rectory Farm, Cambridge Road, Littlebury,
Saffron Walden, Essex CB11 4TL
Tel: 01799 522464

176 Stockbrook Farm Shop & Nursery

Stock Road, Stock, Ingatestone, Essex CM4 9PQ
Tel: 01277 840046
Specialities: Farm Shop

177 Straight Mile Nursery Gardens

Ongar Road, Pilgrims Hatch, Brentwood,
Essex CM15 9SA
Tel: 01277 374439
Specialities: Landscape Gardening, Annuals/Biennials/Bedding

178 Sturmer Nurseries

Church Walk, Sturmer, Haverhill, Essex CB9 7XD
Tel: 01440 706416

179 Summerhill Nurseries & Garden Centre

Pipps Hill Road North, Crays Hill, Billericay,
Essex CM11 2UJ
Tel: 01268 280699

180 Sweet Hedges Nursery

31 Queens Road, Frinton-on-Sea, Essex CO13 9BL
Tel: 01255 671532
Specialities: Annuals/Biennials/Bedding, Shrubs

181 Thurrock Garden Centre

South Road, South Ockendon, Essex RM15 6DU
Tel: 01708 851991

182 Timag Nursery

2 Westwood Cottages, Colne Engaine, Colchester,
Essex CO6 2JH
Tel: 01787 473541
Specialities: Annuals/Biennials/Bedding, Herbaceous Perennials

183 Timmermans Nurseries

Arterial Road, West Horndon, Brentwood,
Essex CM13 3TB
Tel: 01277 811723

184 Todd's Botanics

West Street, Coggeshall, Colchester, Essex CO6 1NT
Tel: 01376 561212
Specialities: Rare and Unusual Plants, Hardy Plants, Exotics

185 Tom Green Plants

Elder St, Wimbish, Saffron Walden, Essex CB10 2XA
Tel: 01799 516529
Specialities: Annuals/Biennials/Bedding, Shrubs, Herbaceous Perennials

186 Tomlins Garden Centre

Mascalls Lane, Brentwood, Essex CM14 5LR
Tel: 01277 212243
Specialities: Floristry, Annuals/Biennials/Bedding

187 Town & Country Nursery

Tilbury Rd, West Horndon, Brentwood,
Essex CM13 3LS
Tel: 01277 811563

188 Treasure Holt Plants

Burrs Road, Clacton-on-Sea, Essex CO15 4SU
Tel: 01255 814838
Specialities: Annuals/Biennials/Bedding, Herbaceous Perennials

189 Tylers Cross Nursery

Epping Road, Nazeing, Waltham Abbey, Essex EN9 2DH
Tel: 01992 892362
Specialities: Annuals/Biennials/Bedding

190 Viliage Nursery

56 Well Lane, Stock, Ingatestone, Essex CM4 9LZ
Tel: 01277 841412
Specialities: Annuals/Biennials/Bedding, Shrubs

191 Vincenzo Rodia

67 Paternoster Hill, Waltham Abbey, Essex EN9 3JY
Tel: 01992 713321

192 Wallace C Co

Broads Green, Great Waltham, Chelmsford, Essex CM3 1DS
Tel: 01245 360413

193 Warley Rose Gardens

Great Warley, Codham Hall Lane, Brentwood, Essex CM13 3JS
Tel: 01277 221966
Specialities: Rose

194 Westview Nurseries

Main Road, Bicknacre, Chelmsford, Essex CM3 4HN
Tel: 01245 324171
Specialities: Annuals/Biennials/Bedding, Shrubs

195 Wheatley Farm Nursery

London Road, Rayleigh, Essex SS6 9ES
Tel: 01268 783111

196 White House Nurseries

Main Rd, Alresford, Colchester, Essex CO7 8DH
Tel: 01206 827360
Specialities: Annuals/Biennials/Bedding

197 Whittaker's Nursery

Heath Road, Orsett, Grays, Essex RM16 3AP
Tel: 01375 374455
Specialities: Annuals/Biennials/Bedding, Herbaceous Perennials

198 Wickham Place Farm

Station Road, Wickham Bishops, Witham, Essex CM8 3JB
Tel: 01621 891282

199 Writtle College Garden Centre

Lordship Road, Writtle, Chelmsford, Essex CM1 3RR
Tel: 01245 422011

200 Writtle Road Nursery

7 Writtle Road, Chelmsford, Essex CM1 3BL
Tel: 01245 265655
Specialities: Giftware, Floristry

201 Wyevale Garden Centre (Braintree)

Cressing Road, Braintree, Essex CM77 8DH
Tel: 01376 553043

202 Wyevale Garden Centre (Chelmsford)

Homelands Retail Park, Cuton Hall Lane, Springfield, Chelmsford, Essex CM2 5PX
Tel: 01245 466466

203 Wyevale Garden Centre (Colchester)

342 London Road, Stanway, Colchester, Essex CO3 8LT
Tel: 01206 213050

204 Wyevale Garden Centre (Rayleigh)

Eastwood Road, Rayleigh, Essex SS6 7LU
Tel: 01702 527331

GLOUCESTERSHIRE

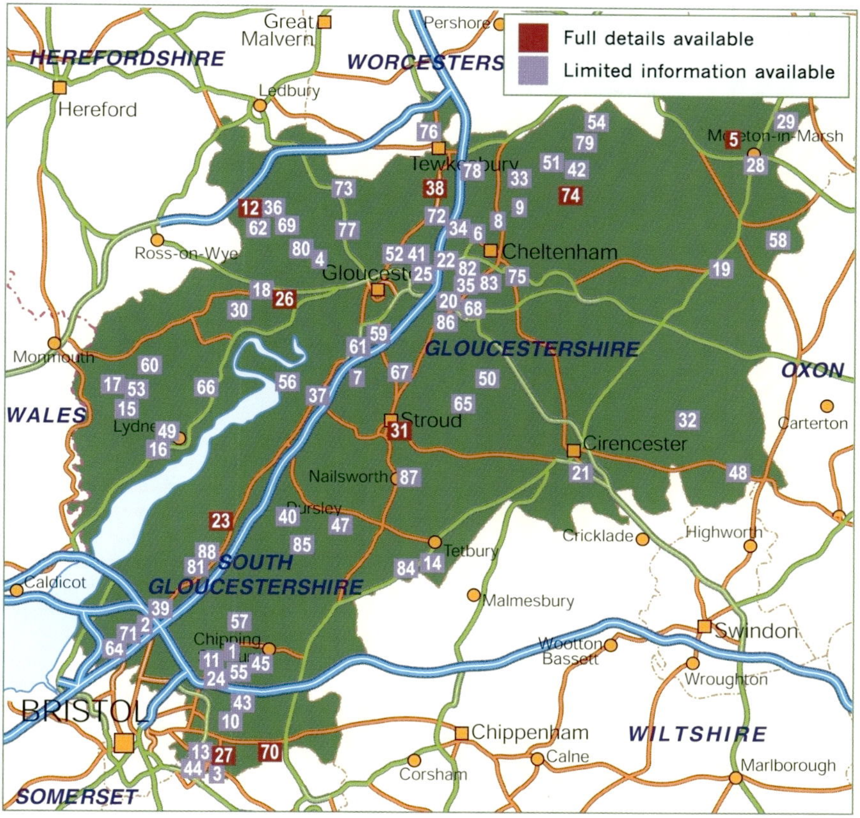

1 Advantage Plants

Heath Farm, Ram Hill, Coalpit Heath, Bristol, Gloucestershire BS36 2TX
Tel: 01454 776789
Specialities: Hanging Baskets

2 Almondsbury Garden Centre

Over Lane, Almondsbury, Bristol, Gloucestershire BS32 4BP
Tel: 01454 457300
Specialities: Fencing, Annuals/Biennials/Bedding, Aquatics

3 Avon Aquatics & Garden Centre

Bath Road, The Park, Willsbridge, Bristol, Gloucestershire BS30 6EE
Tel: 01179 325629
Specialities: Water Products and Features

4 Barbers Bridge Nurseries

Tibberton Corner, Barbers Bridge, Rudford, Gloucester, Gloucestershire GL2 8DX
Tel: 01452 790443

5 Batsford Plant Centre & Arboretum

Batsford, Moreton-in-Marsh, Gloucestershire GL56 6AB
Tel: 01386 700409
Fax: 01386 701148
e-mail: nursery@batsfordfoundation.co.uk
website: www.batsarb.co.uk

Specialities:
Japanese Maples; Herbaceous Perennials; Trees, Shrubs, Bamboo

Products and Services:

- Credit Cards Accepted
- Display Gardens
- Fruit Trees
- Garden Products
- Information Desk
- Large Car Park
- Plants
- Restaurant/Coffee Shop
- Seeds and Bulbs
- Trees

Directions:

Batsford Plant Centre is signposted on the A44, 1 mile west of Moreton-in-Marsh

Description:

Batsford Garden Centre is a specialist plantsman's garden showcasing many rare and unusual trees, shrubs, alpine plants, herbaceous and perennials plus a full complement of garden sundries, compost mulches, terracotta and glazed pots, seeds and gifts. The Centre offers specialist advice on its stock of Ferns, Bamboos, Japanese Maples and Magnolias. The adjacent Batsford Arboretum covers 56 acres and is one of the largest private collections of trees and shrubs in the country.

It was originally designed in the late 1800s by Algernon Freeman Mitford, later Lord Redesdale, who was influenced in his plantings by his love of the landscapes of China and Japan which developed during his diplomatic postings there earlier in the century. He imported the Buddha and other bronze statues for the arboretum and introduced features such as the Japanese Rest House, Hermit's Cave, rockeries and a 600-metre long artificial stream.

6 The Bedding Plant Centre

Tewkesbury Road, Uckington, Cheltenham, Gloucestershire GL51 9SL
Tel: 01242 680366
Specialities: Annuals/Biennials/Bedding

7 Blooms of Bressingham

Bath Road, Haresfield, Stonehouse, Gloucestershire GL10 3DP
Tel: 01452 721081

8 Blooms of Bressingham

Evesham Road, Prestbury, Cheltenham, Gloucestershire GL50 4SJ
Tel: 01242 672560

9 Bonsai World of Cheltenham

Two Hedges Road, Woodmancote, Cheltenham, Gloucestershire GL52 9PT
Tel: 01242 674389
Specialities: Trees

10 Brecklands

Siston Lane, Bristol, Gloucestershire BS30 5LX
Tel: 01179 610554
Specialities: Herbaceous Perennials, Herbs, Alpines/Rock Plants

11 C S Lockyer

Lansbury, Henfield Road, Coalpit Heath, Bristol, Gloucestershire BS36 2UZ
Tel: 01454 772219

12 Chamberlain's Plant Centre

Ledbury Road, Newent, Gloucestershire GL18 1DL
Tel: 01531 828488

Specialities:

Bedding plants; Seasonal trees and shrubs; Specimen trees and shrubs

Description:

In the height of the bedding season, Chamberlain's offers acres of bedding plants, simply take your pick from the 1000s of them directly from the growing beds – you can't buy any fresher than this! Also offered is a quality mix of seasonal trees and shrubs, along with specimen plants.

Directions:

From Newent Town Centre traffic lights take the B4215 (Bridge Street) to Ledbury. The garden centre will be found on your right hand side after approx 1 mile.

13 Chief Trading Post Ltd

Barry Road, Oldland Common, Bristol, Gloucestershire BS30 6QY
Tel: 01179 323112
Specialities: Water Products and Features, Ferns, Aquatics

14 Close Nursery

Shipton Moyne, Tetbury, Gloucestershire GL8 8PJ
Tel: 01666 505021

15 Coinros Nursery

Clements End, Sling, Coleford,
Gloucestershire GL16 8LN
Tel: 01594 562610
Specialities: Herbaceous Perennials, Shrubs, Trees

16 Coinros Park

New Office, Lydney Park Estate, Lydney,
Gloucestershire GL15 6BU
Tel: 01594 842634
Specialities: Herbaceous Perennials, Shrubs, Trees

17 Coleford Garden Centre

Lambsquay Road, Milkwall, Coleford,
Gloucestershire GL16 8QA
Tel: 01594 832700
Specialities: Annuals/Biennials/Bedding, Herbaceous Perennials, Shrubs

18 Country Garden Centre

Ross Road, Huntley, Gloucester,
Gloucestershire GL19 3EY
Tel: 01452 830229
Specialities: Shrubs, Herbaceous Perennials

19 Countrywide Farmers Plc

Station Rd, Bourton-on-The-Water, Cheltenham,
Gloucestershire GL54 2EP
Tel: 01451 820551
Specialities: Pet Products, Annuals/Biennials/Bedding, Garden Furniture

20 Dawn Nurseries

Shurdington Road, Shurdington, Cheltenham,
Gloucestershire GL51 4UD
Tel: 01242 862877

21 Dobbies Garden World

Siddington, Siddington, Cirencester,
Gloucestershire GL7 6EU
Tel: 01285 884300

22 Dundry Nurseries

Bamfurlong Lane, Cheltenham,
Gloucestershire GL51 6SL
Tel: 01452 713124
Specialities: Annuals/Biennials/Bedding, Shrubs, Garden Products

23 Eastwood Garden Plant Centre

Eastwood Park, Falfield, Wotton-under-Edge,
Gloucestershire GL12 8DA
Tel: 01454 260288
e-mail: neildavidevans@btinternet.com
website: www.eastwoodgardencentre.co.uk
Specialities: Cafe, Fruit Trees, Bedding Plants

24 Elmwood Nursery

Westerleigh Road, Henfield, Bristol,
Gloucestershire BS36 2UP
Tel: 01179 561495

25 Evergreen Nurseries

Parton Road, Churchdown, Gloucester,
Gloucestershire GL3 2JB
Tel: 01452 713506
Specialities: Conifers, Annuals/Biennials/Bedding

26 Fairview Nursery & Garden Shop

Chapel Lane, Birdwood, Churcham,
nr Gloucester GL2 8AR
Tel: 01452 750436

Specialities:

Geraniums & Fuchsias; Hanging Baskets & Patio Plants; Vegetable Plants

Products and Services:

- Seasonal Displays
- Credit Cards Accepted
- Fruit Trees
- Garden Furniture
- Garden Products
- Gift Vouchers
- Giftware
- Houseplants
- Information Desk
- Large Car Park
- Plant Guarantee
- Plants
- Seeds and Bulbs
- Trees

Directions:

Fairview Nursery is just off the A40, 6 miles west of Gloucester, signposted to Churcham then Birdwood.

27 Fonthill Garden Centre

Bath Road, Bitton, Bristol BS30 6HX
Tel: 0117 932 3110
website: www.fonthill-lakeside.com

SEE ADVERTISEMENT PAGE 96

Specialities:

Trees & Shrubs, Bedding Plants,

Products and Services:

- Barbeques
- Books & Stationery
- Seasonal Displays
- Clothing
- Credit Cards Accepted
- Disabled Facilities
- Display Gardens
- Dried & Artificial Flowers
- Fencing
- Floristry
- Fruit Trees
- Garden Furniture
- Garden Products
- Giftware
- Greenhouses and Sheds
- Houseplants
- Information Desk
- Large Car Park
- Pet Products
- Plant Guarantee
- Plants
- Restaurant/Coffee Shop
- Seeds and Bulbs
- Trees

Directions:

Fonthill Garden Centre is conveniently situated on the A431, half way between Bath and Bristol in the village of Bitton.

28 Fosseway Garden Centre

Stow Road, Moreton-in-Marsh,
Gloucestershire GL56 0DS
Tel: 01608 651757

29 Four Counties Nursery and Garden Centre

Todenham, Moreton-in-Marsh,
Gloucestershire GL56 9PN
Tel: 01608 650522

30 G B Keyse

Blaisdon Nurseries, Blaisdon, Longhope,
Gloucestershire GL17 0AH
Tel: 01452 830359
Specialities: Annuals/Biennials/Bedding, Bulbs, Vegetables

31 The Garden Shop

30 High Street, Stroud, Gloucestershire GL5 1AJ
Tel: 01453 758432
e-mail: trudiapple@aol.com

Specialities:

Bulb Displays; Garden Sundries; Giftware

Products and Services:

- Fencing
- Floristry
- Fruit Trees
- Garden Products
- Giftware
- Houseplants
- Plants
- Seeds and Bulbs

Directions:

The shop is located at the top of the High Street with parking close by.

32 Gardens & Plants

Hatherop Castle Gardens, Hatherop, Nr Cirencester,
Gloucestershire GL7 3NA
Tel: 01285 750363
Specialities: Trees, Herbaceous Perennials, Annuals/Biennials/Bedding

33 Gotherington Nurseries

Gretton Road, Gotherington, Cheltenham, Gloucestershire GL52 9QY
Tel: 01242 676722
Specialities: Annuals/Biennials/Bedding, Hanging Baskets, Herbaceous Perennials

34 Green Acre Nursery

Tewkesbury Road, Uckington, Cheltenham, Gloucestershire GL51 9SW
Tel: 01242 680610

35 Greenway Nurseries

Shurdington Road, Shurdington, Cheltenham, Gloucestershire GL51 4TT
Tel: 01242 862248

36 Hazelfield Garden Centre

Greentree Leisure Ltd, Birches Lane, Newent, Gloucestershire GL18 1DN
Tel: 01531 820941
Specialities: Annuals/Biennials/Bedding, Herbaceous Perennials, Shrubs

37 Highfield Garden World

Bristol Road, Whitminster, Gloucester, Gloucestershire GL2 7PB
Tel: 01452 741444

38 Hoo House Nursery

Gloucester Road, Tewkesbury, Gloucestershire GL20 7DA
Tel: 01684 293389
e-mail: plants@hoohouse.co.uk
website: www.hoohouse.co.uk
Illustrated brochure available on website

Specialities:
Hardy perennials and Alpines

Products and Services:

- Free generous gardening advice
- Nearly 1000 unusual and easily grown varieties
- Plant assurance

Directions:
The nursery is on the A38, 2 miles south of Tewkesbury. Turn towards Bishops Cleeve at the Odessa Inn and the nursery is 100yards on the right. 5 mins from the M5.

Description:
Established in 1986, **Hoo House Nursery** is owned and run by Julie Ritchie, who grows her plants in peat-free compost. The nursery holds the National Collections of Balloon Flowers and Border Gentians which can be seen growing, with many of the other plants, in the garden. Occasionally Julie has been persuaded to do mail order, though a visit is much to be preferred.

Enjoy browsing the pictures on the website.

Open:
Everyday 10am-5pm; Sunday 11am-5pm

39 Hortham Nurseries

47 Hortham Lane, Almondsbury, Bristol, Gloucestershire BS32 4JJ
Tel: 01454 613468

40 Hunts Court Garden & Nursery

North Nibley, North Nibley, Dursley, Gloucestershire GL11 6DZ
Tel: 01453 547440
Specialities: Rose, Geranium, Penstemon

41 Hurrans Garden Centre

Cheltenham Road East, Churchdown, Gloucester, Gloucestershire GL3 1AB
Tel: 01452 712232

42 Ireley Hardy Plants

Broadway Road, Winchcombe, Cheltenham, Gloucestershire GL54 5NY
Tel: 01242 604160

43 J Pollitt

St. Aldams Ash Farm, Westerleigh Road, Pucklechurch, Bristol, Gloucestershire BS16 9PY
Tel: 01179 374212

44 Jarretts Garden Centre

Bath Rd, Willsbridge, Bristol, Gloucestershire BS30 6ES
Tel: 01179 323716
Specialities: Aquatics, Restaurant/Coffee Shop

45 Kemps Plants

Westerleigh Road, Westerleigh, Bristol, Gloucestershire BS37 8QH
Tel: 01454 322123

46 Kiftsgate Court Gardens

Kiftsgate Court, Chipping Campden, Gloucestershire GL55 6LN
Tel: 01386 438777

47 The Lavender Garden

Ashcroft Nurseries, Kingscote, Tetbury, Gloucestershire GL8 8YF
Tel: 01453 860356
Specialities: Buddleja (Buddleia), Lavender

48 Lechlade Garden Centre

Warrens Cross, Lechlade, Gloucestershire GL7 3DP
Tel: 01367 254100

49 Lydney Garden Centre

15 High Street, Lydney, Gloucestershire GL15 5DP
Tel: 01594 842121
Specialities: Annuals/Biennials/Bedding, Herbaceous Perennials, Water Products and Features

50 Miserden Nursery

Miserden, Stroud, Gloucestershire GL6 7JA
Tel: 01285 821638
Specialities: Herbaceous Perennials

51 Nomads Fuschsias

Evesham Road, Greet, Cheltenham, Gloucestershire GL54 5BX
Tel: 01242 604304
Specialities: Fuchsia

52 Norton Garden Centre

Tewkesbury Road, Down Hatherley, Gloucester, Gloucestershire GL2 9PU
Tel: 01452 730852
Specialities: Garden Furniture, Annuals/Biennials/Bedding, Giftware

53 The Nursery

Little Drybrook, Coleford, Gloucestershire GL16 8LP
Tel: 01594 563266
Specialities: Annuals/Biennials/Bedding

54 Oak Piece Farm Nurseries

Stanway Road, Stanton, Broadway, Gloucestershire WR12 7NQ
Tel: 01386 584077
Specialities: Ornamentals

55 Oakley Green Nurseries

Westerleigh Hill Road, Westerleigh, Bristol, Gloucestershire BS37 8QZ
Tel: 01179 708040

56 Pan-Global Plants

The Walled Garden, The Green, Frampton On Severn, Gloucester, Gloucestershire GL2 7EX
Tel: 01452 741641
Specialities: Exotics

57 Parkers Aquatic Garden Centre

Wotton Road, Iron Acton, Bristol BS37 9XA
Tel: 01454 228761
Specialities: Aquatics, Garden Furniture, Garden Machinery

58 Pebbly Hill Farm Nursery

Icomb Rd, Bledington, Chipping Norton, Gloucestershire OX7 6XJ
Tel: 01608 659851
Specialities: Annuals/Biennials/Bedding, Herbaceous Perennials

59 Pound Farm Shop & Plant Centre

Stroud Road, Whaddon, Gloucester, Gloucestershire GL4 0UG
Tel: 01452 523936
Specialities: Annuals/Biennials/Bedding, Hanging Baskets, Herbaceous Perennials

60 Pygmy Pinetum Garden Nursery

Cannop Crossroads, Cannop, Coleford, Gloucestershire GL16 7EQ
Tel: 01594 833398
Specialities: Herbaceous Perennials, Shrubs, Trees

61 Quality Fuchsias

Sunnyholm, Naas Lane, Quedgeley, Gloucester, Gloucestershire GL2 2SD
Tel: 01452 721713
Specialities: Fuchsia, Geranium, Begonia

62 Roses Country Fayre

52 Ledbury Road, Newent, Gloucestershire GL18 1DL
Tel: 01531 821242
Specialities: Rose, Hanging Baskets, Annuals/Biennials/Bedding

63 Sabin's Nurseries

Oakham Lodge, Ebrington, Chipping Campden, Gloucestershire GL55 6NL
Tel: 01386 593510
Specialities: Annuals/Biennials/Bedding, Hanging Baskets

64 Sandy's Nursery

Over Lane, Almondsbury, Bristol,
Gloucestershire BS32 4DD
Tel: 01454 615076

65 Selsley Herb Nursery

Hyhedge Lane, Bisley, Stroud,
Gloucestershire GL6 7AN
Tel: 01452 770073
Specialities: Herbs

66 Severn View Garden Centre

Newnham Road, Blakeney, Gloucestershire GL15 4AE
Tel: 01594 516777
Specialities: Annuals/Biennials/Bedding, Shrubs, Herbaceous Perennials

67 Shady Plants

Upper Doreys Mill, Edge, Nr Stroud,
Gloucestershire GL6 6NF
Tel: 01452 812459
Specialities: Exotics, Hosta

68 Shurdington Nurseries

Little Shurdington, Shurdington, Cheltenham,
Gloucestershire GL51 4TX
Tel: 01242 863738
Specialities: Trees, Shrubs, Herbaceous Perennials

69 Smith's Clematis Nurseries

36 Ford House Road, Newent,
Gloucestershire GL18 1LQ
Tel: 01531 820842
Specialities: Clematis

70 Special Plants

Greenway Lane, Cold Ashton, Chippenham,
Gloucestershire SN14 8LA
Tel: 01225 891686
e-mail: derry@specialplants.net
website: www.specialplants.net
Specialities: Geranium, Seeds, Rockery Plants, Rare and Unusual Plants

71 Springwood Nurseries

62a Over Lane, Almondsbury, Bristol,
Gloucestershire BS32 4BW
Tel: 01454 616110
Specialities: Alpines/Rock Plants, Herbaceous Perennials, Shrubs

72 Stanboro Lane Nurseries

Stanboro Lane, Elmstone Hardwicke, Cheltenham,
Gloucestershire GL51 9TN
Tel: 01242 681115
Specialities: Shrubs, Garden Products

73 Staunton Garden Centre

Staunton Cross, Ledbury Road, Staunton,
Gloucester, Gloucestershire GL19 3QA
Tel: 01452 840612
Specialities: Annuals/Biennials/Bedding, Shrubs

74 Sudeley Castle Roses & Gardens

Sudeley Castle, Winchcombe,
Gloucestershire GL54 5JD
Tel: 01242 602308

SEE ADVERTISEMENT PAGE 100

Specialities:

Award-winning Gardens; David Austin Roses; Gardening Events

Products and Services:

- Credit Cards Accepted
- Disabled Facilities
- Display Gardens
- Garden Products
- Giftware
- Houseplants
- Information Desk
- Large Car Park
- Plants
- Restaurant/Coffee Shop
- Seeds and Bulbs

Directions:

Sudeley Castle is about 1 mile south of Winchcombe, off the B4632, and widely signposted.

75 Telling and Coates

64A Church Street, Charlton Kings, Cheltenham,
Gloucestershire GL53 8AS
Tel: 01242 514472
Specialities: Alpines/Rock Plants, Heathers

76 Tewkesbury Garden Centre

Worcester Road, Mythe Road, Tewkesbury,
Gloucestershire GL20 6EB
Tel: 01684 293103

77 Thuya Alpine Nursery

Glebelands, Gloucester Road, Hartpury, Gloucester,
Gloucestershire GL19 3BW
Tel: 01452 700548
Specialities: Alpines/Rock Plants

78 Tinpenny Plants

Tinpenny Farm, Fiddington, Tewkesbury,
Gloucestershire GL20 7BJ
Tel: 01684 292668
Specialities: Hosta, Iris, Hemerocallis (Daylily)

79 Toddington Garden Centre

Toddington, Toddington, Cheltenham,
Gloucestershire GL54 5DT
Tel: 01242 621314
Specialities: Climbers, Trees, Rose

80 Trioscape Garden Centre

Highleadon, Highleadon, Newent,
Gloucestershire GL18 1HQ
Tel: 01452 790550
Specialities: Annuals/Biennials/Bedding, Shrubs, Garden Furniture

81 Tytherington Road Nursery

Tytherington Road, Grovesend Thornbury, Bristol,
Gloucestershire BS35 3TT
Tel: 01454 281300
Specialities: Hanging Baskets, Annuals/Biennials/Bedding

82 Valley Roundabout Nurseries

Badgeworth Road, The Reddings, Cheltenham,
Gloucestershire GL51 6SJ
Tel: 01452 713102

83 Waoku Nurseries

Kidnappers Lane, Cheltenham,
Gloucestershire GL53 0NP
Tel: 01242 513401
Specialities: Exotics, Restaurant/Coffee Shop, Bamboos

84 Westonbirt Plant Centre

Forest Enterprise, Westonburt Aboretum, Westonbirt,
Tetbury, Gloucestershire GL8 8QS
Tel: 01666 880544
Specialities: Acer, Trees

85 Wotton Farm Shop

Bradley Raod, Wotton-under-Edge,
Gloucestershire GL12 7DT
Tel: 01453 521546
Specialities: Annuals/Biennials/Bedding, Hanging Baskets, Trees

86 Wyevale Garden Centre (Brockworth)

Shurdington Road, Brockworth, Gloucester,
Gloucestershire GL3 4PU
Tel: 01452 862334

87 Wyevale Garden Centre (Nailsworth)

Waterside Garden Centre, Avening Road, Nailsworth,
Stroud, Gloucestershire GL6 0BS
Tel: 01453 833989

88 Wyevale Garden Centre (Thornbury)

Milbury Heath, Milbury Heath, Wotton-under-Edge,
Gloucestershire GL12 8QH
Tel: 01454 412247

HAMPSHIRE

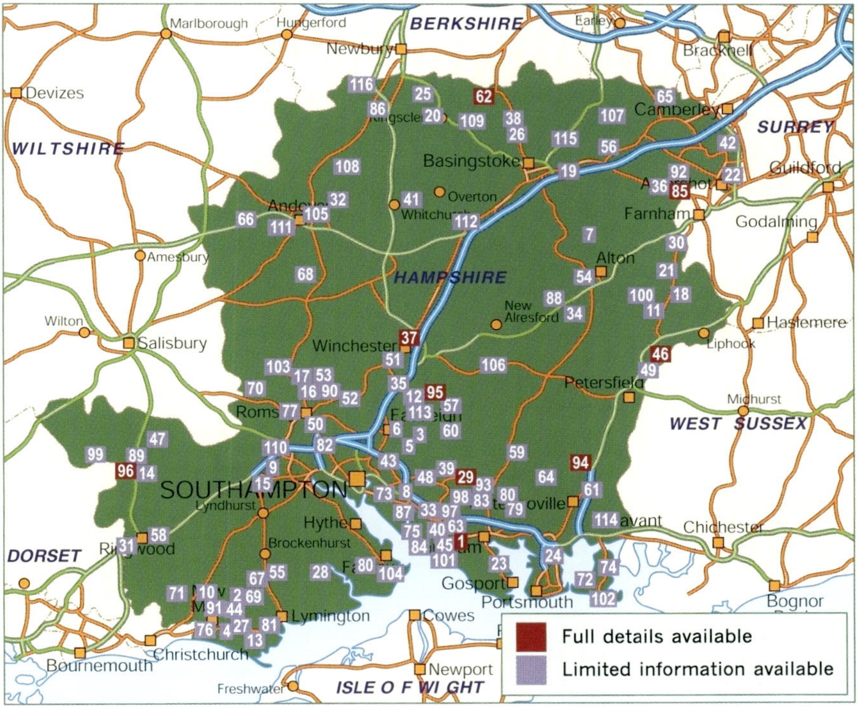

1 Abbey Garden Centre

Mill Lane, Titchfield, Fareham, Hampshire PO15 5RB
Tel/Fax: 01329 842225
e-mail: info@abbeygardencentre.co.uk
website: www.abbeygardencentre.co.uk

SEE ADVERTISEMENT PAGE 102

Specialities:

Seasonal Plants & Shrubs; Bedding Plants; Garden Furniture & BBQs

Description:

The Centre stocks thousands of different ranges for both the garden and indoors, along with a huge selection of gardening accessories. It also has a fully licensed restaurant offering a wide range of food throughout the day, every day.

Directions:

Titchfield village is south of the A27 about 5 miles west of Fareham. The Centre is to the north of the village

2 Agars Nursery

Agars Lane, Hordle, Lymington, Hampshire SO41 0FL
Tel: 01590 683703

3 Allington Nursery

Allington Lane, Fair Oak, Eastleigh,
Hampshire SO50 7DD
Tel: 023 8060 0182
Specialities: Aquatics, Garden Furniture, Restaurant/Coffee Shop

4 Apple Court

Hordle Lane, Hordle, Lymington,
Hampshire SO41 0HU
Tel: 01590 642130
Specialities: Garden Products

5 Arturi's Garden Centre

Allington Lane, Fair Oak, Eastleigh, Hampshire SO50 7DE
Tel: 023 8060 2234
Specialities: Restaurant/Coffee Shop, Trees, Shrubs

6 Ashtrim Landscape Centre

1 Bishopstoke Road, Eastleigh, Hampshire SO50 6AD
Tel: 023 8061 2385
Specialities: Landscape Gardening

7 Avenue Nurseries Garden Centre

The Avenue, Lasham, Alton, Hampshire GU34 5SU
Tel: 01256 381648
Specialities: Restaurant/Coffee Shop, Greenhouses and Sheds, Annuals/Biennials/Bedding

8 Barbara Price

1 Pewitt Hill, Bursledon, Southampton, Hampshire SO31 8BL
Tel: 023 8040 3775
Specialities: Herbaceous Perennials, Shrubs

9 Barrow Hill Nurseries

Barrow Hill Road, Copythorne, Southampton, Hampshire SO40 2PH
Tel: 023 8081 4078
Specialities: Trees, Shrubs

10 Bashley Plant Centre

Bashley Common Road, New Milton, Hampshire BH25 5SG
Tel: 01425 612442

11 Blackmoor Plant Centre

Blackmoor, Liss, Hampshire GU33 6BP
Tel: 01420 488822
Specialities: Window Boxes, Air Plants

12 Brambridge Park Garden Centre

Kiln Lane, Brambridge, Eastleigh, Hampshire SO50 6HT
Tel: 01962 713707

13 Braxton Gardens

Lymore Lane, Milford On Sea, Lymington, Hampshire SO41 0TX
Tel: 01590 642008
Specialities: Garden Products, Restaurant/Coffee Shop, Giftware

14 Brookfields Nursery

Stuckton, Fordingbridge, Hampshire SP6 2HE
Tel: 01425 656779
Specialities: Annuals/Biennials/Bedding, Hanging Baskets

15 Cadnam Garden Centre

Southampton Road, Cadnam, Southampton, Hampshire SO40 2NB
Tel: 023 8081 2240
Specialities: Giftware, Garden Furniture, Annuals/Biennials/Bedding

16 Cedar Nurseries

Sandy Lane, Romsey, Hampshire SO51 0PD
Tel: 01794 368375

17 Choice Plant Nurseries

Stockbridge Rd, Timsbury, Romsey, Hampshire SO51 0NB
Tel: 01794 368895
Specialities: Annuals/Biennials/Bedding

18 CMS Water Features

11 Montrose Close, Whitehill, Bordon, Hampshire GU35 9RG
Tel: 07919 116890
Specialities: Water Products and Features

19 Conkers Garden Centre

London Road, Hatch, Basingstoke, Hampshire RG24 7JL
Tel: 01256 840515
Specialities: Restaurant/Coffee Shop, Annuals/Biennials/Bedding, Herbaceous Perennials

20 Cottismore Garden Centre

Newbury Road, Kingsclere, Newbury, Hampshire RG20 4SY
Tel: 01635 298368

21 Country Market

Malthouse & Osbourne Farms, Main Road, Kingsley, Bordon, Hampshire GU35 9LW
Tel: 01420 472486

22 Covers

Unit 6, Blackwater Park, Aldershot, Hampshire GU12 4PQ
Tel: 01252 320554

23 Covers

Gunners Way, Gosport, Hampshire PO12 4DW
Tel: 023 9260 4400

24 Covers

110 Norway Road, Portsmouth, Hampshire PO3 5FT
Tel: 023 9267 1900

25 The Darling Buds of May Nursery

Newbury Rd, Headley, Thatcham, Hampshire RG19 8LA
Tel: 01635 269308
Specialities: Annuals/Biennials/Bedding, Shrubs, Herbaceous Perennials

26 Elm Park Garden Centre

Aldermaston Road, Pamber End, Tadley, Hampshire RG26 5QW
Tel: 01256 850587
Specialities: Rose

27 Everton Nurseries Garden Centre

Farmers Walk, Everton, Lymington, Hampshire SO41 0JZ
Tel: 01590 642155
Specialities: Garden Products

28 Fairweather's Garden Centre

High Street, Beaulieu, Hampshire SO42 7YB
Tel: 01590 612307
Specialities: Garden Products, Giftware, Restaurant/Coffee Shop

29 Family Trees

Sandy Lane, Shedfield, Hampshire SO32 2HQ
Tel: 01329 834812

Specialities:

Trained Fruit Trees & Full Standards; Young Trees; Old Roses

Description:

Family Trees was founded in 1980 to supply more mature and freshly dug trees to gardeners and professional landscapers. It is open between October 15th and April 30th on Tuesdays, Wednesdays, Fridays and Saturdays from 9.30am to 12.30pm for the sale of trees and collection of any orders. Send a SAE 9''X6'' to request a catalogue.

Directions:

From the A334, about 2 miles west of Wickham, turn right into Sandy Lane. Follow it round two bends and then take first left to Family Trees.

30 Forest Lodge Garden Centre

Denys E Head Ltd, Farnham Road, Holt Pound, Farnham, Hampshire GU10 4LD
Tel: 01420 23275

31 Frampton & Sons

48 High Street, Ringwood, Hampshire BH24 1BQ
Tel: 01425 472342
Specialities: Composts

32 The Garden Superstore

Unit 101, The Commercial Centre, Picket Piece, Andover, Hampshire SP11 6RU
Tel: 01264 362255
Specialities: Annuals/Biennials/Bedding

33 Garsons

Fontley Garden Centre, Fontley Road, Titchfield, Fareham, Hampshire PO15 6QX
Tel: 01329 844336
Specialities: Garden Furniture, Annuals/Biennials/Bedding

34 Garthowen Garden Centre & Nursery

Alton Lane, Four Marks, Alton, Hampshire GU34 5AJ
Tel: 01962 773225
Specialities: Annuals/Biennials/Bedding

35 George Beckett Nurseries

Otterbourne Road, Compton, Winchester, Hampshire SO21 2RT
Tel: 01962 713732
Specialities: Trees, Shrubs

36 Goddards Nurseries

Goddards Farm, Bowling Alley, Crondall, Farnham, Hampshire GU10 5RJ
Tel: 01252 851145

37 The Goodlife Home & Garden Centre

Springvale Road, Headbourne Worthy, Winchester, Hampshire SO23 7LD
Tel: 01962 889000 Fax: 01962 889400
e-mail: enquiries@goodlifegardencentre.co.uk

SEE ADVERTISEMENT PAGE 105

Specialities:

Plants & Shrubs; Food Hall; Gift Shop

Description:

In addition to its healthy and vigorous plants – grown on reclaimed watercress beds – the Goodlife Centre also stocks a fine collection of pots and ornamental wrought iron garden furniture. Also on site is a Food Hall with a wonderful range of food from all around the world, and a Gift Shop filled with elegant and unusual ornaments and home-ware.

Products and Services:

- Seasonal Displays
- Credit Cards Accepted
- Disabled Facilities
- Farm Shop
- Garden Furniture
- Garden Products

Directions:

From junction 9 of the M3 take the A34 towards Newbury and bear off to the right alongthe A33. Take first left along the B3047 and just after the A34 flyover turn right into Bedfield Lane. At the T junction turn left and the centre is on the left.

38 Greenacres Landscapes & Nursery

Aldermaston Rd, Pamber Green, Tadley, Hampshire RG26 3AF
Tel: 01256 850470
Specialities: Trees, Shrubs, Hydroponics

39 Hambrook Garden Landscape Centre

Wangfield Lane, Curdridge, Southampton, Hampshire SO32 2DA
Tel: 01489 779998
Specialities: Shrubs, Trees, Landscape Gardening

40 Hambrooks Garden Centre

135 Southampton Road, Titchfield, Fareham, Hampshire PO14 4PR
Tel: 01489 572285
Specialities: Annuals/Biennials/Bedding, Herbaceous Perennials

41 Hardys Cottage Garden Plants

Priory Lane, Freefolk Priors, Freefolk, Whitchurch, Hampshire RG28 7NJ
Tel: 01256 896533
Specialities: Annuals/Biennials/Bedding, Herbaceous Perennials, Cottage Garden Plants

42 Hart Canna

27 Guildford Rd West, Farnborough, Hampshire GU14 6PS
Tel: 01252 514421
Specialities: Landscape Gardening, Paving

43 Haskins Garden Centre

Mansbridge Road, West End, Southampton,
Hampshire SO18 3HW
Tel: 023 8047 2324

44 Heatherleigh Nursery

Everton Road, Hordle, Lymington,
Hampshire SO41 0FD
Tel: 01425 614777

45 Heisei Acre Nurseries

St. Margarets Lane, Fareham, Hampshire PO14 4BL
Tel: 01329 846024
Specialities: Japanese

46 Highfield Hollies

Highfield Farm, Hatch Lane, Liss,
Hampshire GU33 7NH
Tel: 01730 892372 Fax: 01730 894853
e-mail: louise@highfieldhollies.com
website: www.highfieldhollies.com

Specialities:

Hollies

Description:

Highfield Hollies is the country's only specialist farm growing a wide range of hollies for all purposes. The nursery stocks everything from large specimen trees and bushes for instant impact, through topiary shapes, hedging plants, collectors rarities and crenatas from Japan for rockeries and miniature plantings.

Directions:

From the A3, take the B3006 into Liss. At roundabout in village turn right then immediately right again. Continue up Rake Road 1 mile then turn right into Hatch Lane, by letter box in wall. Highfield Farm is on the right.

47 Hill Garth Nursery

Hill Garth, Woodgreen Road, Godshill, Fordingbridge,
Hampshire SP6 2LP
Tel: 01425 657619
Specialities: Cottage Garden Plants, Rare and Unusual Plants

48 Hillier Garden Centre Botley

Woodhouse Lane, Botley, Southampton, Hampshire SO30 2EZ
Tel: 01489 782306

49 Hillier Garden Centre Liss

Farnham Road, Liss, Hampshire GU33 6LJ
Tel: 01730 892196

50 Hillier Garden Centre Romsey

Botley Road, Romsey, Hampshire SO51 8ZL
Tel: 01794 513459

51 Hillier Garden Centre Winchester

Romsey Road, Winchester, Hampshire SO22 5DN
Tel: 01962 842288

52 Hillier Nurseries Ltd

Ampfield House, Winchester Road, Ampfield, Romsey, Hampshire SO51 9PA
Tel: 01794 368733

53 Hillier Plant Centre Braishfield

Braishfield, Jermyns Lane, Ampfield, Romsey, Hampshire SO51 0QA
Tel: 01794 368407

54 Hillside Nursery

Basingstoke Road, Beech, Alton, Hampshire GU34 4AB
Tel: 01420 80544
Specialities: Annuals/Biennials/Bedding

55 Hollybush Garden Centre

Lymington Road, Brockenhurst, Hampshire SO42 7UF
Tel: 01590 622839
Specialities: Annuals/Biennials/Bedding

56 Hook Garden Centre

Reading Road, Rotherwick, Hook, Hampshire RG27 9DB
Tel: 01256 768700
Specialities: Landscape Gardening

57 Humbees of Marwell Ltd

Thompsons Lane, Owslebury, Winchester, Hampshire SO21 1JH
Tel: 01962 777531
Specialities: Shrubs, Herbaceous Perennials

58 In-Excess UK

Forest Corner, Ringwood, Hampshire BH24 3HW
Tel: 01425 473113

59 J & J Nurseries

Tower Farm, Station Road, Soberton, Southampton, Hampshire SO32 3QU
Tel: 01489 877177
Specialities: Annuals/Biennials/Bedding, Herbaceous Perennials, Landscape Gardening

60 Job's Cottage Nursery

Durley Hall Lane, Durley, Southampton, Hampshire SO32 2AN
Tel: 01489 860456
Specialities: Garden Products

61 Keydell Nurseries

Havant Road, Horndean, Waterlooville, Hampshire PO8 0UU
Tel: 023 9259 3839

62 Lakeside Garden Centre

Kinsclere Road, Brimpton Common, Hampshire RG7 4RT
Tel: 0118 981 4138 Fax: 0118 981 0075
website: www.lakeside-gardencentre.co.uk

Specialities:

Knowledgeable Friendly Staff; Local Delivery Service

Products and Services:

- Aquatics
- Barbeques
- Conservatories
- Seasonal Displays
- Credit Cards Accepted
- Fencing
- Fireworks
- Garden Furniture
- Garden Products
- Large Car Park
- Paving
- Pet Products
- Plant Guarantee
- Plants
- Seeds and Bulbs
- Trees

Directions:

From Basingstoke take the A339 towards Newbury. At Kingsclere, turn right towards Aldermaston on the B3051 and follow the road for about 4 miles. The Garden Centre is on the left

63 Leydene Gardens Nursery

33 Titchfield Grange, Segensworth Road, Fareham, Hampshire PO15 5AR
Tel: 01329 843899
Specialities: Annuals/Biennials/Bedding, Pot Plants

64 Lincoln Green Nurseries

Willow House, Hambledon Road, Denmead, Waterlooville, Hampshire PO7 6EU
Tel: 023 9226 3515
Specialities: Hanging Baskets, Shrubs, Herbaceous Perennials

65 Lindum Nurseries

Eversley, Hook, Hampshire RG27 0NJ
Tel: 01189 733138
Specialities: Annuals/Biennials/Bedding

66 Lineway Partners Ltd

Weyhill Nurseries, Amesbury Road, Weyhill, Andover, Hampshire SP11 8ED
Tel: 01264 771714
Specialities: Giftware, Landscape Gardening

67 Lodore

Mount Pleasant Lane, Lymington, Hampshire SO41 8LS
Tel: 01590 682577
Specialities: Lilium (Lily)

68 Longstock Park Nursery

Longstock Park, Stockbridge, Hampshire SO20 6EH
Tel: 01264 810894
Specialities: Trees, Shrubs, Clematis

69 Lymington Plant Centre

Pitmore Lane, Pennington, Lymington, Hampshire SO41 8LL
Tel: 01590 682611

70 Macgregors Plants for Shade

Carters Clay Road, Lockerley, Romsey, Hampshire SO51 0GL
Tel: 01794 340256
Specialities: Shade Loving Plants

71 Macpennys Nurseries

154 Burley Road, Bransgore, Christchurch, Hampshire BH23 8DB
Tel: 01425 672348

72 Manor Road Nurseries

Manor Rd, Hayling Island, Hampshire PO11 0QL
Tel: 023 9246 3539
Specialities: Annuals/Biennials/Bedding, Herbaceous Perennials

73 Mayfield Nursery

Mayfield Park, Weston Lane, Southampton, Hampshire SO19 0QZ
Tel: 023 8044 7743
Specialities: Garden Products

74 Meadow Farm Nursery

Woodgaston Lane, Hayling Island, Hampshire PO11 0RL
Tel: 023 9246 1570
Specialities: Garden Products, Giftware

75 Medina Nursery

46-50 Brook Lane, Warsash, Southampton, Hampshire SO31 9FG
Tel: 01489 584476
Specialities: Hanging Baskets, Annuals/Biennials/Bedding

76 Milford Supplies Garden

112 Gore Road, New Milton, Hampshire BH25 6SJ
Tel: 01425 627659

77 Mill Water Gardens

Mead Mill, Mill Lane, Romsey, Hampshire SO51 8EQ
Tel: 01794 513444
Specialities: Water Products and Features

78 Mount Folly Nurseries

Southwick Road, North Boarhunt, Fareham, Hampshire PO17 6JH
Tel: 01329 832294
Specialities: Cottage Garden Plants, Fuchsia, Garden Furniture

79 Mud Island Nurseries

Southwick Road, North Boarhunt, Fareham, Hampshire PO17 6JF
Tel: 01329 834407
Specialities: Annuals/Biennials/Bedding

80 New Forest Plants Ltd

Unit 7 Graham Norris Business Centre, 114-118 Hampton Lane, Blackfield, Southampton, Hampshire SO45 1UB
Tel: 023 8089 9949
Specialities: Garden Products

81 Otter Nurseries Lymington

Milford Rd, Everton, Lymington, Hampshire SO41 0JD
Tel: 01590 678679
Specialities: Annuals/Biennials/Bedding

82 Paddock Plants

The Paddock, Upper Toothill Road, Rownhams, Southampton, Hampshire SO16 8AL
Tel: 023 8073 9912
Specialities: Herbaceous Perennials, Shrubs, Ferns

83 Park Place Farm Nursery

Titchfield Lane, Wickham, Fareham, Hampshire PO17 5HB
Tel: 01329 834991

84 The Patio Centre

69 Brook Lane, Warsash, Southampton,
Hampshire SO31 9HU
Tel: 01489 579119
Specialities: Paving

85 Peacocks Garden World

Farnham Road (A287), Ewshot, Farnham,
Hampshire GU10 5BB
Tel: 01252 850895
e-mail: info@peacocksgardenworld.com
website: www.peacocksgardenworld.com

SEE ADVERTISEMENT PAGE 109

Specialities:

Hanging Baskets; Geraniums; Fuchsias

Description:

Peacocks Garden World has been a family-run nursery for 30 years. It offers a very wide range of geraniums, fuchsias and 2 to 12 litre filled pots. It specialises in traditional, mossed, conical grass and rattan hanging baskets. The nursery also stocks a comprehensive selection of terracotta or glazed pots, as well as garden statuary. Joyce and Ian Gardner provide friendly service and advice, and free local deliveries.

Directions:

Peacocks Garden World is on the A287 about 4 miles northwest of Farnham.

86 Penwood Nurseries Ltd

Penwood, Newbury, Hampshire RG20 9EW
Tel: 01635 254366
Specialities: Annuals/Biennials/Bedding, Herbaceous Perennials, Shrubs

87 Petandra Nursery

Hamble Lane Farm, Hamble Lane, Bursledon,
Southampton, Hampshire SO31 8GQ
Tel: 023 8040 6875

88 Phoenix Perennial Plants

Paice Lane, Medstead, Alton, Hampshire GU34 5PR
Tel: 01420 560695
Specialities: Garden Products, Herbaceous Perennials

89 Pixies Plant Centre

Pixies Farm, Stuckton Road, Fordingbridge,
Hampshire SP6 1AR
Tel: 01425 657909
Specialities: Annuals/Biennials/Bedding

90 Pococks Roses

Jermyns Lane, Ampfield, Romsey,
Hampshire SO51 0QA
Tel: 01794 367500
Specialities: Rose

91 Redcliffe Garden Centre

Bashley Road, New Milton, Hampshire BH25 5RY
Tel: 01425 619691
Specialities: Water Products and Features, Pots, Annuals/Biennials/Bedding

92 Redfields Garden Nursery

Ewshot Lane, Church Crookham, Fleet,
Hampshire GU52 8UB
Tel: 01252 624444
Specialities: Garden Products, Giftware, Restaurant/Coffee Shop

93 Redhill Nursery

Winchester Road, Wickham, Fareham,
Hampshire PO17 5HE
Tel: 01329 833185
Specialities: Annuals/Biennials/Bedding, Cut Flowers

94 Rumsey Gardens

117 Drift Road, Clanfield, Waterlooville,
Hampshire PO8 0PD
Tel: 023 9259 3367
website: www.rumsey-gardens.co.uk
Specialities: Organic Fruit Trees, Ferns

95 Sandyfields Nurseries

103 Main Road, Colden Common, Winchester,
Hampshire SO21 1TB
Tel: 01962 712218 Fax: 01962 717057
e-mail: info@sandyfieldsnurseries.co.uk website: www.sandyfieldsnurseries.co.uk

Specialities:

Hanging Baskets & Containers; Perennials

Products and Services:

- Seasonal Displays
- Credit Cards Accepted
- Farm Shop
- Fruit Trees
- Garden Products
- Houseplants
- Large Car Park
- Pet Products
- Plants
- Seeds and Bulbs
- Trees

Directions:

From Exit 11 of the M3, take the B3335 towards Twyfold. Continue through Twyford to the next village, Colden Common. The nurseries are located on the left hand side as you go towards Fair Oak.

Description:

Sandyfields is a traditional family nursery, owned and run by the Hazzard family who have been

commercial growers in the county of Hampshire for two centuries. (One of their forefathers was amongst the first to grow the strawberry commercially in England) The last three generations have been Nurserymen, growing and selling top quality garden plants of all descriptions, from annual bedding plants to trees. The firm is still operated by three family members, supported by a small loyal team of helpers, and all sales are from the nursery. The business moved to the present site in 1955 and formed the company, so 2005 was their Golden Anniversary year under the Sandyfields banner. The nurseries have become renowned over the past couple of decades for their perennials and patio plants, and especially for the tubs and hanging baskets that they build to customers orders, and which adorn many of the local prize-winning hostelries and business premises, as well as private dwellings.

96 Scent Sational Plants

Frog Lane, Bowerwood Road, Fordingbridge,
Hampshire SP6 1BN
Tel: 01425 653010

Specialities:

National Collection of Bamboos on Display.

Description:

This superb plant nursery is home to the National Collection of Bamboos which offers more than 30 varieties for sale. Also stocks a large selection of shrubs, climbers, lasting-beauty perennials, geraniums, fuchsias, basket plants, a variety of bedding plants and hanging baskets.

Directions:

From Junction 1 of the M27 take the B3078 to Fordingbridge. Cross bridge over River Avon and at the mini-roundabout, turn left, continue along High Street. Take the first left towards Alderholt on the B3078. Continue along Provost Street, turn left into Frog Lane. Entrance to the car park on the right.

97 Silver Springs Nurseries

Fontley Road, Titchfield, Fareham, Hants PO15 6QX
Tel: 01329 842114
Specialities: Restaurant/Coffee Shop, Annuals/Biennials/Bedding, Herbaceous Perennials

98 Solent Nurseries

Fontley Road, Titchfield, Fareham, Hants PO15 6QS
Tel: 01329 846359
Specialities: Annuals/Biennials/Bedding, Herbaceous Perennials

99 Spring Meadow Nurseries

South End, Damerham, Fordingbridge, Hampshire SP6 3HW
Tel: 01725 518573
Specialities: Rhododendron (Inc Azalea), Camellia

100 Springfields Nursery

Oakhanger, Bordon, Hampshire GU35 9JD
Tel: 01420 472528
Specialities: Annuals/Biennials/Bedding

101 St Margarets Nurseries

St. Margarets Lane, St. Margarets Lane, Fareham, Hampshire PO14 4BL
Tel: 01329 846006
Specialities: Garden Products, Fuchsia

102 The Terracotta Pot Shop

4a Bembridge House, Bembridge Drive, Hayling Island, Hampshire PO11 9LU
Tel: 023 9263 7590
Specialities: Pots

103 Test Valley Nursery

Stockbridge Road, Timsbury, Romsey, Hampshire SO51 0NG
Tel: 01794 368881
Specialities: Herbaceous Perennials

104 The Tree Field

Ashville, Badminston Drove, Fawley, Southampton, Hampshire SO45 1BW
Tel: 023 8089 7188
Specialities: Trees

105 Vigo Nursery

72 Vigo Road, Andover, Hampshire SP10 1HP
Tel: 01264 323893
Specialities: Shrubs, Trees

106 Water Meadow Nursery

Water Meadows, Cheriton, Alresford, Hampshire SO24 0QB
Tel: 01962 771895
Specialities: Garden Products

107 Whitewater Nursery Ltd

Hound Green, Hook, Hampshire RG27 8LQ
Tel: 01189 326487
Specialities: Herbaceous Perennials, Shrubs, Trees

108 Winterbourne Nursery

Stoke, Andover, Hampshire SP11 0NP
Tel: 01264 738808
Specialities: Hanging Baskets, Pot Plants

109 Wolverton Plants Ltd

Wolverton Common, Wolverton, Tadley, Hampshire RG26 5RU
Tel: 01635 298453

110 Wreford Ltd

Romsey Road, Ower, Romsey, Hampshire SO51 6AF
Tel: 023 8081 3776
Specialities: Trees, Topiary, Hedging

111 Wyevale Garden Centre (Andover)

Salisbury Road, Andover, Hampshire SP11 7DN
Tel: 01264 710551

112 Wyevale Garden Centre (Basingstoke)

Winchester Road, North Waltham, Basingstoke, Hampshire RG25 2DJ
Tel: 01256 397155

113 Wyevale Garden Centre (Fair Oak)

Winchester Road, Fair Oak, Eastleigh, Hampshire SO50 7HD
Tel: 023 8060 0392

114 Wyevale Garden Centre (Havant)

Bartons Road, Havant, Hampshire PO9 5NA
Tel: 023 9245 6200

115 Wyevale Garden Centre (Sherfield on Loddon)

Wildmoor Lane, Sherfield-on-Loddon, Hook, Hampshire RG27 0JD
Tel: 01256 882239

116 Yew Tree Garden Centre

Hatt Common, Newbury, Hampshire RG20 0NG
Tel: 01635 255250
Specialities: Annuals/Biennials/Bedding, Landscape Gardening, Garden Furniture

HEREFORDSHIRE

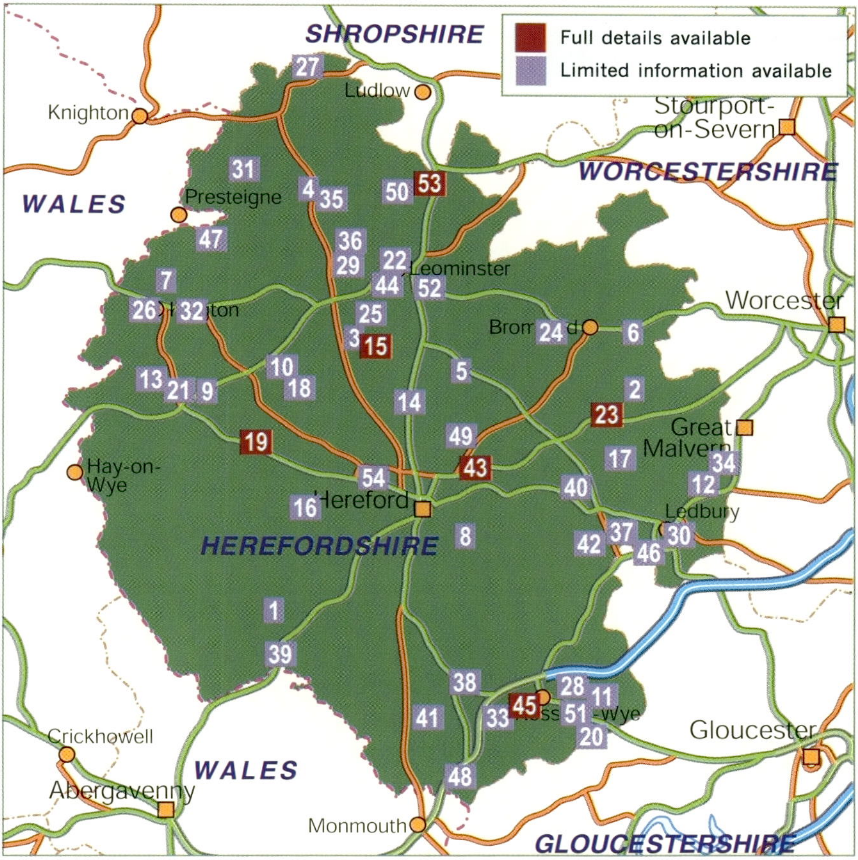

1 Abbey Dore Court Garden

Abbey Dore Court, Abbeydore, Hereford, Herefordshire HR2 0AD
Tel: 01981 240419
Specialities: Herbaceous Perennials

2 Acton Beauchamp Roses

Acton Beauchamp, Worcester, Herefordshire WR6 5AE
Tel: 01531 640433
Specialities: Rose

3 Aulden Farm

Aulden, Leominster, Herefordshire HR6 0JT
Tel: 01568 720129
Specialities: Iris, Lilium (Lily)

4 Aymestrey House

Aymestrey, Leominster, Herefordshire HR6 9ST
Tel: 01568 708342

5 Baches Bargain Warehouse

The Moor, Bodenham, Hereford, Herefordshire HR1 3HT
Tel: 01568 797319
Specialities: Annuals/Biennials/Bedding, Shrubs

6 The Bannut Garden & Nursery

The Bannut, Bromyard Road, Bringsty, Worcester, Herefordshire WR6 5TA
Tel: 01885 482206

7 Beech Grove Nurseries

Rushcock, Beechgrove, Kington,
Herefordshire HR5 3RH
Tel: 01544 231422
Specialities: Herbaceous Perennials, Shrubs

8 Bulmers Nursery

Field Farm, Hampton Bishop, Hereford,
Herefordshire HR1 4JP
Tel: 01432 356906

9 Bulmers Orchard Nursery

Lower Newton Farm, Kinnersley, Hereford,
Herefordshire HR3 6QB
Tel: 01544 327808
Specialities: Annuals/Biennials/Bedding

10 Carrs Special Organic Produce

The Ley Weobley, Weobley, Hereford,
Herefordshire HR4 8QR
Tel: 01544 318321
Specialities: Organic, Organic Edible Plantrs

11 Castrees Garden Plants

Bromsash, Ross-on-Wye, Herefordshire HR9 7PL
Tel: 01989 750315
Specialities: Annuals/Biennials/Bedding, Shrubs

12 Caves Folly Nurseries

Evedine Lane, Colwall Green, Malvern,
Herefordshire WR13 6DU
Tel: 01684 540631
Specialities: Grasses, Herbaceous Perennials, Organic Compost

13 Chennels Gate Gardens & Nursery

Eardisley, Hereford, Herefordshire HR3 6LT
Tel: 01544 327288
Specialities: Annuals/Biennials/Bedding, Herbaceous Perennials

14 Country Homes and Gardens

Wellington, Hereford, Herefordshire HR4 8BB
Tel: 01432 839000
Specialities: Annuals/Biennials/Bedding, Shrubs

15 Credale Nursery

Upper Hill, Leominster, Herefordshire HR6 0JZ
Tel: 01568 720476
e-mail: nursery@credale.fsnet.co.uk

Specialities:

Japanese Maples (Acers); Hellebores; Dierama

Products and Services:

- Plants
- Trees

Directions:

From Hereford take the A4110 Canon Pyon Road. After 7 miles at Bush Bank turn right sign posted Upper Hill. After 2 miles turn right as you enter Upper Hill. Credale Nursery is 100 yards on the left hand side.

16 Danmek Nursery

Madley, Hereford, Herefordshire HR2 9NT
Tel: 01981 251635
Specialities: Shrubs, Annuals/Biennials/Bedding

17 Fabulous Fuchsias

The Martins, Bosbury, Ledbury,
Herefordshire HR8 1HE
Tel: 01531 640298
Specialities: Fuchsia

18 Flowers of the Field Nursery

Field Farm, Weobley, Hereford,
Herefordshire HR4 8QJ
Tel: 01544 318262
Specialities: Ornamentals, Flower, Herbaceous Perennials

19 Garden Express @ Oakchurch

Staunton on Wye, Herefordshire HR4 7NE
Tel: 01981 500100

SEE ADVERTISEMENT PAGE 113

Specialities:

Specialist Garden Centre, Unusual Plants, Food Hall

Products and Services:

- Barbeques
- Christmas Trees
- Farm Food Hall
- Fireworks
- Fruit Trees
- Garden Furniture
- Garden Design
- Garden Products
- Giftware
- Large Car Park
- Paving
- Plants
- Restaurant/Coffee Shop
- Trees & Shrubs
- Water Features

Directions:

Garden Express is attached to Oakchurch Farm Shop, which is 10 miles west of Hereford on the main Brecon Road (A438).

20 Green's Leaves

Leba Orchard, Lea Bailey, Ross-on-Wye, Herefordshire HR9 5TY
Tel: 01989 750303
Specialities: Rare and Unusual Plants, Herbaceous Perennials, Grasses

21 Hadley Rill Ltd

Hadley Rill, Church Road, Eardisley, Hereford, Herefordshire HR3 6NN
Tel: 01544 327438

22 Hintons Country & Garden

17-19 Broad Street, Leominster, Herefordshire HR6 8DB
Tel: 01568 612277

23 Hop Pocket Plant Centre

Bishops Frome, Worcester, Herefordshire WR6 5BT
Tel: 01531 640121
e-mail: trishweston@dsl.pipex.com
Specialities: Grasses, Perennials, Stoneware, Terracotta, Soft Fruit

24 Hopkiln Nursery

The Green Lane, Bromyard, Herefordshire HR7 4RZ
Tel: 01885 488961
Specialities: Climbers, Annuals/Biennials/Bedding

25 Ivycroft Plants

Upper Ivington, Leominster, Herefordshire HR6 0JN
Tel: 01568 720344
Specialities: Annuals/Biennials/Bedding, Shrubs

26 J & P Turner

Arrow Mills, Kington, Herefordshire HR5 3DU
Tel: 01544 230536
Specialities: Herbaceous Perennials, Alpines/Rock Plants, Aquatics

27 Jungle Giants Bamboo Specialists

Ferney, Onibury, Craven Arms, Herefordshire SY7 9BJ
Tel: 01584 856200
Specialities: Bamboos

28 Kingstone Cottage Plants

Weston Under Penyard, Ross-on-Wye, Herefordshire HR9 7PH
Tel: 01989 565267
Specialities: Dianthus (Carnations/Pinks)

29 Lawton Hall Herbs

Lawton Hall, Eardisland, Leominster, Herefordshire HR6 9AX
Tel: 01568 709215
Specialities: Herbs

30 Ledbury's Secret Garden Centre

58 The Homend, Ledbury, Herefordshire HR8 1BT
Tel: 01531 636087

31 Lingen Nursery and Garden Centre

Lingen, Bucknell, Herefordshire SY7 0DY
Tel: 01544 267720
Specialities: Herbaceous Perennials, Garden Products

32 Lyonshall Nurseries

Lyonshall, Kington, Herefordshire HR5 3LN
Tel: 01544 340214
Specialities: Herbaceous Perennials, Seasonal Plants

33 The Market Garden

8 Gloucester Road, Ross-on-Wye, Herefordshire HR9 5BU
Tel: 01989 564609

34 Old Court Nurseries

The Picton Garden, Walwyn Road, Colwall, Malvern, Herefordshire WR13 6QE
Tel: 01684 540416
Specialities: Aster (Michaelmas Daisy)

35 The Old Vicarage Nursery

Lucton, Leominster, Herefordshire HR6 9PN
Tel: 01568 780538
Specialities: Seasonal Plants, Herbaceous Perennials

36 Ornamental Tree Nurseries

Broomy Hill Gardens, Cobnash Kingsland, Leominster, Herefordshire HR6 9QZ
Tel: 01568 708016
Specialities: Trees, Ornamentals

37 Overseas Farm Nurseries

Little Marcle, Ledbury, Herefordshire HR8 2JY
Tel: 01531 670377
Specialities: Seasonal Plants

38 Pengethley Nurseries & Garden Centre

Peterstow, Ross-on-Wye, Herefordshire HR9 6LL
Tel: 01989 730284

39 Pontrilas Garden Centre

Old Brickyard, Pontrilas, Hereford, Herefordshire HR2 0DJ
Tel: 01981 240485

40 Poppinger Nurseries

Upper Poppinger, Woodend, Ledbury, Herefordshire HR8 2RR
Tel: 01432 890295
Specialities: Herbaceous Perennials

41 Premier Plant Centre

Three Ashes, Hereford, Herefordshire HR2 8LX
Tel: 01981 580100
Specialities: Annuals/Biennials/Bedding, Shrubs

42 R E Griffiths

Hatsford, Aylton, Ledbury, Herefordshire HR8 2QJ
Tel: 01531 670275

43 Radway Bridge Nurseries & Garden Centre

Worcester Road, Whitestone, nr Hereford HR1 3RY Tel: 01432 65009
e-mail: info@hereford-garden-centre.co.uk
website: www.hereford-garden centre.co.uk

Specialities:

Herbaceous Plants; Trees & Shrubs; Pots & Containers

Products and Services:

- Barbeques
- Seasonal Displays
- Credit Cards Accepted
- Cut Flowers
- Disabled Facilities
- Dried & Artificial Flowers
- Floristry
- Fruit Trees
- Garden Design
- Garden Products
- Gift Vouchers
- Giftware
- Greenhouses and Sheds
- Houseplants
- Information Desk
- Large Car Park
- Mail Order
- Paving
- Pet Products
- Plant Guarantee
- Plants
- Restaurant/Coffee Shop
- Seeds and Bulbs
- Trees

Directions:

On the A4103 Hereford to Worcester road, about 3 miles east of Hereford.

44 Rainbow Street Nurseries

Rainbow Street, Leominster, Herefordshire HR6 8DQ
Tel: 01568 615648

45 Ross Garden Store

The Engine Shed, Station Approach, Ashburton,
Ross-on-Wye, Herefordshire HR9 7BW
Tel: 01989 568999 Fax: 01989 568157
e-mail: sales@rossgardenstore.co.uk
website: www.rossgardenstore.co.uk

Specialities:

Specimen Trees & Shrubs; Seasonal Plants; Bedding Plants

Products and Services:

- Barbecues
- Credit Cards Accepted
- Cut Flowers
- Floristry
- Garden Design
- Garden Furniture
- Garden Products
- Large Car Park
- Plants
- Seeds and Bulbs
- Trees

Directions:

From the A40, take the Ross-on-Wye exit towards Ross, then take the first right and The Garden Store will be found on your right.

Description:

In April 2006, Ross Garden Store opened its doors for the first time, located on the outskirts of the beautiful market town of Ross-on-Wye. Run by a local couple, the "team" at the store have many years of horticultural experience between them and aim to offer quality plants and good service at value for money prices. The store is housed within an old Engine Shed, formerly of the Great Western Railway and designed by Brunel, a perfect and unique setting for a garden centre. The store provides a wide range of seasonal garden plants which are, by definition, looking good for the time of year, and a range of garden requisites that complement these. The Store is the main stockist of "Kettler" quality garden furniture in the area. There are 3 trained florists on site – all floral requests can be catered for. Also offered is a full garden landscape and design service and plants can be sourced in quantity for trade customers.

46 Rushfields of Ledbury

Ross Road, Ledbury, Herefordshire HR8 2LP
Tel: 01531 632004
Specialities: Grasses, Hosta, Geranium

47 Stansbatch Nursery

Stansbatch, Leominster, Herefordshire HR6 9LH
Tel: 01544 267474
Specialities: Shrubs, Herbaceous Perennials

48 Symonds Yat Garden Centre

Whitchurch, Ross-on-Wye, Herefordshire HR9 6DA
Tel: 01600 891177
Specialities: Garden Products, Trees

49 Three Acre Nurseries

Wyatt Road, Cross Keys, Hereford,
Herefordshire HR1 3NP
Tel: 01432 820471
Specialities: Annuals/Biennials/Bedding, Shrubs, Herbaceous Perennials

50 Tunnel Lane Nursery

Orleton, Ludlow, Herefordshire SY8 4HY
Tel: 01568 780378
Specialities: Shrubs, Herbaceous Perennials

51 Waterpump Plants

Waterpump Farm, Ryeford, Ross-on-Wye,
Herefordshire HR9 7PU
Tel: 01989 750177
Specialities: Aquatics, Rare and Unusual Plants, Geranium

52 West Eaton Nurseries

Stoke Prior Lane, Leominster,
Herefordshire HR6 0NA
Tel: 01568 615757
Specialities: Herbaceous Perennials, Shrubs

53 Woodstock Terracotta & Garden Centre

Brimfield, nr Ludlow, Herefordshire SY8 4NY
Tel: 01584 711445
e-mail: goodwmart@aol.com
website: www.lotsapots.co.uk

Specialities:

Terracotta Pots; Tree Ferns & Palms; Exotic Plants

Description:

With more than 10,000 pots of every size, shape and colour in stock, Woodstock Terracotta offers the largest selection of pots to be found in the Midlands. Also within the well laid out site is a restaurant and coffee shop.

Directions:

Woodstock Terracotta is on the A49, 4 miles south of Ludlow

54 Wyevale Garden Centre (Hereford)

Kings Acre Road, Hereford, Herefordshire HR4 0SE
Tel: 01432 266261

HERTFORDSHIRE

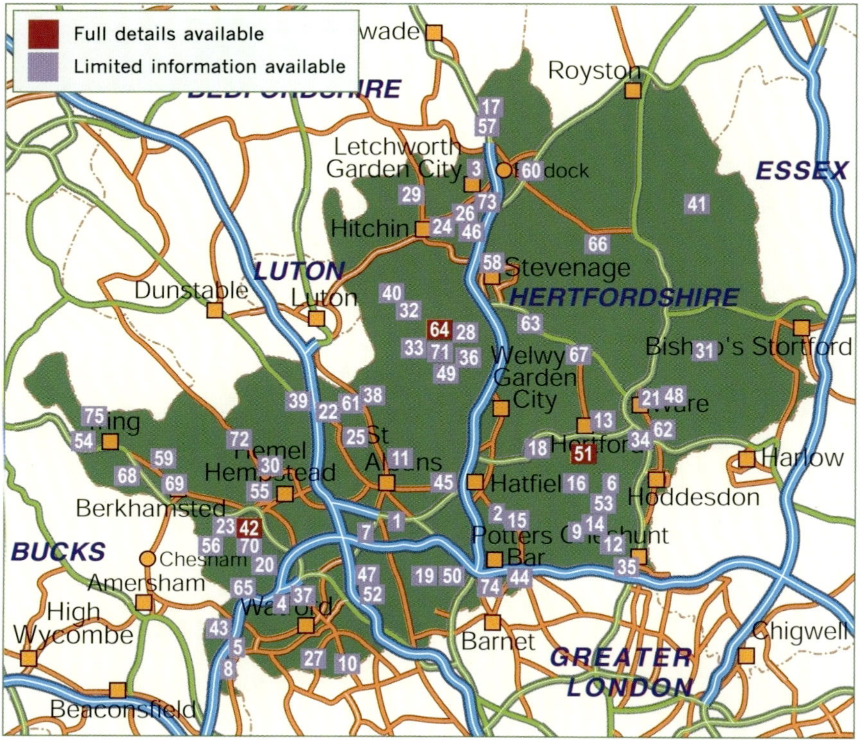

1 Aylett Nurseries

North Orbital Road, St. Albans,
Hertfordshire AL2 1DH
Tel: 01727 822255
Specialities: Restaurant/Coffee Shop, Garden Furniture, Dahlia

2 Bell Bar Nursery

Bulls Lane, North Mymms, Hatfield,
Hertfordshire AL9 7BB
Tel: 07724 064629
Specialities: Annuals/Biennials/Bedding, Shrubs

3 Bickerdikes Garden Centre

Norton Road, Letchworth Garden City,
Hertfordshire SG6 1AG
Tel: 01462 673333
Specialities: Garden Products

4 Blacketts Nursery

Rouse Barn Lane, Chandlers Cross, Rickmansworth,
Hertfordshire WD3 4NB
Tel: 01923 265743
Specialities: Shrubs, Herbaceous Perennials

5 Brambles Nursery

24 Denham Way, Maple Cross, Rickmansworth,
Hertfordshire WD3 9RN
Tel: 01923 779551
Specialities: Landscape Gardening, Lawn, Annuals/Biennials/Bedding

6 Brookfield Nursery

Wormley West End, Broxbourne,
Hertfordshire EN10 7QN
Tel: 01992 462045
Specialities: Shrubs, Herbaceous Perennials

7 Burston Rose & Garden Centre

North Orbital Road, St. Albans,
Hertfordshire AL2 2DS
Tel: 01727 832444
Specialities: Rose

8 C & P Better Homes Garden Centre

Old Uxbridge Road, West Hyde, Rickmansworth,
Hertfordshire WD3 9XP
Tel: 01895 824787

9 C G Edward (Goffs Oak) Ltd

104 Cuffley Hill, Cuffley Hill, Goffs Oak,
Waltham Cross, Hertfordshire EN7 5EY
Tel: 01707 873166

10 Carpenders Park Nursery

Little Oxhey Lane, Watford, Hertfordshire WD19 5BA
Tel: 020 8420 1959
Specialities: Herbaceous Perennials, Garden Furniture, Restaurant/Coffee Shop

11 Carpenters Nursery

106 St. Albans Road, Sandridge, St. Albans,
Hertfordshire AL4 9LJ
Tel: 01727 853340

12 Crescent Nursery

Barrow Lane, Cheshunt, Waltham Cross,
Hertfordshire EN7 5LY
Tel: 01992 636776

13 Daisy Roots Nursery

8 Gosselin Road, Hertford, Hertfordshire SG14 3LG
Tel: 01992 582401
Specialities: Annuals/Biennials/Bedding, Shrubs

14 Dileto Bros

Tina Nursery, Goffs Lane, Goffs Oak, Waltham Cross,
Hertfordshire EN7 5EP
Tel: 01707 872652

15 Dutch Nursery & Garden Centre

Great North Road, Brookmans Park, Hatfield,
Hertfordshire AL9 6ND
Tel: 01707 653372
Specialities: Annuals/Biennials/Bedding, Restaurant/Coffee Shop, Garden Products

16 Fletchers Garden Centre

White Stubbs Lane, Bayford, Hertford,
Hertfordshire SG13 8QA
Tel: 01992 511611
Specialities: Annuals/Biennials/Bedding, Garden Furniture, Garden Products

17 Foxhollow Nursery

Hinxworth, Baldock, Hertfordshire SG7 5EZ
Tel: 01462 733138
Specialities: Annuals/Biennials/Bedding, Shrubs

18 Garden & Pet Direct

Birchall Lane, Cole Green, Hertford,
Hertfordshire SG14 2NR
Tel: 01707 336928
Specialities: Annuals/Biennials/Bedding, Pet Products

19 The Garden Nursery

39 London Road, Shenley, Radlett,
Hertfordshire WD7 9ER
Tel: 01923 854955
Specialities: Annuals/Biennials/Bedding, Shrubs

20 Garden Scene

Chapel Croft, Chipperfield, Kings Langley,
Hertfordshire WD4 9EG
Tel: 01923 262492
Specialities: Herbaceous Perennials, Annuals/Biennials/Bedding

21 Garden Services

Bowling Road, Ware, Hertfordshire SG12 7EE
Tel: 01920 466645

22 Godly's Roses

Redding Lane, Redbourn, St. Albans,
Hertfordshire AL3 7PS
Tel: 01582 792255
Specialities: Rose

23 Greenbloom

Balis, Chipperfield Road, Bovingdon, Hemel
Hempstead, Hertfordshire HP3 0JN
Tel: 01442 831249
Specialities: Trees, Bonsai

24 Grove Road Garden Centre

Grove Road, Hitchin, Hertfordshire SG5 1SE
Tel: 01462 451519

25 Hammonds End Nursery

Hammonds End Lane, Harpenden,
Hertfordshire AL5 2AY
Tel: 01582 460099
Specialities: Annuals/Biennials/Bedding, Trees

26 Harkness Roses

Cambridge Road, Hitchin, Hertfordshire SG4 0JT
Tel: 01462 420402
Specialities: Rose

27 Hazeltone Gardens Ltd

Sandy Lane, Northwood, Hertfordshire HA6 3HB
Tel: 020 8421 5977

28 Highfield Nurseries

Slip Lane, Old Knebworth, Knebworth,
Hertfordshire SG3 6QG
Tel: 01438 812524
Specialities: Annuals/Biennials/Bedding, Herbaceous Perennials

29 Hillcrest Nursery

Bedford Road, Holwell, Hitchin,
Hertfordshire SG5 3RT
Tel: 01462 712000
Specialities: Herbaceous Perennials, Trees, Shrubs

30 Hillier Garden Centre Hemel Hempstead

Leighton Buzzard Road, Piccotts End,
Hemel Hempstead, Hertfordshire HP1 3BA
Tel: 01442 242637

31 Hopleys Plants Ltd

High Street, Much Hadham,
Hertfordshire SG10 6BU
Tel: 01279 842509
Specialities: Annuals/Biennials/Bedding, Shrubs

32 Ivor A Sansom

Lilley Bottom Road, Whitwell, Hitchin,
Hertfordshire SG4 8JP
Tel: 01438 871371

33 King Acre Landscaping Materials

High St, Codicote, Hitchin, Hertfordshire SG4 8XA
Te : 01438 822050
Specialities: Landscape Gardening

34 Leafy Oak Nursery

Hertford Rd, Great Amwell, Ware,
Hertfordshire SG12 9RR
Tel: 01920 463573
Specialities: Annuals/Biennials/Bedding, Herbaceous Perennials

35 Little Greenhouse

Oylers Farm Aio, Great Cambridge Road,
Waltham Cross, Hertfordshire EN8 8EU
Tel: 01992 634446
Specialities: Trees, Shrubs

36 Little Orchard Nurseries

Rabley Heath Road, Welwyn, Hertfordshire AL6 9UA
Tel: 01438 821696
Specialities: Herbaceous Perennials, Shrubs

37 Low Cost Nursery

Hempstead Road, Watford, Hertfordshire WD17 3NB
Tel: 01923 235679
Specialities: Annuals/Biennials/Bedding, Herbaceous Perennials

38 LW Plants

23 Wroxham Way, Harpenden,
Hertfordshire AL5 4PP
Tel: 01582 768467

39 Majestic Trees Ltd

Chequers Meadow, Chequer Hill, Flamstead,
St. Albans, Hertfordshire AL3 8ET
Tel: 01582 843881
Specialities: Shrubs, Trees

40 Manor Gardens Nursery

Church Rd, Kings Walden, Hitchin,
Hertfordshire SG4 8JU
Tel: 01438 871556
Specialities: Annuals/Biennials/Bedding, Herbaceous Perennials

41 Martins Plants Ltd

The Old Bell, Anstey, Buntingford,
Hertfordshire SG9 0DA
Tel: 01763 849496
Specialities: Annuals/Biennials/Bedding, Herbaceous Perennials, Shrubs

42 The Mediterranean Nursery

Kenwood Farm Nurseries, Flaunden Lane, Bovingdon,
Hertfordshire HP3 0RL
Tel: 01442 832482 or 07976 388432
e-mail: enquiries@themediterraneannursery.co.uk
website: www.themediterraneannursery.co.uk

SEE ADVERTISEMENT BELOW

Specialities:

Hardy Mediterranean Plants and Trees; Bedding Plants; Salad Crops in Season

Description:

This is a family-run business specialising in hardy Mediterranean plants. Their range also includes shrubs, perennials, bedding in season and some topiary. Landscape gardeners and garden designers are welcome and are offered a wide range of architectural plants ranging in sizes from 30ft bamboo to alpines. There is a free local delivery service.

Directions:

Bovingdon is on the B4505, about 4.5 miles northeast of Chesham. In the village, turn right into Chipperfield Road and then right into Flaunden Lane. The nursery is on the right hand side.

43 Nacarat

41 Carpenters Wood Drive, Chorleywood, Rickmansworth, Hertfordshire WD3 5RN
Tel: 01923 286362

44 Naybur Bros Ltd

Potters Bar Nursery, The Ridgeway, Potters Bar, Hertfordshire EN6 5QS
Tel: 01707 658444

45 Notcutts Garden Centre

Hatfield Road, Smallford, St. Albans, Hertfordshire AL4 0HN
Tel: 01727 853224

46 Pioneer Nursery

Baldock Lane, Willian, Letchworth Garden City, Hertfordshire SG6 2AE
Tel: 01462 675858
Specialities: Annuals/Biennials/Bedding, Shrubs

47 Plants Direct

Roundbush Lane, Watford, Hertfordshire WD25 8BQ
Tel: 01923 850809

48 Priorswood Clematis Nursery

Priorswood, Widbury Hill, Ware, Hertfordshire SG12 7QH
Tel: 01920 461543
Specialities: Clematis

49 Pulmore Nursery

St. Albans Road, Codicote, Hitchin, Herts SG4 8UY
Tel: 01438 820672
Specialities: Annuals/Biennials/Bedding, Shrubs

50 Ridge Nurseries

Cross Oaks Lane, Ridge, Potters Bar, Hertfordshire EN6 3LH
Tel: 01707 650133
Specialities: Annuals/Biennials/Bedding, Herbaceous Perennials

51 Riverside Garden Centre

Lower Hatfield Road, Hertford, Herts. SG12 8XX
Tel: 01992 501502

SEE ADVERTISEMENT BELOW

Specialities:

Thai Restaurant; Extensive Pet Supply Area; Comprehensive Range of Garden Furniture

Description:

This family owned garden centre carries a large stock of high quality barbecues, garden furniture, tools, animal feeds and all sorts of gifts. There's an extensive kitchen department and a fully licensed riverside restaurant.

Directions:

Riverside Garden Centre is on the B158 on the southeast outskirts of Hertford.

52 Round Bush Nursery

Roundbush Lane, Aldenham, Watford, Hertfordshire WD25 8BQ
Tel: 01923 857265
Specialities: Seasonal Plants, Shrubs

53 Rovaand Nurseries

Crouch Lane, Goffs Oak, Waltham Cross, Hertfordshire EN7 6TL
Tel: 01707 874426
Specialities: Annuals/Biennials/Bedding, Shrubs

54 Rowan Nursery

Thrift Cottage, Park Road, Tring, Hertfordshire HP23 6BU
Tel: 01442 891213
Specialities: Annuals/Biennials/Bedding, Herbaceous Perennials, Shrubs

55 Seeds by Size

45 Crouchfield, Hemel Hempstead,
Hertfordshire HP1 1PA
Tel: 01442 251458
Specialities: Seeds

56 Shantock Nurseries

Shantock Lane, Bovingdon, Hemel Hempstead,
Hertfordshire HP3 0NG
Tel: 01442 834754
Specialities: Shrubs, Herbaceous Perennials

57 Springtime Nursery

Great North Road, Stotfold, Hitchin,
Hertfordshire SG5 4BL
Tel: 01462 835833

58 Stevenage Garden Centre

Gravely Road, Stevenage, Hertfordshire SG1 4AH
Tel: 01438 312660

59 Sunnyside Nursery

Upper Allotment Site, New Road, Northchurch,
Berkhamsted, Hertfordshire HP4 1NJ
Tel: 07743 552154
Specialities: Vegetables, Shrubs

60 Tapps Garden Centre

Wallington Road, Baldock, Hertfordshire SG7 6RS
Tel: 01462 896302
Specialities: Herbaceous Perennials, Garden Products, Seasonal Displays

61 Townsend Nursery

Townsend Lane, Harpenden, Hertfordshire AL5 2RH
Tel: 01582 713083
Specialities: Herbaceous Perennials, Shrubs

62 Van Hage Garden Centre

Amwell Hill Great Amwell, Ware,
Hertfordshire SG12 9RP
Tel: 01920 870811

63 The Van Hage Garden Company

Farm House, Bragbury Lane, Stevenage,
Hertfordshire SG2 8TJ
Tel: 01438 811777

64 Vanstone Park

Vanstone Park of Codicote, Hitchin Road, Codicote,
Hertfordshire SG4 8TH
Tel: 01438 820412
e-mail: info@vanstone-gardencentre.co.uk
website:www.vanstoneparkgardencentre.co.uk

SEE ADVERTISEMENT PAGE 122

Specialities:

Plants, Giftware and Outdoor Garden Furniture

Description:

Family-owned for more than 30 years, Vanstone Park provides a great day out for all the family. Located in beautiful parkland with a miniature railway at weekends and is stocked with everything a gardener could wish for. Its Garden Shop features a fantastic range of gifts and furniture.

Directions:

Vanstone Park is on the B656 Hitchin to Welwyn road, 1 mile north of Codicote and 6 miles south of Hitchin.

65 Wards Nurseries

Dawes Lane, Sarratt, Rickmansworth,
Hertfordshire WD3 6BQ
Tel: 01923 263237
Specialities: Annuals/Biennials/Bedding, Herbaceous Perennials

66 Warren Nursery

Warren Lane, Cottered, Buntingford,
Hertfordshire SG9 9QG
Tel: 01763 281386
Specialities: Grasses, Herbaceous Perennials

67 Watton Nursery

Ware Road, Watton At Stone, Hertford,
Hertfordshire SG14 3RA
Tel: 01920 830944
Specialities: Annuals/Biennials/Bedding, Herbaceous Perennials

68 Wood View Nurseries

Tinkers Lane, Wigginton, Tring,
Hertfordshire HP23 6JB
Tel: 01442 865722
Specialities: Floristry, Specimen Plants

69 Wood's of Berkhamsted Garden Centre

The Old Ironworks, High Street, Berkhamsted, Hertfordshire HP4 1BJ
Tel: 01442 863159

70 Wyevale Garden Centre (Chipperfield)

Tower Hill, Chipperfield, Kings Langley, Hertfordshire WD4 9LH
Tel: 01442 834364

71 Wyevale Garden Centre (Codicote)

High Street, Codicote, Hitchin, Hertfordshire SG4 8XA
Tel: 01438 820433

72 Wyevale Garden Centre (Hemel Hempstead)

Broadwater Garden Centre, Dagnall Road, Great Gaddesden, Hemel Hempstead, Hertfordshire HP1 3BW
Tel: 01442 231284

73 Wyevale Garden Centre (Hitchin)

Cambridge Road, Hitchin, Hertfordshire SG4 0JT
Tel: 01462 434027

74 Wyevale Garden Centre (Potters Bar)

Dancers Hill Road, Barnet, Hertfordshire EN5 4RP
Tel: 020 8440 4734

75 Wyevale Garden Centre (Tring)

Bulbourne Road, Tring, Hertfordshire HP23 5HF
Tel: 01442 891393

ISLE OF WIGHT

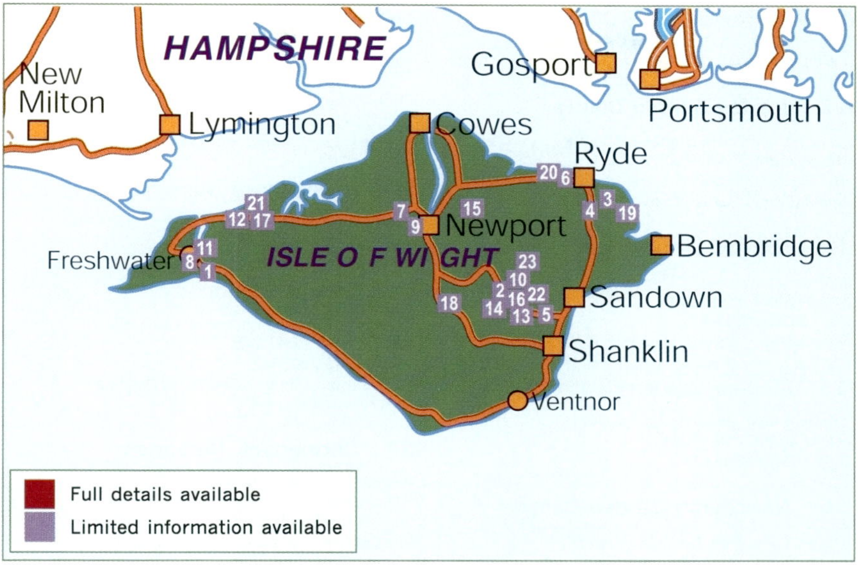

1 Afton Park Nursery & Orchard

Newport Road, Freshwater, Isle of Wight PO40 9XR
Tel: 01983 755774

2 Arreton Garden & Aquatic Centre

Sandown Rd, Hale Common, Newport, Isle of Wight PO30 3AR
Tel: 01983 865885

3 Bullen Mead Nursery

Bullen Road, Ryde, Isle of Wight PO33 1QE
Tel: 01983 568518
Specialities: Annuals/Biennials/Bedding, Herbaceous Perennials

4 Busy Bee Nurseries

Brading Road, Ryde, Isle of Wight PO33 1QG
Tel: 01983 811096
Specialities: Annuals/Biennials/Bedding, Herbaceous Perennials

5 Dalverton Garden Centre

Newport Road, Sandown, Isle of Wight PO36 9PJ
Tel: 01983 868602
Specialities: Garden Products

6 Duffetts Garden Store

12 John St, Ryde, Isle of Wight PO33 2PY
Tel: 01983 563254

7 Forest View Nursery

Forest Road, Newport, Isle of Wight PO30 5NB
Tel: 01983 530675

8 Freshwater Trees & Shrubs

Soake Farm House, Queens Road, Freshwater, Isle of Wight PO40 9ES
Tel: 01983 756464
Specialities: Trees, Shrubs

9 Guys Limited

115 Pyle Street, Newport, Isle of Wight PO30 1XA
Tel: 01983 522094

10 Harbors Lake Nursery

Harbors Lake Lane, Newchurch, Sandown, Isle of Wight PO36 0LY
Tel: 01983 866301

11 Honnor & Jeffrey Ltd

Afton Road, Freshwater, Isle of Wight PO40 9UH
Tel: 01983 752870

12 Jayne Jose Nursery

Main Road, Bouldnor, Yarmouth,
Isle of Wight PO41 0XN
Tel: 01983 761286
Specialities: Shrubs, Conifers, Herbaceous Perennials

13 Jubilee Garden Centre

Newport Road, Branstone, Sandown,
Isle of Wight PO36 0LT
Tel: 01983 865562
Specialities: Garden Furniture, Bulbs

14 Margaret Nurseries

Macketts Lane, Hale Common, Newport,
Isle of Wight PO30 3AS
Tel: 01983 865239
Specialities: Shrubs, Herbaceous Perennials

15 Medina Garden Centre

Staplers Road, Wootton Bridge, Ryde,
Isle of Wight PO33 4RW
Tel: 01983 883430

16 Newchurch Garden Centre

Watery Lane, Newchurch, Sandown,
Isle of Wight PO36 0LX
Tel: 01983 865292

17 Ningwood Nurseries

Main Road, Ningwood, Newport,
Isle of Wight PO30 4NW
Tel: 01983 760376
Specialities: Annuals/Biennials/Bedding

18 Organa Glow

Unit 1 Pritchetts Way, Rookley, Ventnor,
Isle of Wight PO38 3LT
Tel: 01983 721122

19 Priory Nurseries

Priory Drive, Seaview, Isle of Wight PO34 5EA
Tel: 01983 613795
Specialities: Shrubs, Herbaceous Perennials

20 Ryde House Nursery

Binstead Road, Ryde, Isle of Wight PO33 3NF
Tel: 01983 565650

21 Sarahs Plants

Cranmore Avenue, Cranmore, Yarmouth,
Isle of Wight PO41 0XR
Tel: 01983 760386
Specialities: Annuals/Biennials/Bedding, Herbaceous Perennials

22 Springbank Nurseries

Winford Road, Winford, Sandown,
Isle of Wight PO36 0JX
Tel: 01983 865444
Specialities: Herbaceous Perennials, Shrubs

23 Sunnycrest Nursery

Wackland Lane, Newchurch, Sandown,
Isle of Wight PO36 0NB
Tel: 01983 868024

KENT

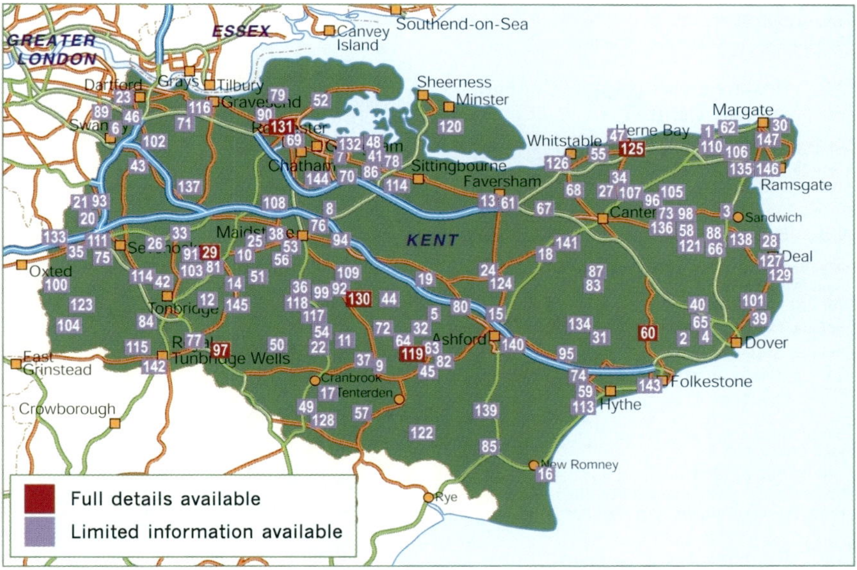

1 Albion Road Garden Centre

2-6 Albion Road, Birchington, Kent CT7 9DN
Tel: 01843 848178

2 Alkham Valley Garden Centre

Alkham Valley Road, Alkham, Dover, Kent CT15 7EW
Tel: 01303 893351

3 Archers Low Nursery

Ash Road, Sandwich, Kent CT13 9JB
Tel: 01304 613150
Specialities: Shrubs, Herbaceous Perennials, Rose

4 Artisan Nursery Garden Design

Le Ferne Farm Cottage, Abbey Road, Hougham, Dover, Kent CT15 7DQ
Tel: 01303 221953
Specialities: Landscape Gardening

5 Bamboo Nursery

The Pond House, Station Road, Pluckley, Ashford, Kent TN27 0QX
Tel: 01233 840209
Specialities: Trees, Shrubs

6 Bee Jays Garden Supplies

London Road, Swanley, Kent BR8 7AP
Tel: 01322 664827

7 Berengrave Nursery

61 Berengrave Lane, Rainham, Gillingham, Kent ME8 7NL
Tel: 01634 388898
Specialities: Rose

8 Better Baskets

1 Black Cottages, Lidsing Road, Boxley, Maidstone, Kent ME14 3EL
Tel: 07941 828268
Specialities: Hanging Baskets

9 Biddenden Plant Centre

Sissinghurst Rd, Biddenden, Ashford, Kent TN27 8EJ
Tel: 01580 292860
Specialities: Garden Products

10 Bijou Nurseries

385 Tonbridge Rd, Wateringbury, Maidstone, Kent ME18 5PG
Tel: 01622 812278
Specialities: Garden Products

11 Brickwall Cottage Nursery

1 Brickwall Cottages, Frittenden, Cranbrook, Kent TN17 2DH
Tel: 01580 852425
Specialities: Herbaceous Perennials

12 Broadview in Hadlow College Garden Centre

Hadlow College, Tonbridge Road, Hadlow, Tonbridge, Kent TN11 0AL
Tel: 01732 853211

13 Brogdale Orchards Ltd

Brogdale Road, Ospringe, Faversham, Kent ME13 8RZ
Tel: 01795 535286
Specialities: Fruit Trees, Giftware, Display Gardens

14 Brookside Nursery & Garden Centre

Seven Mile Lane, East Peckham, Tonbridge, Kent TN12 5JG
Tel: 01622 871250
Specialities: Restaurant/Coffee Shop, Water Products and Features, Pet Products

15 Bybrook Barn Garden Centre

Cemetery Lane, Kennington, Ashford, Kent TN24 9JZ
Tel: 01233 631959

16 Capel's Garden Centre

Mountfield Road, New Romney, Kent TN28 8XU
Tel: 01797 366899
Specialities: Drought Resistant Plants

17 Charity Farm Countrystore

Swattenden Lane, Cranbrook, Kent TN17 3PS
Tel: 01580 713189
Specialities: Herbaceous Perennials, Shrubs

18 Chilham Farm Shop and Plant Centre

Canterbury Road, Chilham, Canterbury, Kent CT4 8DX
Tel: 01227 730348

19 Church Hill Cottage Gardens

Charing Heath, Ashford, Kent TN27 0BU
Tel: 01233 712522
Specialities: Display Gardens

20 Coolings Nurseries Ltd

Main Rd, Knockholt, Sevenoaks, Kent TN14 7LJ
Tel: 01959 534386
Specialities: Garden Furniture

21 Coolings Nurseries Ltd

Rushmore Hill, Knockholt, Sevenoaks, Kent TN14 7NN
Tel: 01959 532269

22 The Cottage Garden

Cranbrook Road, Staplehurst, Kent TN12 0EU
Tel: 01580 891312
Specialities: Herbaceous Perennials, Geranium

23 Country Cottage Compost

6 Raeburn Avenue, Dartford, Kent DA1 3BG
Tel: 01322 277410

24 Crispin Nursery

Faversham Road, Challock, Ashford, Kent TN25 4BQ
Tel: 01233 740898
Specialities: Landscape Gardening

25 Cromar Nursery

39 Livesey Street, Wateringbury, Maidstone, Kent ME18 5BQ
Tel: 01622 812380
Specialities: Trees

26 Crown Point Nursery

Sevenoaks Road, Ightham, Nr Sevenoaks, Kent TN15 0HB
Tel: 01732 810694
Specialities: Rhododendron (Inc Azalea), Trees, Climbers

27 Davids Exotic Plants

37 Shalloak Road, Broad Oak, Canterbury, Kent CT2 0QH
Tel: 01227 711897
Specialities: Exotics

28 Deal Garden Centre

48 Queen Street, Deal, Kent CT14 6HD
Tel: 01304 373069

29 Downderry Nursery

Pillar Box Lane, Hadlow, Tonbridge,
Kent TN11 9SW
Tel: 01732 810081 Fax: 01732 811398
e-mail: info@downderry-nursery.co.uk
website: www.downderry-nursery.co.uk

Specialities:

Lavender; Rosemary

Description:

Downderry is home to the most comprehensive National Plant Collections® of lavender and rosemary, and as the country's premier lavender nursery, set in a beautiful and peaceful walled garden, we offer superb choice, unrivalled expertise and a personal service. Open May-Oct. 10-5 except Mondays.

Directions:

From M20 at J4 or from Tonbridge turn off A26 NE of Hadlow at Downderry Lavender tourist sign then follow small red rosette tourist signs, taking 3rd right after 1.5 miles and then 1st right.

30 East Northdown Farm Nursery

Friends Growers, East Northdown, Margate,
Kent CT9 3TS
Tel: 01843 862060

31 Farthing Common Plant Centre

Stone Street, Farthing Common, Lyminge,
Folkestone, Kent CT18 8DH
Tel: 01303 863438
Specialities: Shrubs, Trees

32 Farthing Common Plant Centre

Heber, Chambers Green Road, Pluckley, Ashford,
Kent TN27 0RH
Tel: 01233 840793

33 Forward Nurseries

Borough Green Road, Ightham, Sevenoaks,
Kent TN15 9JA
Tel: 01732 884726
Specialities: Hedging, Garden Products

34 Foxhill Garden Centre

Calcott Hill, Sturry, Canterbury, Kent CT3 4ND
Tel: 01227 713012

35 French Street Nurseries

French Street, Westerham, Kent TN16 1PW
Tel: 01959 563366

36 G & S Smallholding

Wheelers Lane, Linton, Maidstone, Kent ME17 4BN
Tel: 01622 744273

37 Garden Crafts

Sissinghurst Road, Biddenden, Ashford,
Kent TN27 8DQ
Tel: 01580 292070
Specialities: Ornamentals

38 Garden Flair

The Old Hermitage, Hermitage Lane, Maidstone,
Kent ME16 9NT
Tel: 01622 727900

39 Garden Plants

Windy Ridge, Victory Road, St. Margarets-at-Cliffe,
Dover, Kent CT15 6HF
Tel: 01304 853225
Specialities: Penstemon, Herbaceous Perennials

40 Gardener's Corner

Church Road, Coldred, Dover, Kent CT15 5AQ
Tel: 01304 831102

41 Gardening World Ltd

Lower Hartlip Rd, Newington, Sittingbourne,
Kent ME9 7SZ
Tel: 01795 841199
Specialities: African Crafts Centre

42 Gate House Nursery

Coldharbour Lane, Hildenborough, Tonbridge,
Kent TN11 9LE
Tel: 01732 832180
Specialities: Annuals/Biennials/Bedding, Hanging Baskets

43 Gorse Hill Nursery

Gorse Hill, Farningham, Dartford, Kent DA4 0JU
Tel: 01322 864704
Specialities: Landscape Gardening

44 Grafty Garden Centre

Crumps Corner, Headcorn Road, Grafty Green, Maidstone, Kent ME17 2AT
Tel: 01622 858800
Specialities: Pet Products, Water Products and Features

45 Greenways Garden Centre

Ashford Road, Bethersden, Ashford, Kent TN26 3LF
Tel: 01233 820526

46 Hawley Garden Centre & Nursery

Hawley Road, Dartford, Kent DA2 7RB
Tel: 01322 224108

47 Herne Bay Garden Centre Ltd

2 Avenue Road, Herne Bay, Kent CT6 8TB
Tel: 01227 362770

48 The Hideaway Nursery

105 Chaffes Lane, Upchurch, Sittingbourne, Kent ME9 7BD
Tel: 01634 231693
Specialities: Garden Products, Garden Design, Landscape Gardening

49 High Banks Nurseries

Gills Green, Cranbrook, Kent TN18 5AG
Tel: 01580 754492
Specialities: Garden Products

50 Highgrove Garden Centre

Staplehurst Road, Marden, Tonbridge, Kent TN12 9NB
Tel: 01622 832299

51 Highwinds Nursery

West Pike Fish Farm, Pike Fish Lane, Laddingford, Maidstone, Kent ME18 6BH
Tel: 01892 730554
Specialities: Hedging

52 Homeleigh Garden Centre

Homeleigh, Ratcliffe Highway, Hoo, Rochester, Kent ME3 8QD
Tel: 01634 250235
Specialities: Farm Shop, Pet Products, Fruit Bushes

53 Homewood Orchard Garden Centre

Farleigh Lane, Maidstone, Kent ME16 9LX
Tel: 01622 727103
Specialities: Hanging Baskets

54 Iden Croft Herbs

Frittenden Road, Staplehurst, Tonbridge, Kent TN12 0DH
Tel: 01580 891432
Specialities: Herbs, Giftware, Restaurant/Coffee Shop

55 Johnsons Nursery

Thanet Way, Chestfield, Whitstable, Kent CT5 3JD
Tel: 01227 793763

56 Keepers Nursery

Gallants Court, Gallants Lane, East Farleigh, Maidstone, Kent ME15 0LE
Tel: 01622 726465
Specialities: Fruit Trees

57 The Laurels Nursery

Dingleden, Benenden, Cranbrook, Kent TN17 4JU
Tel: 01580 240463
Specialities: Shrubs, Trees

58 Layham Garden Centre and Nursery

Lower Road, Staple, Canterbury, Kent CT3 1LH
Tel: 01304 813267
Specialities: Rose

59 Longacre Nursery

St. Marys Road, West Hythe, Hythe, Kent CT21 4NU
Tel: 01303 265444
Specialities: Annuals/Biennials/Bedding

60 MacFarlane's Nursery & Garden Centre

Swingfield, nr Dover, Kent CT15 7HX
Tel/Fax: 01303 844244
website: www.macfarlanesgardens.co.uk

Specialities:

Mature Trees; Shrubs; Herbaceous

Products and Services:

- Aquatics
- Books and Stationery
- Butterfly Centre
- Seasonal Displays
- Log Cabins

- Credit Cards Accepted
- Disabled Facilities
- Fencing
- Fruit Trees
- Garden Furniture
- Garden Products
- Giftware
- Greenhouses and Sheds
- Houseplants
- Information Desk
- Large Car Park
- Paving
- Plants
- Restaurant/Coffee Shop
- Seeds and Bulbs
- Trees

Directions:

From the A2 take the A260 Folkestone exit. The nursery is on this road, on the left.

61 Macknade Natural Garden Centre

Selling Road, Faversham, Kent ME13 8XF
Tel: 01795 531213
Specialities: Lawn

62 The Mad Potter Garden Centre

Station Road, Westgate-on-Sea, Kent CT8 8QY
Tel: 01843 835723
Specialities: Palms

63 Madrona Nursery

Pluckley Road, Bethersden, Ashford,
Kent TN26 3DD
Tel: 01233 820100
Specialities: Grasses, Shrubs, Ferns

64 Maltmans Nurseries

Maltmans Hill, Smarden, Ashford, Kent TN27 8RD
Tel: 01233 770307
Specialities: Garden Products

65 Manor View Nursery

Lower Road, Temple Ewell, Dover, Kent CT16 3DY
Tel: 01304 822541

66 Martin's Nursery

Poison Cross, Statenborough Lane, Eastry,
Sandwich, Kent CT13 0DL
Tel: 01304 611262
Specialities: Annuals/Biennials/Bedding, Shrubs, Herbaceous Perennials

67 Maytree Nursery

Coriolan, Courtenay Road, Dunkirk, Faversham,
Kent ME13 9LH
Tel: 01227 750353

68 Meadow Grange Nursery

Honey Hill, Blean, Whitstable, Kent CT5 3BP
Tel: 01227 471205

69 Medway Hydroponics

7a Cuxton Road, Rochester, Kent ME2 2BT
Tel: 01634 735444
Specialities: Garden Products, Hydroponics

70 Meresborough Nursery Ltd

Meresborough Road, Rainham, Gillingham,
Kent ME8 8PP
Tel: 01634 231639
Specialities: Geranium, Chrysanthemum, Fuchsia

71 Millbrook Garden Company

Station Road, Southfleet, Gravesend,
Kent DA13 9PA
Tel: 01474 331135

72 Mower Plant Services

Sunpatch Farm, Headcorn Road, Smarden, Ashford,
Kent TN27 8PJ
Tel: 01622 890046
Specialities: Garden Products

73 Nash Nurseries

Nash, Ash, Canterbury, Kent CT3 2JU
Tel: 01304 812250
Specialities: Organic Edible Plantrs

74 Newingreen Nurseries

Cydonia, Ashford Road, Newingreen, Hythe,
Kent CT21 4JD
Tel: 01303 260863

75 Newlands Nursery

Goathurst Common, Ide Hill, Sevenoaks,
Kent TN14 6DA
Tel: 01732 750591

76 Notcutts Garden Centre

Newnham Court, Bearsted Road, Weavering,
Maidstone, Kent ME14 5LH
Tel: 01622 739944

77 Notcutts Garden Centre

Tonbridge Road, Pembury, Tunbridge Wells,
Kent TN2 4QN
Tel: 01892 822636

78 Oaklands Nurseries

24 London Road, Newington, Sittingbourne,
Kent ME9 7NR
Tel: 01795 842173
Specialities: Garden Products

79 Oakleigh Farm Nursery

Oakleigh Farm Nursery, Buckland Road, Higham, Rochester, Kent ME3 7HY
Tel: 01634 222987
Specialities: Hanging Baskets

80 Oakover Plants

Maidstone Road, Hothfield, Ashford, Kent TN26 1AR
Tel: 01233 712424
Specialities: Trees, Shrubs

81 The Old Walled Garden

Oxen Hoath Road, Hadlow, Tonbridge, Kent TN11 9SS
Tel: 01732 810012
Specialities: Shrubs, Climbers

82 Oldbury Nurseries

Brissenden Green, Bethersden, Ashford, Kent TN26 3BJ
Tel: 01233 820416
Specialities: Fuchsia

83 The Orchard Nurseries

Stone Street, Petham, Canterbury, Kent CT4 5PR
Tel: 01227 700375
Specialities: Alpines/Rock Plants, Conifers

84 P & J Hutton

Hayesden Herb & Honey Farm, Upper Hayesden Lane, Tonbridge, Kent TN11 8AA
Tel: 01732 353421

85 P W Waters

George Street, Brenzett, Romney Marsh, Kent TN29 9UN
Tel: 01797 344418
Specialities: Annuals/Biennials/Bedding, Cut Flowers

86 Paradise Barn Nursery

Paradise Farm Barn, Lower Hartlip Road, Hartlip, Sittingbourne, Kent ME9 7SU
Tel: 01795 843577

87 Petham Garden Centre

Stone Street, Petham, Canterbury, Kent CT4 5PW
Tel: 01227 700449

88 Pettet's Nursery

Drainless Road, Eastry, Sandwich, Kent CT13 0EA
Tel: 01304 613869

89 Phoenix Nursery

Stable Lane, Off Vicarage Road, Bexley, Dartford, Kent DA5 2AW
Tel: 01322 521933

90 The Plant Centre

Waterland, Dillywood Lane, Higham, Rochester, Kent ME3 7NT
Tel: 01634 723123
Specialities: Aquatics

91 Plaxtol Nurseries

Plaxtol, Sevenoaks, Kent TN15 0QR
Tel: 01732 810550
Specialities: Climbers, Herbaceous Perennials, Ferns

92 Pleasant View Garden Centre

Plough Wents Road, Chart Sutton, Maidstone, Kent ME17 3SA
Tel: 01622 842042
Specialities: Garden Products, Giftware, Greenhouses and Sheds

93 Polhill Garden Centre

London Road, Badgers Mount, Sevenoaks, Kent TN14 7BD
Tel: 01959 534212
Specialities: Garden Products, Giftware, Restaurant/Coffee Shop

94 The Potted Garden Nursery Ltd

Ashford Road, Bearsted, Maidstone, Kent ME14 4NH
Tel: 01622 737801

95 Potten Farm Nursery

Potten Farm, Main Road, Sellindge, Ashford, Kent TN25 6EQ
Tel: 01303 814422
Specialities: Climbers

96 Preston Nursery

The Street, Preston, Nr Wingham, Canterbury, Kent CT3 1ED
Tel: 08000 351169
Specialities: Trees

97 Primavera

Inc. Prima Pots and Kings Toll Nursery
Maidstone Road, Matfield, Kent, TN12 7HA
Tel: 01892 824474
e-mail: prima@kingstoll.fsnet.co.uk
website: www.esotico.co.uk

Specialities:
Mediterranean Plants; Peat-Free Plants; Terracotta

Products and Services:

- Credit cards accepted
- Seeds and Bulbs
- Cut Flowers
- Fruit Trees
- Houseplants
- Large Car Park

Directions:
From the A21, about 5 miles east of Tunbridge Wells, take the B2160 towards Matfield. Primavera is about 200 yards along on the left.

Description:
When Nick Carrannante launched Prima Pots in 1987, with a few pallets of Italian terracotta, he endeavoured to supply customers "the highest quality at the lowest price."

Today this philosophy remains with more that 1000 varieties . Lavish hand-thrown terracotta from Tuscany, the evocative glazes of the far east to the latest composite materials in modern styles. Sized from 3cm up to over 100cm, truly a pot lovers paradise. This is complemented by an excellent range of stoneware planters, statues and birdbaths.

The family business acquired Kings Toll Nursery ten years ago and the Mediterranean influence spills over to the plant area. Now boasting a wonderful selection of exotics - olive, banana, citrus, bamboo, palm, bougainvillaea and more, maiking it a must for the contemporary gardener. However, with a large range of bedding, herbs and vegetables all grown on site in peat-free compost, the traditional eco-friendly gardener is far from forgotten. Helpful knowledgeable staff are at hand for advice or just to carry purchases to the car. You are sure to be delighted with a visit to PRIMAVERA.

Opening times Monday to Saturday from 9am to 5pm; Sunday 10am to 4pm; (Closed - on a Monday, November - February)

98 R Minter & Son

Sandilane Nurseries, Durlock Road, Ash, Canterbury, Kent CT3 2HU
Tel: 01304 812352

99 Ralphs Farm Shop & Nursery

Marlpit Farm, Wierton Road, Boughton Monchelsea, Maidstone, Kent ME17 4JW
Tel: 01622 743851

100 Ring Lodge Nursery

Main Road, Edenbridge, Kent TN8 6SN
Tel: 01732 862930
Specialities: Trees, Cut Flowers, Shrubs

101 Ringwould Alpines

Dover Rd, Ringwould, Deal, Kent CT14 8HG
Tel: 01304 360034
Specialities: Alpines/Rock Plants

102 Riverside Fuchsias

Gravel Road, Sutton At Hone, Dartford, Kent DA4 9HQ
Tel: 01322 863891
Specialities: Fuchsia

103 Riverside Nurseries

Hollys Cottage, Hamptons Road, Shipbourne, Tonbridge, Kent TN11 9RQ
Tel: 01732 811449

104 Roger Platts Garden Design and Nursery

Hartfield Road, Edenbridge, Kent TN8 5NH
Tel: 01732 863318

105 Rookston Nurseries

The Street, Stourmouth, Canterbury, Kent CT3 1HX
Tel: 01227 721257

106 Rosemary Nursery

Manston Road, Margate, Kent CT9 4LT
Tel: 01843 823282

107 Rosewood Daylilies

70 Deansway Avenue, Sturry, Canterbury, Kent CT2 0NN
Tel: 01227 711071
Specialities: Hemerocallis (Daylily)

108 Rosie's Garden Plants

Rochester Road, Aylesford, Kent ME20 7EB
Tel: 01622 715777
Specialities: Geranium, Aster (Michaelmas Daisy), Buddleja (Buddleia)

109 Rumwood Nurseries

Sutton Road, Langley, Maidstone, Kent ME17 3ND
Tel: 01622 861477
Specialities: Shrubs, Trees, Rose

110 The Secret Garden

Nursery & Plant Centre, Quex Park, Birchington, Kent CT7 0BB
Tel: 01843 847916
Specialities: Trees, Herbaceous Perennials, Orchids

111 Select Garden & Pet Centre

Main Road, Sundridge, Sevenoaks, Kent TN14 6ED
Tel: 01959 562400
Specialities: Specimen Plants, Palms, Bonsai

112 Shillinghurst Nursery

Oad Street, Borden, Sittingbourne, Kent ME9 8LA
Tel: 01795 842446

113 Solway Nurseries

Burmarsh Road, Hythe, Kent CT21 4NJ
Tel: 01303 260977

114 Spadework Nurseries

Riding Lane, Hildenborough, Tonbridge, Kent TN11 9LP
Tel: 01732 870002

115 Speldhurst Nurseries

Langton Road, Speldhurst, Tunbridge Wells, Kent TN3 0NR
Tel: 01892 862682
Specialities: Garden Products, Garden Design

116 Springhead Nurseries

Watling St, Gravesend, Kent DA11 8AA
Tel: 01474 361370

117 Staplehurst Nursery

Sundridge Nursery, Clapper Lane, Staplehurst, Kent TN12 0JT
Tel: 01580 893607

118 Stilebridge Nursery

Staplehurst Road, Marden, Tonbridge, Kent TN12 9BL
Tel: 01622 832265

119 Stone Green Nursery

Pluckley Road, Bethersden, Ashford, Kent TN26 3ET
Tel: 01233 820998
website: www.stonegreennurseries.co.uk
Specialities: Specimen plants, Acers, Clematis

120 Stones Garden Centre

Halfway Road, Minster On Sea, Sheerness, Kent ME12 3AA
Tel: 01795 580411

121 Summerfield Nurseries

Barnsole Road, Staple, Canterbury, Kent CT3 1LD
Tel: 01304 812549
Specialities: Annuals/Biennials/Bedding, Ponsettias

122 Tenterden Garden Centre

Reading Street, Tenterden, Kent TN30 7HT
Tel: 01233 758510
Specialities: Garden Products

123 Tenterden Garden Centre

Bough Beech Road, Four Elms, Edenbridge, Kent TN8 6NE
Tel: 01732 700032

124 Victoriana Nursery Gardens

Buck Street, Challock, Ashford, Kent TN25 4DG
Tel: 01233 740529
Specialities: Vegetables, Fruit Bushes, Fuchsia

125 Vincent Nurseries & Plant Centre

Eddington Lane, Herne Bay, Kent CT6 5TS
Tel: 01227 375806

Specialities:
Shrubs grown on site; Basket Plants; Bedding Plants

Products and Services:

- Aquatics
- Seasonal Displays
- Credit Cards Accepted
- Cut Flowers
- Disabled Facilities
- Floristry
- Fruit Trees
- Garden Products
- Giftware
- Houseplants
- Information Desk
- Large Car Park
- Plants
- Seeds and Bulbs
- Trees

Directions:
From Canterbury follow the A291 to Herne Bay via Herne village. Cross 3 roundabouts then 2nd left into Eddington Lane.

Description:
Vincent Nurseries is a family business that was established in 1947 by Philip and David Maunder and began by growing vegetables and salad crops. It is now run by Philip's daughter Christine who joined the business in 1983 after a 3 year course at Hadlow College. On her return the business moved from producing for the wholesale market into retailing. The centre covers 5 acres with a mixture of glass and tunnels. It is stocked with a comprehensive range of trees, shrubs, conifers, perennials and fruit bushes as well as seasonal bedding and basket plants. Many of the plants are grown on the nursery and special care and attention is given to the quality of all the plants. A full range of garden sundries, chemicals, composts and teracotta are also stocked.

126 W Carter & Son

34 Borstal Hill, Whitstable, Kent CT5 4LX
Tel: 01227 272903

127 Wale & Son

10-11 The Strand, Walmer, Deal, Kent CT14 7DY
Tel: 01304 374854

128 The Walled Nursery

Water Lane, Hawkhurst, Cranbrook, Kent TN18 5DH
Tel: 01580 752425
Specialities: Garden Products

129 Walmer Nursery

19-21 Dover Road, Walmer, Deal, Kent CT14 7HR
Tel: 01304 375277

130 Warmlake Nurseries

North Street, Sutton Valence, Maidstone, Kent ME17 3LW
Tel: 01622 844000 Fax: 01622 840051

Specialities:
Hardy Perennials; Paving; Pet Foods

Products and Services:
- Seasonal Displays
- Credit Cards Accepted
- Cut Flowers
- Disabled Facilities
- Fruit Trees
- Fuchsias
- Garden Products
- Information Desk
- Large Car Park
- Paving
- Pet Products
- Plant Guarantee
- Plants
- Seasonal Bedding
- Seeds and Bulbs
- Trees

Directions:
From Maidstone take the A274. Join the A229 towards Tenterden. The Nursery is approx.3.5 miles on A274 on the right hand side.

131 Waterlands

Dillywood Lane, Higham, Kent ME3 7NT
Tel: 01634 719889 Fax: 01634 711225
e-mail: waterlandscapes@btconnect.com
website: www.waterlands.org.uk

Specialities:
Aquatics; Ponds; Water Features

Products and Services:
- Aquatics
- Garden Design
- Plants

Directions:
Dillywood Lane is off the B2000 to the south of Higham.

Description:
Ideally located near the A2/M2, Waterlands was established in 1948 and has been supplying the community with unrivalled services and products for gardening and landscaping for more than 50 years. Waterlands has developed an extensive aquatics centre where advice and products are suitable to all. Whether you are looking to install your first water feature, build a new pond, expand or service your current set up Waterlands are on hand to offer expert advice. Not only does Waterlands have a superb aquatics centre but can also supply a wide selection of products from stock including barbecues, hammocks, furniture, plants, water features, waterfalls, ornamental stone, ponds, fresh, tropical and marine fish and possibly the largest stockist of pots in Kent. All the staff at Waterlands have expert knowledge of high quality products on sale and are always willing to offer friendly advice and guidance to all its customers. Why not come and see for yourself our oasis in rural Kent.

132 Waverley Nurseries

622 Lower Rainham Road, Rainham, Gillingham, Kent ME8 7TX
Tel: 01634 232520

133 Westerham Heights Nursery

Main Road, Westerham, Kent TN16 2HW
Tel: 01959 571545

134 Whiteways Bonsai & Nursery

Curties Lane, Stowting, Ashford, Kent TN25 6BS
Tel: 01303 862259
Specialities: Bonsai, Conifers, Shrubs

135 Willowbank Nurseries

High Street, Manston, Nr Ramsgate, Kent CT12 5BQ
Tel: 01843 822973
Specialities: Hanging Baskets, Annuals/Biennials/Bedding, Climbers

136 Wingham Country Market

Shatterling, Shatterling, Canterbury, Kent CT3 1JW
Tel: 01227 720567

137 Woodlands Garden Centre

Ash Lane, Ash, Sevenoaks, Kent TN15 7EG
Tel: 01474 852788
Specialities: Garden Products, Garden Furniture, Giftware

138 Worth Plant Centre

Jubilee Rd, Worth, Deal, Kent CT14 0DT
Tel: 01304 612055

139 Wyevale Garden Centre (Ashford Romney Marsh)

Hamstreet, Ashford, Kent TN26 2QP
Tel: 01233 732988

140 Wyevale Garden Centre (Ashford Willesborough)

Hythe Road, Willesborough, Ashford, Kent TN24 0NE
Tel: 01233 502136

141 Wyevale Garden Centre (Canterbury - Chartham)

Stour Business Park, Chartham, Canterbury, Kent CT4 7HF
Tel: 01227 731033

142 Wyevale Garden Centre (Eridge Road)

Eridge Road, Tunbridge Wells, Kent TN4 8HR
Tel: 01892 515234

143 Wyevale Garden Centre (Folkestone)

Ingles Meadow, Jointon Road, Folkestone, Kent CT20 2RF
Tel: 01303 258100

144 Wyevale Garden Centre (Gillingham)

Elm Court, Capstone Road, Gillingham, Kent ME7 3JQ
Tel: 01634 813778

145 Wyevale Garden Centre (Paddock Wood)

Maidstone Road, Paddock Wood, Tonbridge, Kent TN12 6QJ
Tel: 01892 835777

146 Wyevale Garden Centre (Ramsgate)

Montefiore Avenue, Ramsgate, Kent CT11 8BW
Tel: 01843 592393

147 Young's Nurseries

Sacketts Hill, Broadstairs, Kent CT10 2QS
Tel: 01843 592556
Specialities: Annuals/Biennials/Bedding, Trees

LANCASHIRE

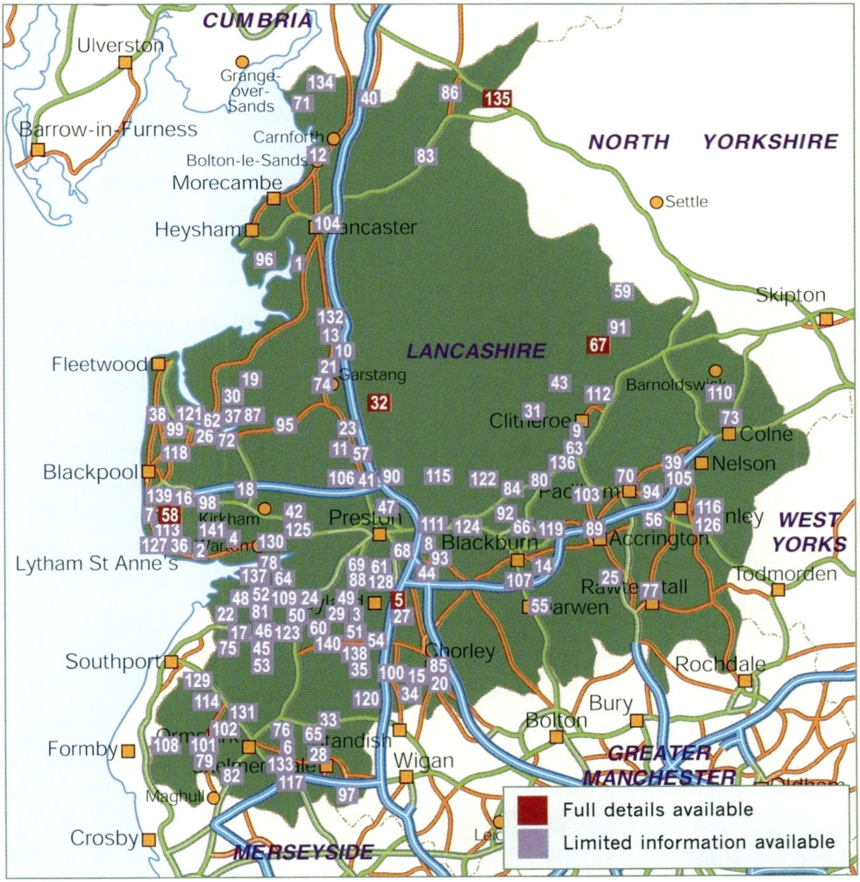

1 Ashton Hall Garden Centre Limited

Ashton With Stodday, Lancaster,
Lancashire LA2 0AJ
Tel: 01524 751767

2 Ashtons Nursery Gardens

Mythop Road, Lytham St. Annes,
Lancashire FY8 4JP
Tel: 01253 736627
Specialities: Annuals/Biennials/Bedding, Shrubs, Trees

3 Ashtons Plant Centre

Longmeanygate, Midge Hall, Leyland,
Lancashire PR26 7TB
Tel: 01772 434233

4 Askew House Plants

Kellamergh Nurseries, Bryning Lane, Warton,
Preston, Lancashire PR4 1TN
Tel: 01772 632506
Specialities: Houseplants

5 Avant Gardens

Wigan Road, Leyland, Preston,
Lancashire PR25 5XW
Tel: 01772 433777 Fax: 01772 436493
e-mail: avantgardens@hotmail.co.uk

Specialities:
Unusual Plant Varieties & Ornaments; Indoor and Outdoor Aquatics

Products and Services:

- Aquatics
- Seasonal Displays
- Clothing
- Conservatories
- Credit Cards Accepted
- Disabled Facilities
- Display Gardens
- Dried and Artificial Flowers
- Fireworks
- Fruit Trees
- Garden Design
- Garden Furniture
- Garden Products
- Giftware
- Greenhouses and Sheds
- Houseplants
- Information Desk
- Large Car Park
- Paving
- Plants
- Restaurant/Coffee Shop
- Seeds and Bulbs
- Trees

Directions:
Avant Gardens are on the A49 east of Leyland, south of Preston and 2 minutes from junction 28 of the M6

Description:
This traditional garden centre stocks just about everything a gardener might ever need. It was established in 1997 by Heather and Juliet Mallinson and their family. Juliet herself is horticulturally trained and her helpful and experienced staff are knowledgeable and ever ready to give advice. The extensive and beautifully maintained site stocks a wide variety of unusual plants and a selection of trees and topiary. There's also a vegetable display area and the concessions on site include paving, spa pools, conservatories and doll's houses. The garden centre is open from 9am to 5.30pm, Monday to Saturday, and from 10am to 4pm on Sunday. The business is a member of the HTA and sells their gift vouchers.

6 Avondale Nurseries (NW) Ltd

Spa Lane, Lathom, Ormskirk, Lancashire L40 6JG
Tel: 01695 50857
Specialities: Annuals/Biennials/Bedding, Homegrown

7 Baguleys Garden Centre

Midgeland Road, Blackpool, Lancashire FY4 5HE
Tel: 01253 762981
Specialities: Garden Products, Trees

8 Banister Hall

Potter Lane, Higher Walton, Preston,
Lancashire PR5 4EN
Tel: 01772 877087

9 Barkers Garden Centres & Florist

Whalley Road, Clitheroe, Lancashire BB7 1HT
Tel: 01200 423521
Specialities: Floristry, Garden Products

10 The Barn Plant Centre & Gift Shop

The Barn, The Square, Scorton, Preston,
Lancashire PR3 1AU
Tel: 01524 793533
Specialities: Herbaceous Perennials, Restaurant/Coffee Shop, Giftware

11 Barton Grange Garden Centre

Garstang Road, Barton, Preston,
Lancashire PR3 5AA
Tel: 01772 864242

12 Bay View Garden Centre

Mill Lane, Bolton Le Sands, Carnforth,
Lancashire LA5 8ET
Tel: 01524 733780

13 Beechgrove Nurseries

Station Lane, Scorton, Preston,
Lancashire PR3 1AN
Tel: 01524 791286
Specialities: Hanging Baskets, Shrubs

14 Beechwood Garden Centre Limited

Roman Road, Blackburn, Lancashire BB1 2LB
Tel: 01254 53310

15 Birkacre Nurseries

Birkacre Road, Chorley, Lancashire PR7 3QL
Tel: 01257 270473
Specialities: Herbaceous Perennials, Annuals/Biennials/Bedding

16 Blaylock Shrubs

Daisyfield, Old House Lane, Blackpool,
Lancashire FY4 5LE
Tel: 01253 695999
Specialities: Shrubs

17 Blundell's Nurseries

68 Southport New Road, Tarleton, Preston, Lancashire PR4 6HY
Tel: 01772 815442
Specialities: Annuals/Biennials/Bedding, Herbaceous Perennials

18 Brackenwood Nurseries

Bradshaw Lane, Greenhalgh, Preston, Lancashire PR4 3HQ
Tel: 01772 686232
Specialities: Seasonal Plants, Herbaceous Perennials

19 Bradshaw Lane Nurseries

Bradshaw Lane, Pilling, Preston, Lancashire PR3 6AX
Tel: 01253 790895
Specialities: Annuals/Biennials/Bedding, Shrubs

20 Bramblewood Nursery & Garden Centre

Wigan Lane, Heath Charnock, Chorley, Lancashire PR7 4DD
Tel: 01257 269600
Specialities: Annuals/Biennials/Bedding, Garden Products

21 Briarcliffe Nurseries

Gubberford Lane, Garstang, Preston, Lancashire PR3 1PS
Tel: 01995 602420
Specialities: Floristry, Shrubs, Trees

22 Briarlee Garden Centre

35-37 Chapel Lane, Banks, Southport, Lancashire PR9 8EY
Tel: 01704 226647
Specialities: Annuals/Biennials/Bedding, Garden Products

23 Brookhouse Nurseries

Garstang Road, Bilsborrow, Preston, Lancashire PR3 0RD
Tel: 01995 640543
Specialities: Annuals/Biennials/Bedding, Shrubs

24 Brooklane Nursery

Brook Lane, Much Hoole, Preston, Lancashire PR4 4RD
Tel: 01772 611629

25 Brookside Garden Centre

Roundhill Road, Accrington, Lancashire BB5 3SW
Tel: 01706 216491
Specialities: Trees, Shrubs, Conifers

26 Burtonwood Nurseries

109 Mains Lane, Poulton-le-Fylde, Lancashire FY6 7LD
Tel: 01253 892404
Specialities: Seasonal Displays, Seasonal Plants

27 Charnock Garden Centre

Wigan Road, Leyland, Lancashire PR25 5DA
Tel: 01772 623350
Specialities: Restaurant/Coffee Shop, Pet Products

28 Cicely's Cottage Garden Plants

43 Elmers Green, Skelmersdale, Lancashire WN8 6SG
Tel: 01695 720790
Specialities: Homegrown, Cottage Garden Plants

29 Claremont Aquatic Nurseries

Clairmont, Cocker Bar Road, Leyland, Lancashire PR26 7TA
Tel: 01772 421860
Specialities: Annuals/Biennials/Bedding, Aquatics, Water Products and Features

30 Cleggs Nurseries

Back Lane, Stalmine, Poulton-le-Fylde, Lancashire FY6 0JN
Tel: 01253 700250
Specialities: Seasonal Plants, Homegrown, Annuals/Biennials/Bedding

31 Clough Bottom Trees & Shrubs

Talbot Bridge, Bashall Eaves, Clitheroe, Lancashire BB7 3NA
Tel: 01254 826285
Specialities: Shrubs, Trees

32 Cobble Hey Farm & Gardens

off Hobbs Lane, Claughton on Brock, Garstang, Preston, Lancashire PR3 0QN
Tel: 01995 602643
e-mail: cobblehey@aol.com
website: www.cobblehey.co.uk

SEE ADVERTISEMENT PAGE 138

Specialities:

Unusual Perennials; Geraniums; Phlox Paniculata

Description:

This award-winning attraction is a working farm with large gardens based at more than 600ft. This venue has something for the whole family, from spectacular views over the Forest of Bowland, to a wide range of walks, rare breed animals, tearooms where all the refreshments are home-produced, a children's play area, wetland conservation area for birds, and a garden which has more than 90 metres of naturally running water through it, along with a plant nursery to tempt you with that unusual plant.

Directions:

As take the A6 towards Garstang you will see an Esso garage called Rogers of Brock with a junction signed for Beacon Fell and Cobble Hey Gardens. Just follow the flower.

33 Cobbs Brow Nurseries

Cobbs Brow Lane, Newburgh, Wigan,
Lancashire WN8 7ND
Tel: 01257 464424
Specialities: Homegrown, Seasonal Plants, Annuals/Biennials/Bedding

34 Coppull Moor Lane Nurseries

Coppull Moor Lane, Coppull, Chorley,
Lancashire PR7 5JH
Tel: 01257 791804
Specialities: Annuals/Biennials/Bedding, Herbaceous Perennials

35 Croston Cactus

43 Southport Road, Eccleston, Chorley,
Lancashire PR7 6ET
Tel: 01257 452555
Specialities: Cacti And Succulents

36 Crowder Nurseries

Division Lane, Blackpool, Lancashire FY4 5DZ
Tel: 01253 763570
Specialities: Shrubs, Bulbs

37 Cuts Lane Nurseries Ltd

Cutts Lane, Out Rawcliffe, Preston,
Lancashire PR3 6SY
Tel: 01253 701196
Specialities: Annuals/Biennials/Bedding

38 Daisybank Garden Centre

Rear of 172 Victoria Road West, Thornton-Cleveleys,
Lancashire FY5 3NE
Tel: 01253 856802
Specialities: Annuals/Biennials/Bedding, Garden Furniture

39 Dawsons Nurseries

Clitheroe Road, Brierfield, Nelson,
Lancashire BB9 5QH
Tel: 01282 615473
Specialities: Annuals/Biennials/Bedding, Shrubs, Conifers

40 Deerslet Nurseries

Burton-in-Kendal, Burton, Carnforth, Lancashire LA6 1JG
Tel: 01524 781777
Specialities: Annuals/Biennials/Bedding, Shrubs

41 Dennys Clematis Nursery

58 Woodplumpton Lane, Broughton, Preston, Lancashire PR3 5JL
Tel: 01772 862331
Specialities: Clematis, Annuals/Biennials/Bedding

42 Dobbies Garden World

Blackpool Road, Clifton, Preston, Lancashire PR4 0XL
Tel: 01772 683844

43 Dove Syke Nursery

Eaves Hall Lane, West Bradford, Clitheroe, Lancashire BB7 3JG
Tel: 01200 428417
Specialities: Annuals/Biennials/Bedding, Shrubs

44 Duckworth House Lane Nurseries

Bankhead Lane, Preston, Lancashire PR5 6YR
Tel: 01772 339616
Specialities: Homegrown, Herbaceous Perennials

45 Dunscar Garden Centre Ltd

116 Southport New Road, Tarleton, Preston, Lancashire PR4 6HY
Tel: 01772 812684

46 Dunscar Nurseries & Garden Centre

188 Blackgate Lane, Tarleton, Preston, Lancashire PR4 6UU
Tel: 01772 812807
Specialities: Annuals/Biennials/Bedding, Garden Products

47 Eastway Nurseries Ltd

58 Coniston Road, Fulwood, Preston, Lancashire PR2 8AY
Tel: 01772 709511

48 Edenbreck Nurseries

139 Station Road, Hesketh Bank, Preston, Lancashire PR4 6SR
Tel: 01772 812574
Specialities: Shrubs, Herbaceous Perennials

49 Eezitill Nurseries

Startley Nook, Long Moss Lane, Whitestake, Preston, Lancashire PR4 4XQ
Tel: 01772 616820
Specialities: Herbaceous Perennials, Shrubs

50 Embley's Nurseries

Liverpool New Road, Much Hoole, Preston, Lancashire PR4 4RL
Tel: 01772 612227

51 Everglades Nurseries

Dawbers Lane, Euxton, Chorley, Lancashire PR7 6EQ
Tel: 01257 450533
Specialities: Herbaceous Perennials, Annuals/Biennials/Bedding

52 Eversley Nursery

10 Granville Avenue, Hesketh Bank, Preston, Lancashire PR4 6AH
Tel: 01772 812538
Specialities: Annuals/Biennials/Bedding, Shrubs

53 Fir Tree Nurseries

Holmeswood Road, Rufford, Ormskirk, Lancashire L40 1TY
Tel: 01704 823334
Specialities: Annuals/Biennials/Bedding, Trees

54 Five Acres Plant Centre

Dawbers Lane, Euxton, Chorley, Lancashire PR7 6EE
Tel: 01257 279024
Specialities: Annuals/Biennials/Bedding, Shrubs

55 Four Seasons Garden Centre

Ashton Road, Darwen, Lancashire BB3 2DX
Tel: 01254 706890
Specialities: Herbaceous Perennials

56 Fresh Fields Enterprises Ltd

Morecambes Lodge, Cemetery Lane, Burnley, Lancashire BB11 5QB
Tel: 01282 832820
Specialities: Homegrown, Annuals/Biennials/Bedding, Vegetables

57 Garden Centre Plants

Barton Grange, Barton Lane, Barton, Preston, Lancashire PR3 5AU
Tel: 01772 863531

58 The Garden Place

Cropper Road, Marton, Blackpool
Lancashire P74 5LB
Tel: 01253 699987

Specialities:

Bedding Plants; Coffee Shop/Restaurant; Garden Furniture

Description:

The Garden Place has recently been upgraded and now offers a full range of quality plants, garden furniture and sundries. It also has an excellent licensed restaurant and coffee shop serving very good home-made cakes and a selection of lunches.

Directions:

From Exit 4 of the M55, continue westwards on the dual carriageway to the roundabout. Turn left and The garden Place is signposted up Cropper Road.

59 Gardenmakers

Coars Farm, Wigglesworth, Skipton,
Lancashire BD23 4SN
Tel: 01729 840848

60 Glenroyde

Glenroyd, North Road, Bretherton, Leyland,
Lancashire PR26 9AY
Tel: 01772 600337

61 Glenview Nurseries

Aspley House, Brook Lane, Whitestake, Preston,
Lancashire PR4 4LA
Tel: 01772 311049
Specialities: Shrubs, Seasonal Plants

62 Hambleton Garden Centre

Bank View Industrial Estate, Shard Lane, Hambleton,
Lancashire FY6 9BX
Tel: 01253 702212
Specialities: Garden Furniture, Fencing, Composts

63 Hansons Garden Centre Limited

Whalley Road, Barrow, Clitheroe,
Lancashire BB7 9BA
Tel: 01254 822145
Specialities: Restaurant/Coffee Shop, Greenhouses and Sheds, Garden Furniture

64 The Hawthornes

Marsh Road, Hesketh Bank, Preston,
Lancashire PR4 6XT
Tel: 01772 812379
Specialities: Shrubs, Seasonal Plants

65 Heare Nursery

Cobbs Brow Lane, Newburgh, Wigan,
Lancashire WN8 7SF
Tel: 01257 463012

66 Hedging Conifers

Higher Waves Farm, Ramsgreave Drive, Blackburn,
Lancashire BB1 8LL
Tel: 01254 248187
Specialities: Conifers, Hedging

67 Holden Clough Nursery

Holden, Bolton-by-Bowland, nr Clitheroe,
Lancashire BB7 4PF
Tel: 01200 447 615 Fax: 01200 447 197
e-mail: info@holdencloughnursery.co.uk
website: www.holdencloughnursery.co.uk

SEE ADVERTISEMENT PAGE 141

Specialities:

Alpines & Rock Garden Plants; Herbaceous Perennials; Ornamental Shrubs; Dwarf Conifers

Description:

Holden Clough is a traditional 2-acre nursery which has won awards at RHS Shows since 1985. Over the years the range of plants propagated and offered for sale has widened immensely to include many fine herbaceous perennials, shrubs, climbing plants, dwarf conifers and heathers.

Directions:

From the A59 about 4 miles north of Clitheroe, turn left on minor road through Sawley to Holden (approx. 3 miles). At Holden fork left and the nursery is 200 yards on the right.

68 Holland Nurseries

Higher Walton Road, Walton-le-Dale, Preston,
Lancashire PR5 4HD
Tel: 01772 202472
Specialities: Shrubs, Herbaceous Perennials

69 Holme Lea Nursery

Pope Lane, Whitestake, Preston,
Lancashire PR4 4BA
Tel: 01772 615244
Specialities: Herbaceous Perennials

Holden Clough Nursery was established in 1927 by Richard Milne-Redhead and in the early days specialised almost solely in the cultivation of alpine and rock garden plants which proved very popular with gardeners across the north and were soon offered by mail order countrywide. Over the years the range of plants propagated and offered for sale has widened immensely and throughout this time the main criteria of the nursery has been to produce quality plants in a hardy manner with no excessive protection from the elements and at the same time ensure that the range of plants offered is hardy in this North West England Pennine location.

- Credit Cards Accepted
- Display Gardens
- Garden Construction
- Garden Design
- Information Desk
- Nursery Events
- Plant Guarantee
- Plants
- Trees

ALPINES & ROCK GARDEN PLANTS; HERBACEOUS PERENNIALS; ORNAMENTAL SHRUBS

HOLDEN CLOUGH NURSERY

HOLDEN, BOLTON-BY-BOWLAND, NR CLITHEROE, LANCASHIRE BB7 4PF
TEL: 01200 447615 FAX: 01200 447197
E-MAIL: INFO@HOLDENCLOUGHNURSERY.CO.UK
WEBSITE: WWW.HOLDENCLOUGHNURSERY.CO.UK

Opening times:
Monday to Friday; March to October: 9am-5pm
Saturday; All Year: 9am-5pm
At other times by appointment only

70 Huntroyde Nurseries

Whins Lane, Padiham, Burnley,
Lancashire BB12 7QJ
Tel: 01282 770753
Specialities: Shrubs, Herbaceous Perennials

71 Kayes Nursery

Lindeth Road, Silverdale, Carnforth,
Lancashire LA5 0TY
Tel: 01524 702010

72 The Koi Pool Water Gardens

35 Mains Lane, Poulton-le-Fylde,
Lancashire FY6 7LJ
Tel: 01253 896127

73 Lakeside Garden Centre

Skipton Road, Foulridge, Colne,
Lancashire BB8 7NW
Tel: 01282 865650
Specialities: Garden Products, Annuals/Biennials/Bedding

74 Laneside Alpines

74 Croston Road, Garstang, Preston,
Lancashire PR3 1HR
Tel: 01995 605537
Specialities: Alpines/Rock Plants

75 Larkfield Nurseries

Mere Lane, Southport, Lancashire PR9 8DA
Tel: 01704 214150
Specialities: Shrubs, Herbaceous Perennials

76 Lathom View Garden Centre

Beech Grove, Hall Lane, Lathom, Ormskirk,
Lancashire L40 5UN
Tel: 01704 892315
Specialities: Water Products and Features, Herbaceous Perennials, Trees

77 Lea Brook Nursery

Burnley Road, Rossendale, Lancashire BB4 8HH
Tel: 01706 230856
Specialities: Annuals/Biennials/Bedding, Shrubs

78 Lemkin Nurseries

Guide Road, Hesketh Bank, Preston,
Lancashire PR4 6XS
Tel: 01772 811541
Specialities: Annuals/Biennials/Bedding, Herbaceous Perennials

79 Lloyds Nursery

1 Broad Lane, Aughton, Ormskirk,
Lancashire L39 7HS
Tel: 01515 260669
Specialities: Shrubs, Annuals/Biennials/Bedding

80 Longsight Nursery

Longsight Road, Langho, Blackburn,
Lancashire BB6 8AD
Tel: 01254 247075
Specialities: Annuals/Biennials/Bedding, Seasonal Plants

81 Lyndene Garden Centre

143 Southport New Road, Tarleton, Preston,
Lancashire PR4 6HX
Tel: 01772 813611
Specialities: Garden Products, Annuals/Biennials/Bedding

82 Mark Taylor

Springfield Road, Aughton, Ormskirk,
Lancashire L39 6ST
Tel: 01695 424037

83 Mary Green

The Walled Garden, Monteagle Drive, Hornby,
Lancaster, Lancashire LA2 8LD
Tel: 01524 221989

84 Mill House Nursery

Longsight Road, Copster Green, Blackburn,
Lancashire BB1 9EU
Tel: 01254 248333
Specialities: Annuals/Biennials/Bedding, Herbaceous Perennials, Shrubs

85 Millwood Alpines

7 Lime Grove, Chorley, Lancashire PR7 3JA
Tel: 01257 413454
Specialities: Alpines/Rock Plants

86 Mole End Nurseries Ltd

Coach House, Burrow, Carnforth,
Lancashire LA6 2RJ
Tel: 01524 274245
Specialities: Annuals/Biennials/Bedding, Shrubs

87 Moorham Hill Nurseries

Chapel Lane, Out Rawcliffe, Preston,
Lancashire PR3 6TB
Tel: 01253 700776

88 Newgate Nurseries

75 Newgate Lane, Whitestake, Preston,
Lancashire PR4 4JU
Tel: 01772 612336
Specialities: Bamboos, Pot Plants, Tropical

89 Nursery End Garden Centre

Harrington Street, Clayton Le Moors, Accrington,
Lancashire BB5 4DF
Tel: 01254 389611
Specialities: Garden Products, Annuals/Biennials/Bedding

90 Oak Nurseries

Pudding Pie Nook Lane, Goosnargh, Preston,
Lancashire PR3 2JL
Tel: 01772 862828
Specialities: Annuals/Biennials/Bedding, Shrubs

91 Oaktree Nurseries

Settle Road, Bolton By Bowland, Clitheroe,
Lancashire BB7 4NT
Tel: 01200 447431
Specialities: Seasonal Plants, Shrubs

92 Old Dads Farm

Mellor Lane, Mellor, Blackburn,
Lancashire BB2 7EN
Tel: 01254 812434
Specialities: Cut Flowers, Hanging Baskets

93 Old Oak Farm Nurseries

Bells Lane, Hoghton, Preston, Lancashire PR5 0JJ
Tel: 01254 852065
Specialities: Alpines/Rock Plants, Herbaceous Perennials, Shrubs

94 Park Lane Plants Ltd

Ightenhill, Park Lane, Burnley,
Lancashire BB12 0RW
Tel: 01282 454802
Specialities: Annuals/Biennials/Bedding, Shrubs

95 Pasture Views Nursery

Moss Side Lane, Great Eccleston, Preston,
Lancashire PR3 0US
Tel: 01995 671628
Specialities: Annuals/Biennials/Bedding, Shrubs

96 Pearsons Garden Centre

40 Main Street, Overton, Morecambe, Lancashire LA3 3HF
Tel: 01524 858335
Specialities: Garden Products, Pet Products

97 Pimbo Garden Centre

32 Pimbo Lane, Upholland, Skelmersdale, Lancashire WN8 9QQ
Tel: 01695 622601
Specialities: Garden Products, Herbaceous Perennials

98 The Plant Emporium

Preston New Road, Blackpool, Lancashire FY4 5JR
Tel: 01253 764099
Specialities: Annuals/Biennials/Bedding, Shrubs

99 The Plant Place

169-185 Fleetwood Road South, Thornton-Cleveleys, Lancashire FY5 5NS
Tel: 01253 856414
Specialities: Herbaceous Perennials, Shrubs

100 Pole Green Nurseries

Church Lane, Charnock Richard, Chorley, Lancashire PR7 5NH
Tel: 01257 791233
Specialities: Annuals/Biennials/Bedding, Herbaceous Perennials

101 Pop Larkin's Garden & Pet Centre

1 Broad Lane, Aughton, Ormskirk, Lancashire L39 7HS
Tel: 01515 200120
Specialities: Garden Products, Pet Products

102 Primrose Hill Nurseries Ltd

Asmall Lane, Halsall, Ormskirk, Lancashire L39 8RB
Tel: 01704 841121
Specialities: Annuals/Biennials/Bedding, Shrubs

103 Read Garden Centre Ltd

Old Coal Staithe, Read, Burnley, Lancashire BB12 7RT
Tel: 01254 884215
Specialities: Garden Products, Herbaceous Perennials

104 Redwood Garden Centre

Wyresdale Road, Lancaster, Lancashire LA1 3LA
Tel: 01524 542685
Specialities: Landscape Gardening, Annuals/ Biennials/Bedding, Paving

105 Reedley Nursery

Robinson Lane, Brierfield, Nelson, Lancashire BB9 5QS
Tel: 01282 693376
Specialities: Annuals/Biennials/Bedding, Shrubs, Homegrown

106 Ribblesdale Nurseries

Newsham Hall Lane, Woodplumpton, Preston, Lancashire PR4 0AS
Tel: 01772 863081
Specialities: Annuals/Biennials/Bedding, Shrubs, Trees

107 Riverside Garden Centre

Branch Road, Lower Darwen, Darwen, Lancashire BB3 0PR
Tel: 01254 699384

108 Robert Wright & Son

Lord Sefton Way, Liverpool, Lancashire L37 5AA
Tel: 01704 878488
Specialities: Rose, Shrubs, Winter Pansies

109 S Banister

Becconsall Lane, Hesketh Bank, Preston, Lancashire PR4 6RR
Tel: 01772 813203

110 Salterforth Nurseries

Moor Lane, Salterforth, Barnoldswick, Lancashire BB18 5SP
Tel: 01282 850649
Specialities: Shrubs, Herbaceous Perennials

111 Samlesbury Bonsai Nursery

The Boat House, Potters Lane, Samlesbury, Preston, Lancashire PR5 0UE
Tel: 01772 877213
Specialities: Bonsai, Herbaceous Perennials

112 Shackletons Garden Nursery

Clitheroe Road, Chatburn, Clitheroe, Lancashire BB7 4JY
Tel: 01200 441230
Specialities: Annuals/Biennials/Bedding, Shrubs, Trees

113 Singleton House Plants

285 Midgeland Road, Blackpool, Lancashire FY4 5JA
Tel: 01253 762345
Specialities: Homegrown, Pot Plants

114 Sintons' Nurseries

44 Renacres Lane, Halsall, Ormskirk, Lancashire L39 8SE
Tel: 01704 840600
Specialities: Homegrown, Annuals/Biennials/Bedding

115 Spoutfarm Nursery

Preston Road, Longridge, Preston, Lancashire PR3 3BE
Tel: 01772 784010

116 Springwood Nursery & Garden Centre

Springwood Road, Burnley, Lancashire BB10 4HR
Tel: 01282 832271
Specialities: Herbaceous Perennials, Trees, Shrubs

117 Stanley Gate Nursery & Garden Centre

Ormskirk Old Road, Bickerstaffe, Ormskirk, Lancashire L39 0HD
Tel: 01695 725262

118 Stella Maris Nursery

24 Westfield Avenue, Blackpool, Lancashire FY3 7LU
Tel: 01253 392587
Specialities: Herbaceous Perennials

119 Stonehill Nurseries & Garden Centre

Whalley Old Road, Blackburn, Lancashire BB1 4AA
Tel: 01254 53110
Specialities: Annuals/Biennials/Bedding, Garden Products, Herbaceous Perennials

120 Stoney Leach

40 Toogood Lane, Wrightington, Wigan, Lancashire WN6 9PL
Tel: 01257 253105

121 Strikes Burnside Garden Centre

New Lane, Thornton-Cleveleys, Lancashire FY5 5NL
Tel: 01253 821383

122 Stydd

Stoneygate Lane, Ribchester, Preston, Lancashire PR3 3YN
Tel: 01254 878797

123 Tarleton Specimen Plants

Fairview, Gorse Lane, Tarleton, Preston, Lancashire PR4 6LH
Tel: 01772 816879
Specialities: Specimen Plants, Rare and Unusual Plants

124 Tarleton Specimen Plants Ltd

Samlesbury Hall, Preston New Road, Samlesbury, Preston, Lancashire PR5 0UP
Tel: 01254 814245
Specialities: Specimen Plants

125 Tavistock Herb Nursery

Tavistock, Preston Old Road, Clifton, Preston, Lancashire PR4 0ZA
Tel: 01772 683505
Specialities: Herbs

126 Towneley Garden Centre

Deerpark Road, Burnley, Lancashire BB10 4SD
Tel: 01282 424162
Specialities: Garden Products

127 Trebaron Garden Centre

350 Common Edge Road, Blackpool, Lancashire FY4 5DY
Tel: 01253 691368
Specialities: Garden Furniture, Annuals/Biennials/Bedding, Bulbs

128 Trebaron Garden Centre

Chain House Lane, Whitestake, Preston, Lancashire PR4 4LB
Tel: 01772 336664
Specialities: Garden Products, Garden Furniture

129 The Vincents Garden Centre Ltd

300 Southport Road, Southport, Lancashire PR8 5LF
Tel: 01704 880236

130 W & M J Whittle & Son

Preston New Road, Freckleton, Preston, Lancashire PR4 1HN
Tel: 01772 632432

131 W Pye & Sons

Pinfold Nurseries, Smithy Lane, Scarisbrick, Ormskirk, Lancashire L40 8HH
Tel: 01704 840610
Specialities: Homegrown, Herbaceous Perennials, Annuals/Biennials/Bedding

132 W Robinson & Sons Ltd

Sunny Bank, Lancaster Road, Forton, Preston, Lancashire PR3 0BN
Tel: 01524 791210
Specialities: Seeds, Water Products and Features

133 Warbreck Garden Centre

Lyelake Lane, Westhead, Ormskirk, Lancashire L40 6JW
Tel: 01695 722960
Specialities: Annuals/Biennials/Bedding, Garden Products

134 Waterslack Garden Centre

Ford Lane, Silverdale, Carnforth, Lancashire LA5 0UH
Tel: 01524 701255
Specialities: Farm Shop, Restaurant/Coffee Shop

135 Westhouse Nursery

A65 Road, Westhouse, Ingleton, Lancashire LA6 3NJ
Tel: 015242 41878 / 42652
website: www.westhousenursery.co.uk

Specialities:

Fuchsias; Bedding Plants; Alpines

Description:

Located in a beautiful part of the Yorkshire Dales National Park, Westhouse Nursery is stocked with hundreds of lovingly home-grown plants. The heated indoor nursery boasts over 400 varieties of fuchsias, as well as perennials, shrubs, hanging baskets and terracotta pots.

Directions:

The nursery is on the A65, six miles southeast of Kirkby Lonsdale and 13 miles from junction 36 of the M6

136 Whalley Nurseries

Clitheroe Road, Whalley, Clitheroe, Lancashire BB7 9AQ
Tel: 01254 824990
Specialities: Garden Products, Herbaceous Perennials

137 Whitegate Nursery

Chapel Road, Hesketh Bank, Preston, Lancashire PR4 6RX
Tel: 01772 812553
Specialities: Annuals/Biennials/Bedding, Seasonal Plants

138 Whiterigg Alpines

1 Southport Road, Eccleston, Chorley, Lancashire PR7 6ET
Tel: 01257 451021
Specialities: Alpines/Rock Plants

139 Willows Nursery

40 Stockydale Road, Blackpool, Lancashire FY4 5HR
Tel: 01253 691504
Specialities: Annuals/Biennials/Bedding, Shrubs

140 Wyevale Garden Centre (Leyland)

Auldene Garden Centre, Southport Road, Leyland, Lancashire PR26 8LQ
Tel: 01772 600271

141 Wyndcote Bonsai

Lytham Road, Moss Side, Lytham St. Annes, Lancashire FY8 4LZ
Tel: 01253 733458
Specialities: Bonsai

LEICESTERSHIRE

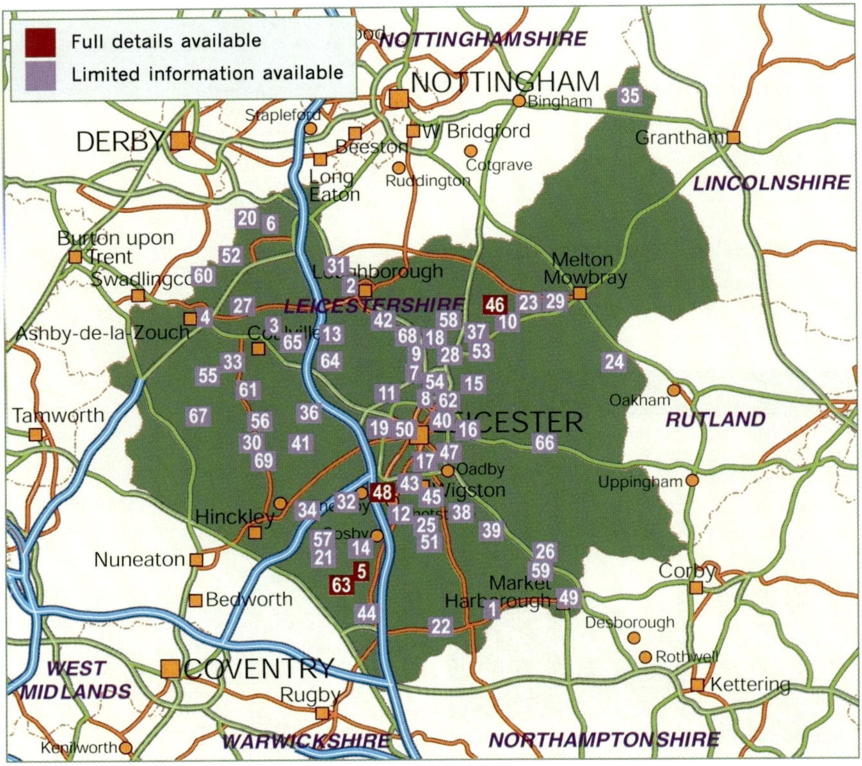

1 A & A Thorp

Bungalow No 5, Main Street, Theddingworth, Lutterworth, Leicestershire LE17 6QZ
Tel: 01858 880496

2 A & C Swift

126 Knightthorpe Road, Loughborough, Leicestershire LE11 5JU
Tel: 01509 239110

3 A & F Nursery

Loughborough Road, Whitwick, Coalville, Leicestershire LE67 5AS
Tel: 01530 834501
Specialities: Annuals/Biennials/Bedding

4 Ashby Aquatics

163 Nottingham Road, Ashby-de-La-Zouch, Leicestershire LE65 1DP
Tel: 01530 560834
Specialities: Water Products and Features

5 Ashby Parva Plant Centre

Ashby Parva, Lutterworth, Leicestershire LE17 5JD
Tel/Fax: 01455 209225
e-mail: parva.nursery@btconnect.com
website: www.ashbyparvaplantcentre.co.uk

SEE ADVERTISEMENT PAGE 147

Specialities:

Planting Plan Advisory Service; Growing Hardy Plants

Description:

The Cobb family started growing and selling plants from their back garden in 1982 – today the site covers more than 5 acres. The family grow over 80% of the plants they sell and have built up an enviable reputation for their range of plant varieties and their comprehensive knowledge and advice.

Directions:

From Junction 20 of the M1 follow signs to Lutterworth. At next roundabout continue straight across onto the A5. At next roundabout turn right. Continue until you reach a T-junction and mini-roundabout. Turn left. Enter Bitteswell. Take your first left then a right-hand turn into Ashby Lane. Follow this road out of Bitteswell, pass the Lutterworth Rugby Football Club on your left, Garden Centresignposted on your left.

6 Bailiffs Mead Nurseries

67 High Street, Castle Donington, Derby, Leicestershire DE74 2PQ
Tel: 01332 810245

7 Bentley Garden Buildings

Loughborough Road, Wanlip, Leicester, Leicestershire LE7 4PN
Tel: 01162 675117
Specialities: Greenhouses and Sheds

8 Birstall Garden Centre

19-35 Sibson Road, Birstall, Leicester, Leicestershire LE4 4DX
Tel: 01162 677091
Specialities: Garden Furniture, Greenhouses and Sheds, Barbeques

9 Brooklea Nursery

975 Loughborough Road, Rothley, Leicester, Leicestershire LE7 7NJ
Tel: 01162 302155

10 Brooksby Plant Centre

Brooksby Melton College, Brooksby, Melton Mowbray, Leicestershire LE14 2LJ
Tel: 01664 850850

11 Brookside Nurseries Ltd

129 Cropston Road, Anstey, Leicester, Leicestershire LE7 7BR
Tel: 01162 364564
Specialities: Annuals/Biennials/Bedding, Hanging Baskets, Floristry

12 Chamney Garden Centre

Lutterworth Road, Blaby, Leicester,
Leicestershire LE8 4DX
Tel: 01162 777020

13 Charley Knoll Nursery

Charley Road, Nanpantan, Loughborough,
Leicestershire LE12 9YE
Tel: 01509 239139

14 Charlie Brown's Garden Centre Ltd

Dunton Road, Broughton Astley, Leicester,
Leicestershire LE9 6NA
Tel: 01455 282310
Specialities: Annuals/Biennials/Bedding, Shrubs, Rose

15 Church View Nursery

Queniborough Road, Barkby, Leicester,
Leicestershire LE7 3QJ
Tel: 01162 695505
Specialities: Herbaceous Perennials

16 Coles (James) & Sons Nurseries

The Nurseries, 624 Uppingham Road, Thurnby,
Leicester, Leicestershire LE7 9QB
Tel: 01162 412115
Specialities: Trees, Shrubs

17 Craighill Nurseries

Craighill Road, Knighton, Leicester,
Leicestershire LE2 3FB
Tel: 01162 707065

18 Derrys Nurseries

Main Street, Cossington, Leicester,
Leicestershire LE7 4UU
Tel: 01509 812815

19 Dominion Road Nursery

171 Dominion Road, Glenfield, Leicester,
Leicestershire LE3 8JB
Tel: 01162 874981

20 Donington Nurseries

Park Lane, Castle Donington, Derby,
Leicestershire DE74 2RS
Tel: 01332 853004

21 Fourwinds Nurseries Limited

Coventry Road, Sharnford, Hinckley,
Leicestershire LE10 3PG
Tel: 01455 221005

22 Gandy's Roses Ltd

Station Road, North Kilworth, Lutterworth,
Leicestershire LE17 6HZ
Tel: 01858 880398
Specialities: Rose

23 The Garden Gallery

Station Lane, Asfordby Hill, Melton Mowbray,
Leicestershire LE14 3SL
Tel: 01664 810108

24 Gates Nurseries & Garden Centre

Somerby Road, Cold Overton, Oakham,
Leicestershire LE15 7QB
Tel: 01664 454309

25 Glebe Garden Centre

Foston Road, Countesthorpe, Leicester,
Leicestershire LE8 5QP
Tel: 01162 771570
Specialities: Annuals/Biennials/Bedding

26 Glenn Stewart Nursery

Melton Road, East Langton, Market Harborough,
Leicestershire LE16 7TG
Tel: 01858 545466

27 Glesmoor Nurseries

Gelsmoor Road, Coleorton, Coalville,
Leicestershire LE67 8JF
Tel: 01530 222693
Specialities: Seasonal Displays

28 Goscote Nurseries Ltd

Syston Road, Cossington, Leicester,
Leicestershire LE7 4UZ
Tel: 01509 812121
Specialities: Herbaceous Perennials, Trees, Shrubs

29 The Grange Garden Centre Limited

Sysonby Grange House, Melton Road, Asfordby Hill,
Melton Mowbray, Leicestershire LE14 3QU
Tel: 01664 812012
Specialities: Rose, Water Products and Features, Garden Furniture

30 Greenacres Nurseries

Ashby Road, Stapleton, Leicester,
Leicestershire LE9 8JE
Tel: 01455 290878

31 Hathern Nurseries

Whatton Road, Hathern, Loughborough,
Leicestershire LE12 5LX
Tel: 01509 646000
Specialities: Exotics

32 Hawgrip Nurseries

Seine Lane, Enderby, Leicester,
Leicestershire LE19 4PD
Tel: 01162 848227
Specialities: Annuals/Biennials/Bedding

33 Heather Lane Nurseries

Heather Lane, Ravenstone, Coalville,
Leicestershire LE67 2AH
Tel: 01530 832101
Specialities: Annuals/Biennials/Bedding

34 Hill View Nurseries

Croft Road, Thurlaston, Leicester,
Leicestershire LE9 7TB
Tel: 01455 888393

35 Homestead Plants

The Homestead, Normanton, Bottesford,
Nottingham, Leicestershire NG13 0EP
Tel: 01949 842745
Specialities: Hardy Plants, Rare and Unusual Plants, Hosta

36 John Smith & Son

Reservoir Road, Thornton, Coalville,
Leicestershire LE67 1AN
Tel: 01530 230331
Specialities: Fuchsia

37 Kayes Garden Nursery

1700 Melton Road, Rearsby, Leicester,
Leicestershire LE7 4YR
Tel: 01664 424578
Specialities: Herbaceous Perennials, Climbers, Grasses

38 Ken Bailey at Wistow

Kibworth Road, Wistow, Leicester,
Leicestershire LE8 0QF
Tel: 01162 592009

39 Kibworth Garden Centre & Nursery

Fleckney Road, Kibworth, Leicester,
Leicestershire LE8 0HJ
Tel: 01162 792754
Specialities: Annuals/Biennials/Bedding

40 Linnaeus Nursery

Cordelia Close, Leicester, Leicestershire LE5 0LE
Tel: 01162 255828
Specialities: Annuals/Biennials/Bedding

41 Lockeymead Farm Nursery

Hunts Lane, Desford, Leicester,
Leicestershire LE9 9LJ
Tel: 01455 822411
Specialities: Exotics, Shrubs, Annuals/Biennials/Bedding

42 Loughborough Road Nurseries

103 Loughborough Road, Quorn, Loughborough,
Leicestershire LE12 8DU
Tel: 01509 416680
Specialities: Annuals/Biennials/Bedding, Herbaceous Perennials, Shrubs

43 Lulworth Plants

28 Gladstone Street, Wigston Magna,
Leicestershire LE18 1AE
Tel: 07814 042889
Specialities: Shade Loving Plants, Exotics

44 Lutterworth Plant& Bonsai Nursery

Guthlaxton Avenue, Lutterworth,
Leicestershire LE17 4ET
Tel: 01455 552211
Specialities: Trees, Annuals/Biennials/Bedding, Bonsai

45 Mozart House Nursery Garden

84 Central Avenue, Wigston,
Leicestershire LE18 2AA
Tel: 01162 889548
Specialities: Grasses, Bamboos, Ferns

46 Oasis in Hoby

Brooksby Road, Hoby, Melton Mowbray,
Leicestershire. LE14 3EA
Tel: 01664 434113
website: www.oasisinhoby.co.uk

SEE ADVERTISEMENT PAGE 151

Specialities:

Perennials, Shrubs, Ferns, Grasses, Bamboo.

Products and Services:

- Credit Cards Accepted
- Large Car Park
- Plant Guarantee
- Plants
- Trees

Directions:

Off the A607 Melton Mowbray to Leicester road near Brooksby.

47 Outer Eden Ltd

Rear of 43-45, Francis Street, Leicester, Leicestershire LE2 2BE
Tel: 01162 708100
Specialities: Garden Furniture

48 Palmers Garden Centre

St Johns, Enderby, Leicestershire LE19 2BS
Tel: 0116 286 3323 Fax: 0116 275 3273
e-mail: info@palmersgardencentre.co.uk
website: www.palmersgardencentre.co.uk

SEE ADVERTISEMENT PAGE 150

Specialities:

Garden Planning & Planting Service; Italian Specimen Plants; Garden & Conservatory Furniture

Products and Services:

- Barbecues
- Books and Stationery
- Children's Play Area
- Seasonal Displays
- Clothing
- Credit Cards Accepted
- Disabled Facilities
- Dried/Artificial Flowers
- Fencing
- Fruit Trees
- Garden Construction
- Garden Design
- Garden Furniture
- Garden Products
- Giftware
- Greenhouses and Sheds
- Houseplants
- Information Desk
- Large Car Park
- Paving
- Pet Products
- Plant Guarantee
- Plants
- Restaurant/Coffee Shop
- Seeds and Bulbs
- Trees

Directions:

From Junction 21 of the M1 turn right on the Outer Ring Road (A563). About ¼ mile along this road, turn right on the roundabout on to the B4114 and follow the brown tourist signs.

49 Park Nurseries

Welland Park Road, Market Harborough, Leicestershire LE16 9DR
Tel: 01858 469889

50 Planet-Hydro Ltd

Unit 11 North Bridge Works, Storey Street, Leicester, Leicestershire LE3 5GR
Tel: 01162 510800
Specialities: Hydroponics

51 Premiere Plants

The Nursery, Leicester Road, Countesthorpe, Leicester, Leicestershire LE8 5QW
Tel: 01162 782611
Specialities: Herbaceous Perennials, Grasses, Annuals/Biennials/Bedding

52 Priory Nurseries

Melbourne Lane, Breedon-on-The-Hill, Derby, Leicestershire DE73 8AT
Tel: 01332 862406
Specialities: Shrubs

53 Rearsby Roses

Melton Road, Rearsby, Leicester,
Leicestershire LE7 4YP
Tel: 01162 601211
Specialities: Rose

54 Rectory Field Nursery

Rectory Road, Wanlip, Leicester,
Leicestershire LE7 4PL
Tel: 01162 674613
Specialities: Herbaceous Perennials, Annuals/Biennials/Bedding

55 S & S Perennials

24 Main Street, Normanton Le Heath, Coalville,
Leicestershire LE67 2TB
Tel: 01530 262250
Specialities: Cyclamen, Narcissus (Daffodil), Bulbs

56 Sage Garden Products & Nursery

19 Main Street, Nailstone, Nuneaton,
Leicestershire CV13 0QE
Tel: 01530 260833

57 Sapcote Nursery & Garden Centre

Hinckley Road, Sapcote, Leicester,
Leicestershire LE9 4LG
Tel: 01455 274049
Specialities: Barbeques, Garden Furniture, Annuals/Biennials/Bedding

58 Seagrave Nurseries

9 King Street, Seagrave, Loughborough,
Leicestershire LE12 7LY
Tel: 01509 816061

59 Seasonal Plants

Langton Road, Foxton, Market Harborough,
Leicestershire LE16 7FG
Tel: 01162 598415

60 Staunton Harold Nurseries & Garden Centre

Melbourne Road, Staunton Harold,
Ashby-de-La-Zouch, Leicestershire LE65 1RU
Tel: 01332 862769

61 Sunnyside Garden Centre Ltd

Leicester Road, Ibstock, Leicestershire LE67 6HL
Tel: 01530 263418
Specialities: Fencing, Annuals/Biennials/Bedding, Herbaceous Perennials

62 Thurmaston Garden Centre

600 Melton Road, Thurmaston, Leicester, Leicestershire LE4 8BB
Tel: 01162 692541
Specialities: Trees, Landscape Gardening, Floristry

63 Ullesthorpe Garden Centre

Lutterworth Road, Ullesthorpe, Leicestershire LE17 5DR
Tel: 01455 202144 Fax: 01455 202585
e-mail: enquiries@ullesthorpegardencentre.co.uk
website: www.ullesthorpegardencentre.co.uk

SEE ADVERTISEMENT PAGE 152

Specialities:

Pets & Aquatics Department

Products and Services:

- Aquatics
- Barbeques
- Books and stationery
- Seasonal Displays
- Clothing
- Credit Cards Accepted
- Disabled Facilities
- Display Gardens
- Dried and Artificial Flowers
- Fencing
- Fruit Trees
- Garden Furniture
- Garden Products
- Giftware
- Houseplants
- Information Desk
- Large Car Park
- Paving
- Pet Products
- Plant Guarantee
- Plants
- Restaurant/Coffee Shop
- Seeds and Bulbs
- Trees

Directions:

5 miles from M6/M69 and approx 2 miles from J20 of M1. Follow brown & white tourist signs for the Garden Centre.

64 Ulverscroft Grange Nursery

Priory Lane, Ulverscroft, Markfield, Leicestershire LE67 9PB
Tel: 01530 243635
Specialities: Herbaceous Perennials

65 Warren Hills Nursery

Warren Hills Cottage, Warren Hills Road, Coalville, Leicestershire LE67 4UY
Tel: 01530 812350

66 Wheats County Garden

Tilton Lane, Billesdon, Leicester, Leicestershire LE7 9FG
Tel: 01162 596248

67 Willow Brook Nurseries

Shadows Lane, Congerstone, Nuneaton, Leicestershire CV13 6NF
Tel: 01827 880305

68 Wyevale Garden Centre (Leicester)

Leicester - Rowena Garden Centre, Loughborough Road, Rothley, Leicester, Leicestershire LE7 7NL
Tel: 01162 376500

69 Wyevale Garden Centre (Woodlands)

Woodlands Nurseries, Ashby Road, Stapleton, Leicester, Leicestershire LE9 8JE
Tel: 01455 291494

LINCOLNSHIRE

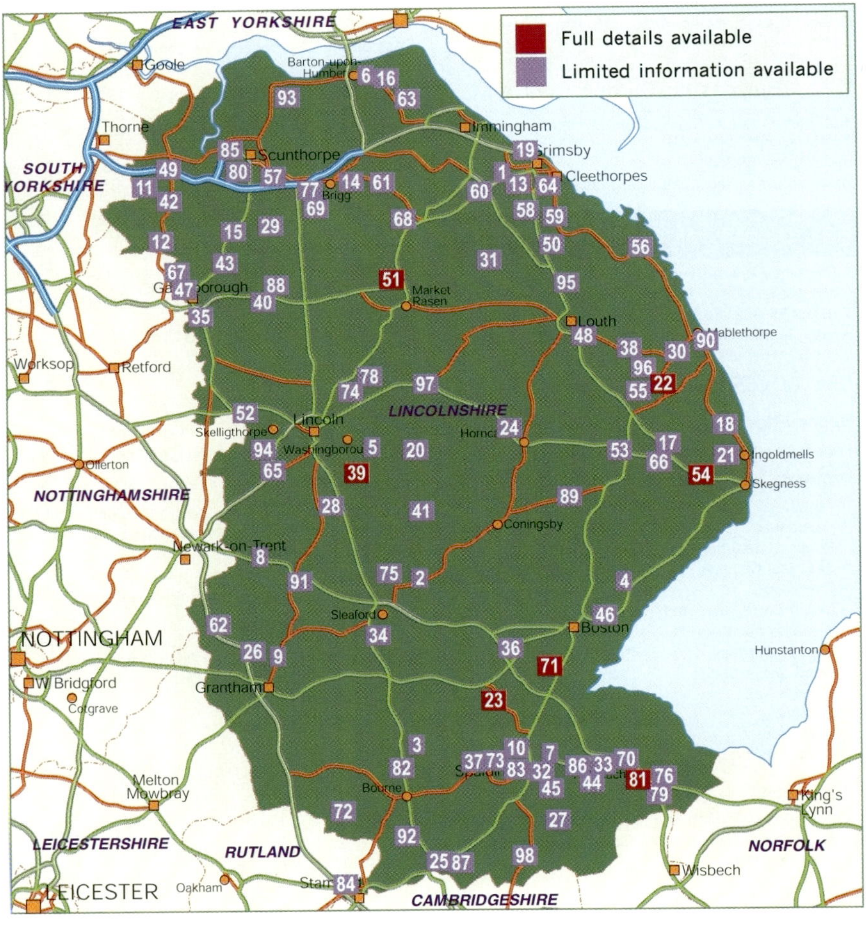

1 Alfords Nurseries

Grimsby Road, Laceby, Grimsby, Lincolnshire DN37 7DY
Tel: 01472 871280

2 Anwick Garden Centre

Anwick, Nr Sleaford, Lincolnshire NG34 9SS
Tel: 01526 832277
Specialities: Garden Products

3 Aveland Trees

Dunsby, Bourne, Lincolnshire PE10 0UB
Tel: 01778 440716
Specialities: Shrubs, Trees

4 B.M Blythe

Westholme Nurseries, The Gride, Old Leake, Boston, Lincolnshire PE22 9SA
Tel: 01205 870202

5 Barn Nursery

Farm Buildings, Five Mile Lane, Washingborough, Lincoln, Lincolnshire LN4 1AF
Tel: 01522 793727

6 Barton Water Gardens

Barrow Road, Barton-upon-Humber, Lincolnshire DN18 6DA
Tel: 01652 660291
Specialities: Aquatics

7 Baytree Nurseries & Garden Centre

High Road, Weston Longville, Spalding, Lincolnshire PE12 6JU
Tel: 01406 370242

8 Beckfield Nursery

High Street, Brant Broughton, Lincolnshire LN5 0QX
Tel: 01400 272493

9 Belton Garden Centre

Belton, Grantham, Lincolnshire NG32 2LN
Tel: 01476 563700
Specialities: Trees, Herbaceous Perennials, Annuals/Biennials/Bedding

10 Birch Grove Garden Centre

Surfleet Road, Pinchbeck, Spalding, Lincolnshire PE11 3XY
Tel: 01775 680490
Specialities: Aquatics, Pet Products

11 Birley's Garden Centre

Sandtoft Road, Belton, Doncaster, Lincolnshire DN9 1PN
Tel: 01427 874397
Specialities: Trees, Garden Furniture, Barbeques

12 Bland Tavern Nurseries

Station Rd, Graiselound Haxey, Doncaster, Lincolnshire DN9 2NJ
Tel: 01427 752267
Specialities: Annuals/Biennials/Bedding

13 Bonsai At Bradley

The Old Chapel, Church Lane, Bradley, Grimsby, Lincolnshire DN37 0AE
Tel: 01472 750707
Specialities: Bonsai

14 Brigg Garden Centre

Bigby High Road, Brigg, Lincolnshire DN20 9HE
Tel: 01652 650484
Specialities: Annuals/Biennials/Bedding, Shrubs, Herbaceous Perennials

15 Bucknell Farm Garden Centre

Bucknall Farm, Scotter Common, Gainsborough, Lincolnshire DN21 3JF
Tel: 01724 764040
Specialities: Bonsai

16 C P Tuplin

Ferry Road, Barrow Haven, Barrow-upon-Humber, Lincolnshire DN19 7HB
Tel: 01469 532662

17 Candlesby Herbs

Cross Keys Cottage, Chalkpit Lane, Candlesby, Spilsby, Lincolnshire PE23 5SF
Tel: 01754 890211
Specialities: Herbs

18 Chapel Garden Centre

Skegness Road, Chapel St. Leonards, Skegness, Lincolnshire PE24 5UB
Tel: 01754 873558

19 Concrete Gardens Patio & Garden Centre

Flour Square, Lockhill, Grimsby, Lincolnshire DN31 3LS
Tel: 01472 352022

20 The Contented Gardener Nursery

The Garden House, Wragby Road, Bardney, Lincoln, Lincolnshire LN3 5XL
Tel: 01798 815571
Specialities: Herbaceous Perennials

21 Corner Farm Nursery

Ingoldmells Rd, Addlethorpe, Skegness, Lincolnshire PE24 4TF
Tel: 01754 610365
Specialities: Restaurant/Coffee Shop

22 Cottage Nurseries

Thoresthorpe, Alford, Lincolnshire LN1 30HX
Tel: 01507 466 968 Fax: 01507 463 409
e-mail: bill@cottagenurseries.net
website: www.cottagenurseries.net

Specialities:

Herbaceous Perennials; Alpines

Products and Services:

- Credit Cards Accepted
- Display Gardens
- Information Desk
- Large Car Park
- Plants
- Seeds and Bulbs
- Trees

Directions:

The nurseries are located on the A1104 Alford to Mablethorpe road, about 1 mile north of Alford. Turn right at the bottom of the hill and the nurseries are 400 yards on the left.

Description:

In its mail order catalogue Cottage Nurseries now lists more than 1000 varieties of plants, some of which are not widely available. New varieties will be added to the list as they become available during the season.

The nursery and garden here have taken on a new image with exciting plants for visitors to see growing in the gardens, many of which are available for purchase. The owners are always pleased to welcome visitors to their family run nursery situated at the edge of the Lincolnshire Wolds near the attractive market town of Alford. The nursery is open every day from 1st March to 30th September from 9am to 5pm and 10am to 3pm, weekends only from 1st October to 28th February. The nursery is closed 18th December until 2nd January. If you wish to visit or collect an order at other times other than the regular opening times please contact the nursery.

23 Cross Roads Nurseries

Main Road, Quadring, Spalding,
Lincs. PE11 4PJ
Tel: 01775 820113 Fax: 01775 821666
e-mail: richard@crossroads@btopenworld.com

Specialities:

Primroses; Bedding Plants; Exotic Plants

Description:

Cross Roads Nurseries stocks an amazing 250,000+ primroses with a wonderful selection of colour from January to March. Also available are exotic plants, virus free cannas, hardy bananas and phormium, plus others from May to July. The nurseries also offer vegetable plants and can source most plants on request. Open from 8am to 12pm, and 1pm to 4pm (Monday to Friday), and from 9am to 12pm on Saturday, the nurseries also welcome visitors at other times by appointment.

Directions:

Cross Roads Nurseries are on the A152 between Gosberton and Donington

24 Crowders Garden Centre

Lincoln Road, Thimbleby, Horncastle,
Lincolnshire LN9 5LZ
Tel: 01507 525252

25 Deeping Gate Tree Nurseries

Northfield Road, Market Deeping, Peterborough,
Lincolnshire PE6 8LG
Tel: 01778 347208
Specialities: Trees

26 Downtown Garden Centre

Gonerby Moor Roundabout, Gonerby Moor,
Grantham, Lincolnshire NG32 2AB
Tel: 01476 512333

27 Dunhorse Nursery

Jekils Bank, Moulton Eaugate, Spalding,
Lincolnshire PE12 0SY
Tel: 01406 380697

28 Ermine Cottage Nurseries

The Heath, Heath Road, Coleby, Lincoln,
Lincolnshire LN5 0AR
Tel: 01522 810396

29 Fair Gardens Plant Centre

Station Road, Kirton Lindsey, Gainsborough,
Lincolnshire DN21 4JR
Tel: 01652 648631

30 Fern Lea Nurseries

Main Rd, Maltby Le Marsh, Maplethorpe,
Lincolnshire LN13 0JP
Tel: 01507 450142
Specialities: Garden Products

31 The Fern Nursery

Grimsby Road, Binbrook, Market Rasen,
Lincolnshire LN8 6DH
Tel: 01472 398092

32 Festival Garden Centre

Springfields Outlet Village, Camel Gate, Spalding, Lincolnshire PE12 6ET
Tel: 01775 760949
Specialities: Garden Furniture, Garden Products, Restaurant/Coffee Shop

33 Fleet Road Nursery & Plant Centre

Fleet Rd, Holbeach Drove, Spalding, Lincolnshire PE12 7AX
Tel: 01406 426016

34 Four Seasons Garden Centre

London Road, Silk Willoughby, Sleaford, Lincolnshire NG34 8NY
Tel: 01529 304970
Specialities: Restaurant/Coffee Shop, Annuals/Biennials/Bedding, Herbaceous Perennials

35 Gainsborough Home & Garden

Lea Rd, Gainsborough, Lincolnshire DN21 1AF
Tel: 01427 612004

36 Glenhirst Cactus Nursery

Station Road, Swineshead, Boston, Lincolnshire PE20 3NX
Tel: 01205 820314
Specialities: Cacti And Succulents

37 Glenside Nurseries

The Rowans, Glenside South, West Pinchbeck, Spalding, Lincolnshire PE11 3NP
Tel: 01775 640520
Specialities: Garden Products

38 Greenways Garden Ornaments

Louth Road, Gayton Top, Alford, Lincolnshire LN13 0NH
Tel: 01507 450336
Specialities: Water Products and Features, Aquatics, Restaurant/Coffee Shop

39 Growing Gifts

Branston Moor Nursery, Moor Lane, Branston, Lincolnshire LN4 1HZ
Tel: 01522 790022
website: www.growinggifts.co.uk
Specialities: Hanging Baskets, Ceramic Planters

40 Hall Farm Nursery

Harpswell, Gainsborough, Lincolnshire DN21 5UU
Tel: 01427 668412
Specialities: Rose, Shrubs, Herbaceous Perennials

41 Holmdale Garden Nursery

High St, Martin, Lincoln, Lincolnshire LN4 3QY
Tel: 01526 378838
Specialities: Garden Products, Display Gardens, Restaurant/Coffee Shop

42 Holmes & Gardens

Belton Road, Epworth, Doncaster, Lincolnshire DN9 1JL
Tel: 01427 872282

43 Hoots Nursery & Plant Centre

Laughton Rd, Blyton, Gainsborough, Lincolnshire DN21 3LQ
Tel: 01427 629285

44 Hurdletree Nurseries

Hurdletree Bank, Holbeach, Spalding, Lincolnshire PE12 8QH
Tel: 01406 540505
Specialities: Shrubs, Trees, Hanging Baskets

45 J Cunnington

7 Swindlers Drove, Spalding, Lincolnshire PE12 6BS
Tel: 01775 712170

46 Johnsons Garden Centre

Wainfleet Road, Fishtoft, Boston, Lincolnshire PE21 9RZ
Tel: 01205 363408

47 Kathleen Muncaster Fuchsias

18 Field Lane, Morton, Gainsborough, Lincolnshire DN21 3BY
Tel: 01427 612329
Specialities: Fuchsia, Display Gardens

48 Kenwick Farmhouse Nursery

Kenwick Road, Kenwick, Louth, Lincolnshire LN11 8NW
Tel: 01507 606469
Specialities: Herbaceous Perennials, Annuals/Biennials/Bedding

49 Kingfisher Nurseries

Double Rivers, Crowle, Scunthorpe, Lincolnshire DN17 4BY
Tel: 01724 710019

50 Lincoln Field

Ludborough Road, North Thoresby, Grimsby, Lincolnshire DN36 5RF
Tel: 01472 840461

51 Lincoln Green Clematis

Sand Lane, Osgodby, Market Rasen,
Lincolnshire LN8 3TE
e-mail: lgn@freenet.co.uk
website: www.clematis-r-us.com

Specialities:

Clematis; Coffee Shop; Educational Talks

Description:

In addition to its specialist clematis plants, the nursery also stocks a wide variety of bulbs, trees, seeds and a range of garden products. It also sells pet products and giftware. Other amenities include a coffee shop, large car park and information desk.

Directions:

From the A46, about 2 ½ miles north of Market Rasen, turn left on minor road to Osgodby. You will see the nursery before you come to the village.

52 Living Stone

Riverside, Skellingthorpe Road, Saxilby, Lincoln,
Lincolnshire LN1 2LR
Tel: 01522 704158

53 Longmate Nurseries

Sausthorpe Road, Partney, Spilsby,
Lincolnshire PE23 4JJ
Tel: 01790 752716

54 Lyndhurst Garden Centre

Skegness Road, Burgh-le-Marsh, nr Skegness,
Lincolnshire PE24 5AA
Tel: 01754 810295
e-mail: lyndhurstgc@tiscali.co.uk
website: lyndhurstgardencentre.co.uk

SEE ADVERTISEMENT PAGE 159

Specialities:

Restaurant & Coffee Shop; All Christmas needs; Landscape Design

Description:

Lyndhurst Garden Centre is a family run business catering for all your gardening needs. Stephen manages the Centre together with his brother, Andrew. Their sister, Susan, has charge of the **Coffee Shop** which features delicious meals as well as home-made cakes. Glenn specializes in **Garden Landscaping**.

Directions:

Lyndhurst Garden Centre is on the A158 Lincoln to Skegness road, about 4 miles west of Skegness

55 Mandy's Plants & Garden Centre

Ailby, Alford, Lincolnshire LN13 0EE
Tel: 01507 466198
Specialities: Fuchsia, Trees

56 Meals Farm Nursery

Marsh Lane, North Somercotes, Louth,
Lincolnshire LN11 7NT
Tel: 01507 358256
Specialities: Annuals/Biennials/Bedding, Shrubs, Trees

57 Mendle Nursery

Holme, Scunthorpe, Lincolnshire DN16 3RF
Tel: 01724 850864
Specialities: Alpines/Rock Plants

58 Millstone Garden Centre

Cheapside, Waltham, Grimsby,
Lincolnshire DN37 0HX
Tel: 01472 824737
Specialities: Garden Products

59 Norburn's Plant Centre

Blenheim House, Station Road, Tetney, Grimsby,
Lincolnshire DN36 5HY
Tel: 01472 815515

60 The Oaklands Nurseries

Barton Street, Laceby, Grimsby,
Lincolnshire DN37 7LF
Tel: 01472 879126

61 The Old Rectory Nurseries

Somerby, Barnetby, Lincolnshire DN38 6EX
Tel: 01643 862363

62 Page Plants Nursery

52 Main Road, Long Bennington, Newark,
Lincolnshire NG23 5DJ
Tel: 01400 281329

63 The Palm Farm & Thornton Hall Gardens

Station Road, Thornton Curtis, Ulceby,
Lincolnshire DN39 6XF
Tel: 01469 531232
Specialities: Exotics, Hardy Garden Palms, Shrubs

64 Pennells Garden Centre

Humberston Road, New Waltham, Grimsby,
Lincolnshire DN36 4RW
Tel: 01472 313600

65 Pennells Garden Centre

Newark Road, South Hykeham, Lincoln,
Lincolnshire LN6 9NT
Tel: 01522 880033

66 Plant Lovers

Candlesby House, Scremby Road, Candlesby,
Spilsby, Lincolnshire PE23 5RU
Tel: 01754 890256

67 Pot Belly Nursery

Walkerith Road, Morton, Gainsborough,
Lincolnshire DN21 3DB
Tel: 01427 610109
Specialities: Aquatics, Shrubs, Annuals/Biennials/Bedding

68 Pottertons Nursery

Moortown Road, Nettleton, Market Rasen,
Lincolnshire LN7 6HX
Tel: 01472 851714
Specialities: Alpines/Rock Plants, Rockery Plants, Dwarf Bulbs

69 R E Smalley

Main Street Sturton, Sturton, Brigg,
Lincolnshire DN20 9DL
Tel: 01652 654335
Specialities: Annuals/Biennials/Bedding, Hanging Baskets, Seasonal Displays

70 R Proctor

192 Fleet Road, Fleet Holbeach, Spalding,
Lincolnshire PE12 8LE
Tel: 01406 423161
Specialities: Cut Flowers

71 Rainbow Landscaping Centre

London Road, Kirton, nr Boston,
Lincolnshire PE20 1PX
Tel: 01205 723555 Fax: 01205 723260
e-mail: sales@rainbowlandscapingcentre.co.uk
web: www.rainbowlandscapingcentre.co.uk

SEE ADVERTISEMENT PAGE 160

Specialities:

Water Features;
Gravels & Granite;
Quality Turf

Description:

This family-owned business specialises in quality water features but also offers a comprehensive stock of other landscaping materials. Also on site is an excellent coffee shop.

Rainbow Landscaping Centre

London Road. Kirton, nr Boston, Lincolnshire PE20 1PX
Tel: 01205 723555 Fax: 01205 723260
e-mail: sales@rainbowlandscapingcentre.co.uk
website: www.rainbowlandscapingcentre.co.uk

Opening Times

8am to 5.30pm, Monday to Saturday
10am-4pm, Sunday

- Fruit Trees
- Garden Construction
- Garden Design
- Garden Furniture
- Garden Machinery
- Garden Products
- Giftware
- Greenhouses and Sheds
- Houseplants
- Barbecues
- Books and Stationery
- Children's Play Equipment
- Seasonal Displays
- Conservatories
- Credit Cards Accepted
- Disabled Facilities
- Dried and Artificial Flowers
- Fencing
- Information Desk
- Large Car Park
- Paving
- Pet Products
- Plants
- Restaurant/Coffee Shop
- Seeds and Bulbs
- Trees

Directions:

From the A17 about 5 miles south of Boston, at Kirton, turn right on the B1397. Stay on this road, going towards Sutterton, and you will see the Centre on your right after about 1 mile.

72 Rasells Nurseries

Station Road, Little Bytham, Grantham, Lincolnshire NG33 4QY
Tel: 01780 410345
Specialities: Herbaceous Perennials, Shrubs, Trees

73 Rose Cottage Water Garden & Spa Centre

Glenside, North Pinchbeck, Spalding, Lincolnshire PE11 3SD
Tel: 01775 710882
Specialities: Water Products and Features, Aquatics

74 Rudies Roots Nurseries

Scothern Road, Nettleham, Lincolnshire LN2 2TU
Tel: 01522 752762
Specialities: Exotics, Hardy Garden Palms, Shrubs

75 Ruskington Garden Centre

Whitehouse Farm, Newton Lane, Ruskington, Sleaford, Lincolnshire NG34 9EB
Tel: 01526 833022
Specialities: Giftware, Annuals/Biennials/Bedding, Herbaceous Perennials

76 S R Luff

Hospital Drove, Little Sutton Long Sutton, Spalding, Lincolnshire PE12 9EL
Tel: 01406 364200

77 Scawby Gardens

The Hall, Vicarage Lane, Scawby, Brigg, Lincolnshire DN20 9LX
Tel: 01652 650515
Specialities: Herbaceous Perennials, Shrubs, Herbs

78 Scothern Nurseries and Garden Centre

8 Dunholme Road, Scothern, Lincolnshire LN2 2UD
Tel: 01673 862297
Specialities: Shrubs, Annuals/Biennials/Bedding, Trees

79 Seagate Irises

Long Sutton By-Pass, Long Sutton, Spalding, Lincolnshire PE12 9RX
Tel: 01406 365138
Specialities: Iris, Display Gardens

80 Silica Lodge Garden Centre

Scotter Road South, Scunthorpe, Lincolnshire DN17 2BN
Tel: 01724 282148

81 Silverwood Garden Centre

Brownsgate, Long Sutton, Lincolnshire PE12 9AD
Tel: 01406 364364
Fax: 01406 365365
e-mail: tyrrelluk@hotmail.com
website: www.tyrrelluk.com

SEE ADVERTISEMENT PAGE 162

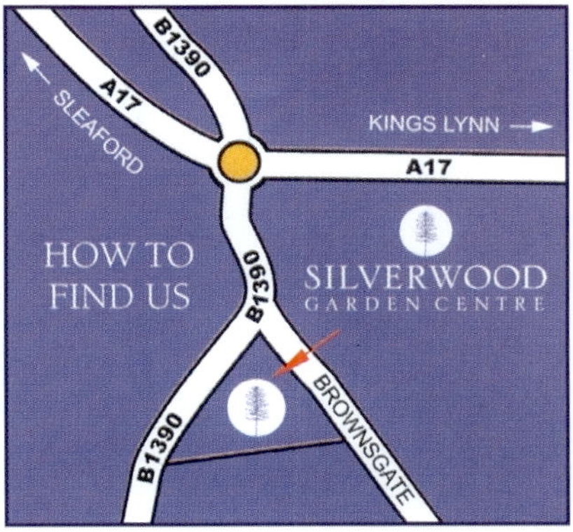

Directions:

From the A17, about 6 miles east of Holbeach, go south on the B1390. At the first left turn into Brownsgate Road. The Nursery is on this road on the right hand side.

82 Southfield Nurseries

Bourne Road, Morton, Bourne, Lincolnshire PE10 0RH
Tel: 01778 570168
Specialities: Cacti And Succulents

83 Spalding Lifestyle & Gardening Centre

210 Spalding Rd, Pinchbeck, Spalding, Lincolnshire PE11 3PB
Tel: 01775 769922

84 Stamford Garden & Leisure Centre

Road End Farm, Toll Bar, Great Casterton, Stamford, Lincolnshire PE9 4BB
Tel: 01780 765656
Specialities: Restaurant/Coffee Shop, Greenhouses and Sheds

85 Stephen H Smith Garden & Leisure

Trent Valley Branch, Doncaster Road, Gunness, Scunthorpe, Lincolnshire DN15 8TE
Tel: 01724 848950
Specialities: Shrubs, Restaurant/Coffee Shop

86 Style Roses

10 Meridian Walk, Holbeach, Spalding, Lincolnshire PE12 7NR
Tel: 01406 424089
Specialities: Rose

87 Swines Meadow Garden Centre

47 Towngate East, Market Deeping, Peterborough, Lincolnshire PE6 8LQ
Tel: 01778 343340
Specialities: Water Products and Features, Seasonal Plants

88 Tower Nurseries

Gibson Road, Hemswell Cliff, Gainsborough, Lincolnshire DN21 5TL
Tel: 01427 668887

89 Townleys Garden Centre

Main Road, Spilsby, Lincolnshire PE23 4BY
Tel: 01790 763301

90 Trusthorpe Garden Centre

Sutton Road, Trusthorpe, Mablethorpe, Lincolnshire LN12 2PH
Tel: 01507 478191

91 The Victorian Kitchen Garden Centre

Fulbeck Garden, Fulbeck, Lincolnshire NG32 3HH
Tel: 01400 273668

92 Waterside Garden Centre

King Street, Baston, Peterborough, Lincolnshire PE6 9NY
Tel: 01778 560000
Specialities: Restaurant/Coffee Shop, Garden Furniture, Aquatics

SILVERWOOD
GARDEN CENTRE

Brownsgate, Long Sutton,
Lincolnshire PE12 9AU
Tel: 01406 364364
Fax: 01406 365365

We pride ourselves in presenting fresh, homecooked food. That's why we have been awarded 3 tulips for excellence! Classic English breakfasts, light lunches of mixed salads, baguettes and sandwiches or more hearty dishes such as steak and ale pie and chicken and broccoli bake. We also offer vegetarian dishes. Or simply settle down by our log fire with tea and a selection of delicious homemade cakes.

Open daily from 10am.

If your lunchtime is limited, why not ring your order through before hand? Sunday lunch from 12 noon - a very popular choice here at Silverwood so we recommend that you book!

April 2008 marks the grand opening of the new and beautiful display garden at Silverwood. Customers can pick up a picnic hamper at the outstanding Tea Room & Restaurant and find a scenic spot in the garden.

A new nursery manager has joined bringing with him many years of experience, busy preparing areas for you to walk around inside and out.

Many varieties are already growing such as bedding, perennials, shrubs, hedging and roses.

We have an exclusively designed working display copper water feature by 'Metal Jacket'

Walk down to another display of paviors, slabs, aggregates and Indian stone opposite our Tea Room and Restaurant.

93 Westshores Nurseries

82 West Street, Winterton, Scunthorpe, Lincolnshire DN15 9QF
Tel: 01724 733940
Specialities: Ornamentals, Herbaceous Perennials

94 Whisby Water and Garden Centre

Whisby Road, Whisby Moor, Lincoln, Lincolnshire LN6 9BY
Tel: 01522 685395

95 Woodlands

Peppin Lane, Fotherby, Louth, Lincolnshire LN11 0UW
Tel: 01507 603586
Specialities: Fuchsia, Climbers

96 Woodthorpe Hall Garden Centre Limited

Woodthorpe Hall, Woodthorpe, Alford, Lincolnshire LN13 0DD
Tel: 01507 450509
Specialities: Garden Products, Restaurant/Coffee Shop

97 The Wragby Maze and Conifer Centre

Bardney Road, Wragby, Lincolnshire LN8 5QZ
Tel: 01673 858660
Specialities: Conifers, Ornamentals

98 Wyevale Garden Centre (Crowland)

Postland Road, Crowland, Peterborough, Lincolnshire PE6 0JB
Tel: 01733 212900

GREATER LONDON

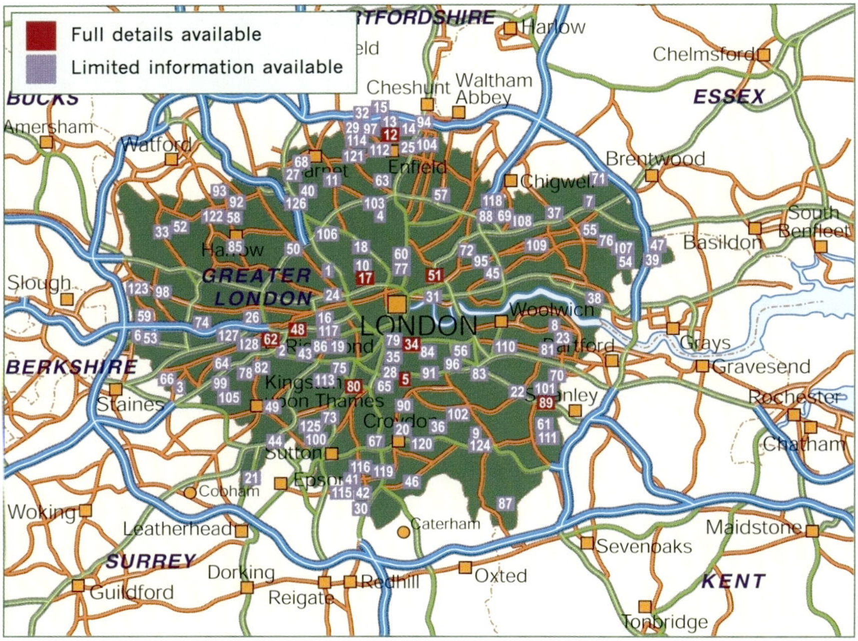

1 Adrian Hall Garden Centres

Hampstead Garden Centre, 161-189 Iverson Road, West Hampstead, London NW6 2RB
Tel: 020 7328 3208

2 Adrian Hall Garden Centres

Sheen Garden Centre, 181-189 Upper Richmond Road West, East Sheen, London SW14 8DU
Tel: 020 8876 3648

3 Adrian Hall Garden Centres

The Garden Centre, Snakey Lane, Feltham, London TW13 7NA
Tel: 020 8751 7600

4 Alexandra Palace Garden Centre

Alexandra Palace, London N22 7BB
Tel: 020 8444 2555

5 Alleyn Park Garden Centre

rear of 77 Park Hall Road, West Dulwich, London SE21 8ES
Tel: 020 8670 7788 Fax: 020 8670 7770
e-mail: info@alleynpark.co.uk
website: www.alleynpark.co.uk

SEE ADVERTISEMENT PAGE 165

Specialities:

Exceptionally Knowledgeable & Friendly Staff; Wide Range of Unusual Plants; Large Trees, including Olives

Description:

In addition to its range of plants, shrubs and trees, the Centre also sells christmas trees, a range of organic products, gifts, tools and irrigation equipment, composts and fertilizers.

Directions:

Close to the junction where Park Hall Road meets Croxted Road, almost opposite Café Rouge

6 Ansell Garden Centre

Holloway Lane, Sipson, West Drayton, London UB7 0AD
Tel: 020 8897 6075
Specialities: Lawn, Garden Furniture, Fencing

7 Bedfords Road Nursery

Lower Bedfords Rd, Romford, London RM1 4DQ
Tel: 01708 733009

8 Belvedere Pet & Garden Centre

6 Albert Road, Belvedere, London DA17 5LJ
Tel: 01322 432482

9 Bencewell Garden Shop

Oakley Road, Bromley, London BR2 8HG
Tel: 020 8462 4221
Specialities: Annuals/Biennials/Bedding, Seeds, Garden Products

10 The Boma Garden Centre

Islip Street, London NW5 2DJ
Tel: 020 7284 4999
Specialities: Landscape Gardening, Garden Design

11 Bridge Nursery

Walfield Avenue, London N20 9PS
Tel: 020 8445 9569
Specialities: Herbaceous Perennials, Shrubs

12 Brown's Garden Centre

Theobalds Park Road, Crews Hill, Enfield, Middlesex EN2 9DG
Tel: 0208 367 1741 Fax: 0208 367 7766
e-mail: info@brownsgardencentre.co.uk
website: www.brownsgardencentre.co.uk

SEE ADVERTISEMENT PAGE 166

Specialities:

Trees and Shrubs; Seeds & Bulbs; Bedding Plants; Gifts

Description:

This friendly family-run centre offers an ever-increasing choice of locally grown top quality plants along with a great selection of trees, shrubs and houseplants sourced from further afield. It also stocks thousands of pots, ornaments, tools and many styles of garden furniture.

Directions:

The nursery is located just five minutes from the A10 (Great Cambridge Road) and junction 25 of the M25.

13 Browns Garden Centre

Cattlegate Road, Enfield, London EN2 9DP
Tel: 020 8363 1383

14 Bulls Cross Nursery

Bulls Cross, Enfield, London EN2 9HE
Tel: 01992 700422

15 By Pass Nurseries

Cattlegate Road, Crews Hill, Enfield, London EN2 9DP
Tel: 020 8367 3377

16 C Rassell Ltd

80 Earls Court Road, London W8 6EQ
Tel: 020 7937 0481
Specialities: Pots, Herbaceous Perennials, Floristry

17 Camden Garden Centre

2 Barker Drive, St Pancras Way, London NW1 OJW
Tel: 0207 387 7080 Fax: 0207 529 8160
e-mail: sales@camdegardencentre.co.uk
website: www.camdengardencentre.co.uk

SEE ADVERTISEMENT PAGE 167

Specialities:

Plants & Products for London; Unusual Plants; Herbs & Vegetables

Description:

This unusual Garden Centre not only supplies a broad range of plants and gardening products but is also a charitable organisation offering employment, training and educational opportunities for trainees such as the long-term unemployed, the homeless, ex-offenders and people with physical and learning disabilities.

Directions:

In Agar Grove 2 turnings left going south down St Pancras Way.

18 Capital Gardens

1 Townsend Yard, Highgate High Street, Highgate Village, London N6 5JF
Tel: 020 8340 1041

19 The Chelsea Gardener

Sydney Street, London SW3 6NR
Tel: 020 7352 5656

20 Cherry Orchard Garden Centre

171 St. James's Road, Croydon, London CR0 2BY
Tel: 020 8689 2625
Specialities: Shrubs, Herbs

21 Chessington Garden Centre

Leatherhead Road, Chessington, London KT9 2NF
Tel: 01372 725638
Specialities: Aquatics, Annuals/Biennials/Bedding, Landscape Gardening

22 Chislehurst Garden Centre

Perry Street, Chislehurst, London BR7 6HA
Tel: 020 8302 2455

23 The City Gardener

150 Long Lane, Bexleyheath, London DA7 5AQ
Tel: 020 8303 2838
Specialities: Garden Design

24 Clifton Nurseries

5a Clifton Villas, London W9 2PH
Tel: 020 7289 6851
Specialities: Floristry, Landscape Gardening, Garden Furniture

25 Clock House Nursery

Forty Hill, Enfield, London EN2 9EU
Tel: 020 8363 1016

26 Concrete Jungle Bonsai

124 South Ealing Road, London W5 4QJ
Tel: 020 8840 7886
Specialities: Bonsai

27 Cottage Garden Nursery

Barnet Road, Barnet, London EN5 3JX
Tel: 020 8441 8829
Specialities: Shrubs, Cottage Garden Plants

28 Croxted Road Garden Centre

Croxted Road, London SE24 9DB
Tel: 020 8674 4366
Specialities: Garden Products

29 Culver Garden Centre

Cattlegate Rd, Enfield, London EN2 9DS
Tel: 020 8366 0701
Specialities: Garden Products

30 Cyan

Jati House, Coulsdon North Industrial Estate, Coulsdon, London CR5 2NR
Tel: 08456 789890
Specialities: Garden Furniture, Ornamentals

31 The Docklands Garden Centre Ltd

244 Arch, Ratcliffe Lane, London E14 7JE
Tel: 020 7790 1146

32 Drayton Nursery

Cattlegate Rd, Crews Hill, Enfield, London EN2 9DJ
Tel: 020 8366 0748

33 Ducks Hill Garden Centre

Ducks Hill Road, Ruislip, London HA4 7TP
Tel: 01895 674286
Specialities: Garden Products

34 Dulwich Garden Centre

20/22 Grove Lane, East Dulwich, London SE22 8EF
Tel: 020 8299 1089 Fax: 020 8693 0335
e-mail: info@dulwichgardencentre.co.uk
website: www.dulwichgardencentre.co.uk

Specialities:

Plants; Bonsai Trees; Turf

Description:

A family concern that offers a friendly, helpful and professional service, Dulwich Garden Centre has been providing the needs of the gardening public for over 24 years. The owners personally travel to Italy each year to source the very best in trees and shrubs.

Directions:

The Centre is located on the A2216, right next to East Dulwich Railway Station.

35 Dulwich Pot & Plant Garden

12B Red Post Hill, London SE21 7BX
Tel: 020 7733 3697
Specialities: Pots

36 Eden Park Garden Centre

Upper Elmers End Road, Beckenham, London BR3 3DE
Tel: 020 8650 2424
Specialities: Annuals/Biennials/Bedding, Shrubs, Climbers

37 Elmstead Nurseries

Whalebone Lane North, Collier Row, Romford, London RM6 5QT
Tel: 020 8597 9020

38 Eve's Eden

Unit 5 New Road, Wennington, Rainham, London RM13 9EB
Tel: 01708 862245

39 Fenlands Nursery

Ockendon Road, North Ockenden, Upminster, London RM14 3PP
Tel: 01708 856364

40 Finchley Nurseries

Burtonhole Lane, London NW7 1AS
Tel: 020 8959 2124
Specialities: Annuals/Biennials/Bedding, Shrubs

41 Flittons Nursery & Plant Centre

51 Woodmansterne Lane, Wallington, London SM6 0SW
Tel: 020 8647 5615
Specialities: Hanging Baskets, Herbaceous Perennials

42 Foxhollow Herb Nursery

73 Lower Pillory Down, Little Woodcote Estate, Carshalton, London SM5 4DD
Tel: 020 8660 0991
Specialities: Herbs

43 Fulham Palace Garden Centre

Bishops Avenue, London SW6 6EE
Tel: 020 7736 2640
Specialities: Garden Products, Restaurant/Coffee Shop

44 Garden Care Supplies

G C S Holdings, Old Kingston Road, Worcester Park, London KT4 7QH
Tel: 020 8337 9922

45 The Garden Centre

39-42 The Shopping Hall, Myrtle Road,
London E6 1HY
Tel: 020 8472 7791

46 Gardencraft

238a Addington Road, South Croydon,
London CR2 8LE
Tel: 020 8651 2302
Specialities: Annuals/Biennials/Bedding, Cut Flowers

47 Gardentime

Clay Tye Rd, Upminster, London RM14 3PL
Tel: 01708 640866

48 Gingko Garden Centre

Ravenscourt Avenue, Hammersmith W6 0SL
Tel: 020 8563 7112 Fax: 0208 5637160
e-mail: info@ginkgogardens.co.uk
website: www.ginkgogardens.co.uk

SEE ADVERTISEMENT BELOW

Specialities:

Landscaping; Plants; Unique Furniture

Description:

Ginkgo Garden Centre sells a wide range of fruit and vegetables as well as a selection of shrubs, flowers and seeds. Furniture, statuary and plants are cleverly displayed whilst keeping the feel of a beautiful garden.

Directions:

Ravenscourt avenue is a cul-de-sac off King Street, Hammersmith, just a 2-minute walk from Ravenscourt Park tube station on the District Line.

49 Green Finger

Park Works, Park Road, Kingston Upon Thames,
London KT2 6BG
Tel: 020 8546 3444

50 Greenhouse Garden and Plant Centre

Birchen Grove, London NW9 8RY
Tel: 020 8905 9189
Specialities: Annuals/Biennials/Bedding, Herbaceous Perennials

Occupying a beautiful location nestled between the Hertford Union Canal and Victoria Park, **Growing Concerns** provides a tranquil retreat in a busy city. There are seats dotted around the site where visitors can sit and enjoy the pergola, plants and pots. The nursery offers a comprehensive range of plants and a friendly plant buying experience. Growing Concerns grew out of a gardening club started by local residents who were being re-housed from flats and houses, many with gardens and balconies for the first. With the community's help, it has grown and grown into a busy garden centre

- Credit Cards Accepted
- Fruit Trees
- Garden Design
- Garden Products
- Houseplants
- Large Car Park
- Plants
- Seeds and Bulbs
- Trees

FOR ALL YOUR GARDEN PLANTS IN A UNIQUE SETTING.....

GROWING CONCERNS GARDEN CENTRE

2 WICK LANE, LONDON E3 2NA
TEL/FAX: 0208 985 3222
E-MAIL: INFO@GRWOINGCONCERNS.ORG
WEBSITE: WWW.GROWINGCONCERNS.ORG

51 Growing Concerns Garden Centre

2 Wick Lane, London E3 2NA
Tel/Fax: 0208 985 3222
e-mail: info@grwoingconcerns.org
website: www.growingconcerns.org

SEE ADVERTISEMENT PAGE 170

Specialities:

Outdoor shrubs & Plants; Friendly Knowledgeable Staff; Unique Setting

Description:

Growing Concerns is a busy, friendly community-based business offering outdoor plants and shrubs in a beautiful setting next to the canal and with its own green oak thatched gazebo. Visitors can partake of excellent coffee and the nursery's famous carrot cake!

Directions:

From the Blackwall Tunnel approach road, exit at Old Ford and go north towards Victoria Park/St Mark's Gate. Growing Concerns is between the Hertford Union canal and Victoria Park, opposite the Top of the Morning pub.

52 Haste Hill Nurseries

Fore Street, Pinner, London HA5 2ND
Tel: 01895 674896

53 Heathrow Garden Centre

Sipson Road, West Drayton, London UB7 0HR
Tel: 020 8897 8893
Specialities: Garden Products, Garden Furniture

54 High House Farm

Ockendon Road, Upminster, London RM14 2DN
Tel: 01708 222181

55 Hornchurch

37 Hazelmere Gardens, Hornchurch, London RM11 2AS
Tel: 01708 472999

56 Hortus

26 Blackheath Village, London SE3 9SY
Tel: 020 8297 9439
Specialities: Landscape Gardening

57 Hurrans Nursery

179 New Road, London E4 9EZ
Tel: 020 8529 1898

58 Jacques Amand International

The Nurseries, Clamp Hill, Stanmore, London HA7 3JS
Tel: 020 8420 7110
Specialities: Bulbs

59 John Train

Harmondsworth Road, West Drayton, London UB7 9JS
Tel: 020 8759 3010
Specialities: Water Products and Features, Herbaceous Perennials, Annuals/Biennials/Bedding

60 John's Garden Centre

175 Stoke Newington Church Street, London N16 0UL
Tel: 020 7275 9494
Specialities: Garden Products

61 Kevington Nurseries Ltd

Little Kevington, Waldens Road, Orpington, London BR5 4EU
Tel: 01689 836505

62 The Kew Gardener

18 Station Parade, Kew Gardens, Surrey TW9 3PZ
Tel: 0208 948 1422 Fax: 0208 332 9630
e-mail: plant@kewgardener.com
website: www.kewgardener.com

SEE ADVERTISEMENT PAGE 172

Specialities:

Unusual Plants, World Class Orchids, Specialist Gardening Service

Products and Services:

- Seasonal Displays
- Credit Cards Accepted
- Garden Construction
- Garden Design
- Garden Maintenance
- Garden Products
- Giftware
- Houseplants
- Information Desk
- Plant Guarantee
- Plants
- Seeds and Bulbs
- Trees
- Christmas Trees

Directions:

Situated next to Kew Gardens tube station (district line). Two minutes walk from the Royal Botanic Gardens, Kew.

63 King Easton Ltd

The Green, Station Road, London N21 3NB
Tel: 020 8886 8783

64 Kneller Garden Supplies

297 Whitton Dene, Isleworth, London TW7 7NE
Tel: 020 8898 7494
Specialities: Garden Products, Garden Furniture

65 Knollys Garden Centre Ltd

Knollys Rd, London SW16 2JJ
Tel: 020 8769 7321

66 Lakeside Garden Centre Ltd

Bedfont Road, Feltham, London TW14 8EB
Tel: 020 8844 2261
Specialities: Annuals/Biennials/Bedding, Herbaceous Perennials

67 Lawmans Garden Centre

28-32 Beddington Lane, Croydon, London CR0 4TB
Tel: 020 8760 0082

68 Leahurst Nursery

36A Galley Lane, Barnet, London EN5 4AJ
Tel: 020 8449 9631
Specialities: Shrubs, Herbaceous Perennials

69 Lee's Garden Centre

14 Mossford Green, Ilford, London IG6 2BX
Tel: 020 8550 8475
Specialities: Garden Products

70 Littleworth Nurseries

Vicarage Road, Bexley, London DA5 2AW
Tel: 01322 550410

71 Long House Plants

Church Road, Noak Hill, Romford, London RM4 1LD
Tel: 01708 371719
Specialities: Annuals/Biennials/Bedding, Shrubs

72 Mini Garden Centre

313B Cann Hall Road, London E11 3NL
Tel: 020 8555 2132

73 Morden Hall Garden Centre

Morden Hall Road, Morden, London SM4 5JD
Tel: 020 8646 3002

74 My Own Garden Centre

16 Burket Close, Southall, London UB2 5NR
Tel: 020 8571 7171

75 Neal's Nurseries

Heathfield Rd, London SW18 2PH
Tel: 020 8874 2037

76 Nelsons Garden Centre

85 North Street, Hornchurch, London RM11 1ST
Tel: 01708 444658

77 North One Garden Centre

The Old Button Factory, 25 Englefield Road, London N1 4EU
Tel: 020 7923 3553
Specialities: Annuals/Biennials/Bedding, Herbaceous Perennials, Giftware

78 The Palm Centre

Ham Central Nursery, Ham Street, Richmond, London TW10 7HA
Tel: 020 8255 6191
Specialities: Exotics, Tropical, Hardy Garden Palms

79 Parkside Garden Centre

154 Denmark Hill, London SE5 8EL
Tel: 020 7738 4240

80 Patio

100 Tooting Bec Road, London SW17 8BG
Tel: 020 8672 2251 Fax: 020 8682 2105

SEE ADVERTISEMENT BELOW

Specialities:

Hand-made Terracotta Pots; Unusual Plants, Shrubs & Herbs

Description:

Patio specialises in hand-made terracotta pots made by potters known to the owners over the years. They come from Tuscany, Provence, Andalusia and Crete. Patio also offers plants, shrubs and herbs from specialist nurseries in the United Kingdom and Italy.

Directions:

Take the A24 south to Tooting Bec underground. Left into the A214 - 400yds on the right is a large white georgian lodge - that's us.

81 Peppers

72 Avenue Road, Bexleyheath, London DA7 4EG
Tel: 020 8303 2195

82 Petersham Nurseries Ltd

Off Petersham Road, River Lane, Richmond, London TW10 7AG
Tel: 020 8940 5230

83 Phoebe's Garden Centre

Mottingham Lane, London SE12 9AW
Tel: 020 8851 8793

84 Plant Nation

2A Upland Road, London SE22 9EE
Tel: 020 8299 0444
Specialities: Annuals/Biennials/Bedding, Herbaceous Perennials

85 Plantation Garden Plant Centre

Kenton Road, Harrow, London HA1 2BW
Tel: 020 8423 2073

86 The Pot Company

340 Fulham Road, London SW10 9UH
Tel: 020 7376 4808
Specialities: Pots

87 Ravensbourne NHS Trust

Horticultural Project, Church Approach, Cudham, Sevenoaks, London TN14 7QF
Tel: 01959 570343

88 Redbridge Garden Centre

Roding Lane North, Woodford Green, London IG8 8LZ
Tel: 020 8550 9000
Specialities: Garden Products, Garden Furniture

89 Ruxley Manor Garden Centre

Maidstone Road, Sidcup, Kent DA14 5BQ
Tel: 020 8300 0084 Fax: 020 8302 3879
e-mail: rmgc@ruxley-manor.co.uk
website: www.ruxley-manor.co.uk

SEE ADVERTISEMENT PAGE 175

Specialities:

Outdoor Plants; Garden Furniture & Barbecues

Products and Services:

- Aquatics
- Barbeques
- Books and stationery
- Seasonal Displays
- Clothing
- Conservatories
- Credit Cards Accepted
- Disabled Facilities
- Display Gardens
- Dried & Artificial Flowers
- Farm Shop
- Fencing
- Fruit Trees
- Garden Design
- Garden Furniture
- Garden Products
- Giftware
- Greenhouses and Sheds
- Houseplants
- Large Car Park
- Paving
- Pet Products
- Plant Guarantee
- Plants
- Restaurant/Coffee Shop
- Seeds and Bulbs
- Trees
- Water features

Directions:

Leave the A20 at Crittall's Corner, take the A223 (Edgington Way). At the roundabout, take 3rd exit onto B2173 (Maidstone Road). Ruxley is 150 metres up on the right hand side.

90 The Secret Garden

Westow Street, London SE19 3AF
Tel: 020 8771 8200

91 Shannon's Garden Centre

99-105 Stanstead Road, Forest Hill, London SE23 1HH
Tel: 020 8291 1507

92 Shaw Trust Horticulture Nurseries Ltd

The Nurseries, Clamp Hill, Stanmore, London HA7 3JS
Tel: 020 8954 4287

93 Shoots Garden Centre and Nursery

Common Road, Stanmore, London HA7 3JF
Tel: 020 8954 4628
Specialities: Aquatics, Restaurant/Coffee Shop, Pet Products

94 Simmons Orchard Side Nurseries

230 Bullsmoor Lane, Enfield, London EN1 4RL
Tel: 01992 763064
Specialities: Annuals/Biennials/Bedding, Herbaceous Perennials

95 Smallholders Pet Garden & Aquatic Centre

113-117 Woodgrange Road, London E7 0EP
Tel: 020 8534 1972
Specialities: Aquatics, Shrubs, Xmas Trees

96 Snapdragon Nursery

266-268 Lee High Road, London SE13 5PL
Tel: 020 8463 0503
Specialities: Annuals/Biennials/Bedding, Floristry

97 Spring Time Nurseries (Enfield) Ltd

Cattlegate Road, Crews Hill, Enfield, London EN2 9EE
Tel: 020 8367 9326

98 Springwell Nursery

Hayes End Road, Hayes, London UB4 8EH
Tel: 020 8573 1479
Specialities: Annuals/Biennials/Bedding, Herbaceous Perennials, Shrubs

99 Squires Garden Centre (Twickenham)

Sixth Cross Road, Twickenham, London TW2 5PA
Tel: 020 8977 9241

100 Stonecot Nurseries

Stonecot Hill, Sutton, London SM3 9HR
Tel: 020 8644 9412
Specialities: Garden Products

101 Stuarts Nurseries

57 North Cray Road, Sidcup, London DA14 5EU
Tel: 020 8300 1933
Specialities: Garden Design, Landscape Gardening

102 Sundridge Park Garden Centre

Plaistow Lane, Bromley, London BR1 3EN
Tel: 020 8466 7002

103 Sunshine Garden Centre

Durnsford Road, London N11 2EL
Tel: 020 8889 4224

104 T Watts & Sons Retail Ltd

249 Hertford Road, Enfield, London EN3 5JJ
Tel: 020 8804 1300

105 Teddington Station Garden Centre

Station Road, Teddington, London TW11 9AA
Tel: 020 8943 5222

106 Temple Fortune Garden Centre

788a Finchley Road, London NW11 7TJ
Tel: 020 8455 5363

107 Tubs & Shrubs

150 Upminster Road, Upminster, London RM14 2RB
Tel: 01708 509306
Specialities: Shrubs

108 W.P Stratford

364 Horns Road, Ilford, London IG6 1BT
Tel: 020 8551 5600

109 Webbs Garden Centre

609 Green Lane, Ilford, London IG3 9RP
Tel: 020 8597 2684
Specialities: Garden Furniture

110 Welling Garden Centre

353 Shooters Hill, Welling, London DA16 3RP
Tel: 020 8856 2933

111 Whitehead's Nursery

Woodside, Skeet Hill Lane, Orpington, London BR5 4HB
Tel: 01689 827503

112 Wildwoods Water Gardens Ltd

Theobalds Park Road, Enfield, London EN2 9BP
Tel: 020 8366 0243
Specialities: Water Products and Features, Aquatics

113 The Wimbledon Garden Centre

366 Merton Road, Southfields, London SW18 5AD
Tel: 020 8874 9962
Specialities: Tropical, Exotics

114 Woldens Garden Centre

Cattlegate Road, Enfield, London EN2 9DX
Tel: 020 8363 6450

115 Wonderland Nursery

69 Little Woodcote Estate, Carshalton, London SM5 4DD
Tel: 020 8668 3133

116 Woodcote Green Nurseries Ltd

Woodmansterne Lane, Wallington, London SM6 0SU
Tel: 020 8647 6838

117 World's End Nurseries

441-457 Kings Road, London SW10 0LR
Tel: 020 7351 3343
Specialities: Annuals/Biennials/Bedding, Herbaceous Perennials

118 Worldwide Pots

638 Chigwell Road, Woodford Green, London IG8 8AF
Tel: 020 8504 9880
Specialities: Pots

119 Wyevale Garden Centre (Croydon Purley Way)

Waddon Way, Croydon, London CR0 4HY
Tel: 020 8688 5117

120 Wyevale Garden Centre (Croydon Shirley)

Wickham Road, Croydon, London CR9 8AG
Tel: 020 8654 3720

121 Wyevale Garden Centre (Enfield)

Cattlegate Road, Enfield, London EN2 9DX
Tel: 020 8367 0422

122 Wyevale Garden Centre (Harrow)

Headstone Lane, Harrow, London HA2 6NB
Tel: 020 8428 3408

123 Wyevale Garden Centre (Hillingdon)

Pield Heath Road, Uxbridge, London UB8 3NP
Tel: 01895 236141

124 Wyevale Garden Centre (Keston)

Oakley Road, Bromley, London BR2 8HD
Tel: 01689 859419

125 Wyevale Garden Centre (Merton)

Lower Morden Lane, Morden, London SM4 4SJ
Tel: 020 8337 7781

126 Wyevale Garden Centre (Mill Hill)

Daws Lane, London NW7 4SL
Tel: 020 8906 4255

127 Wyevale Garden Centre (Osterley)

Windmill Lane, Isleworth, London TW7 5PR
Tel: 020 8847 2468

128 Wyevale Garden Centre (Syon Park)

Syon Park, London Road, Brentford, London TW8 8JG
Tel: 020 8568 0134

GREATER MANCHESTER

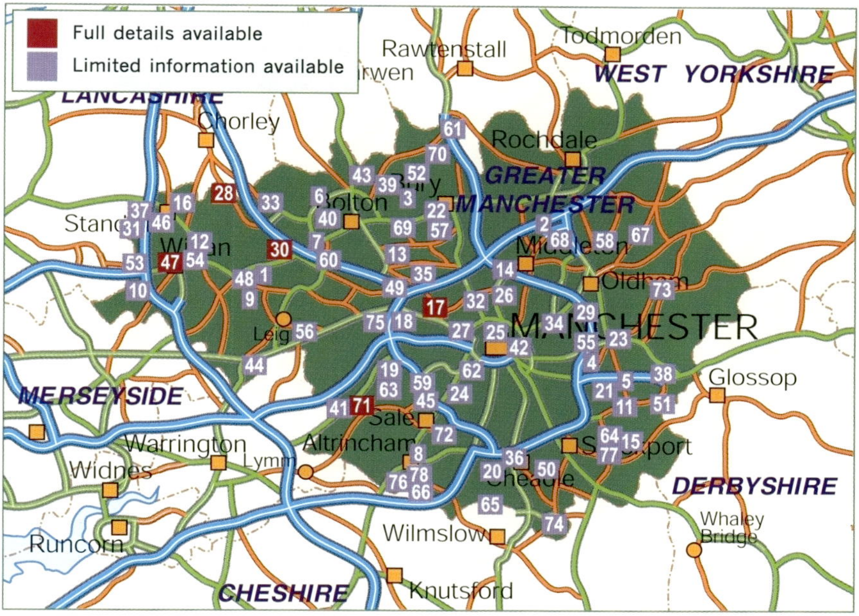

1 Alder Lane Nurseries

10 Alder Lane, Hindley Green, Wigan, Manchester WN2 4ET
Tel: 01942 255141
Specialities: Homegrown, Herbaceous Perennials

2 All In One Garden Centre

Rochdale Road, Middleton, Manchester M24 2RB
Tel: 01706 711711
Specialities: Garden Products, Annuals/Biennials/Bedding

3 Arthur Lane Nurseries

1 Arthur Lane, Ainsworth, Bolton, Manchester BL2 5PW
Tel: 01204 529929
Specialities: Annuals/Biennials/Bedding, Herbaceous Perennials

4 Ashton Park Garden Centre

Lord Sheldon Way, Lapwing Lane, Audenshaw, Manchester M34 5QL
Tel: 01613 701495

5 Barbers Kingston Nursery

Dukinfield Rd, Hyde, Manchester SK14 4PD
Tel: 01613 683008

6 Barrow Bridge

Bazley Street, Bolton, Manchester BL1 7NG
Tel: 01204 844152
Specialities: Annuals/Biennials/Bedding

7 Barton Grange Garden Centre

Wigan Road, Bolton, Manchester BL3 4RD
Tel: 01204 660660
Specialities: Garden Products

8 Battman & Sons

29 Deansgate Lane, Timperley, Altrincham, Manchester WA15 6SQ
Tel: 01619 280556

9 Bickershaw Hall Nurseries

599 Bickershaw Lane, Bickershaw, Wigan, Manchester WN2 4AA
Tel: 01942 866400
Specialities: Herbaceous Perennials, Shrubs

10 Billinge Garden Centre

Winstanley Road, Billinge, Wigan, Manchester WN5 7XG
Tel: 01695 632317
Specialities: Garden Products

11 Bluebell Centre

Bamford Hall, Stockport Road, Hyde, Manchester SK14 5EZ
Tel: 01614 941018
Specialities: Garden Products, Giftware

12 Bonsai Centre

92 Wigan Lane, Wigan, Manchester WN1 2LF
Tel: 01942 244827
Specialities: Bonsai

13 Boscow Nurseries

Boscow Road, Little Lever, Bolton, Manchester BL3 1AH
Tel: 01204 573195
Specialities: Herbaceous Perennials, Seasonal Plants

14 Bowlee Nursery

Heywood Old Road, Middleton, Manchester M24 4GS
Tel: 01616 533624
Specialities: Shrubs, Seasonal Plants

15 Brabyns Garden Nursery

Brabyns Park, Marple Bridge, Stockport, Manchester SK6 5LB
Tel: 01614 497242

16 Bradley Hall Nurseries

Bradley Lane, Standish, Wigan, Manchester WN6 0XQ
Tel: 01257 427085
Specialities: Annuals/Biennials/Bedding, Shrubs

17 Bridge Street Garden Centre

Bridge Street, Swinton, Manchester, Lancs M27 4DU
Tel: 0161 7283737 Fax: 0161 7937017
e-mail: bridgestreetgardencentre@msn.com

Specialities:

Hanging Baskets, Shrubs, Christmas Trees, Chrismas Wreaths

Products and Services:

- Barbeques
- Seasonal Displays
- Cut Flowers
- Display Gardens
- Dried & Artificial Flowers
- Fruit Trees
- Garden Construction
- Garden Furniture
- Garden Design
- Garden Products
- Giftware
- Greenhouses & sheds
- Houseplants
- Large Car Park
- Paving
- Pet Products
- Pick your own
- Plant Guarantee
- Plants
- Seeds and Bulbs
- Trees

Directions:

Off junction 17 M62 just off A6 or A580 East Lanc's Road.

18 Broadoak Park Garden Centre

173 Worsley Road, Worsley, Manchester M28 2SJ
Tel: 01617 943377
Specialities: Garden Products, Annuals/Biennials/ Bedding

19 Brookedge Nurseries

Woodhouse Road, Urmston, Manchester M41 8NU
Tel: 01617 467300
Specialities: Seasonal Plants, Herbaceous Perennials

20 Brookside Nursery

Brookside Road, Gatley, Cheadle, Manchester SK8 4BG
Tel: 01614 283053

21 Broomstair Nurseries

Watson Street, Denton, Manchester M34 3EN
Tel: 01613 364377
Specialities: Annuals/Biennials/Bedding, Shrubs

22 Bury Shed Centre

162a Bolton Road, Bury, Manchester BL8 2NP
Tel: 01617 977372
Specialities: Greenhouses and Sheds

23 Cheshire Hanging Baskets

12 Whitelands Road, Ashton-under-Lyne, Manchester OL6 6UT
Tel: 01613 300986
Specialities: Hanging Baskets

24 Chorlton Nursery

Vicars Road, Manchester M21 9GY
Tel: 01618 818142
Specialities: Shrubs, Herbaceous Perennials

25 The City Garden Centre

30-36 Swan Street, Manchester M4 5JG
Tel: 01618 325834
Specialities: Greenhouses and Sheds, Trees

26 Cleveland Garden Centre

Bank Road, Manchester M8 4QE
Tel: 01617 952244

27 Cozycare Nursery Products

Gloucester House, Cheltenham Street, Salford, Manchester M6 6WY
Tel: 01617 372748

28 Crows Nest Cottage Garden Centre

Crows Nest Farm, Chorley Road, nr Bolton, Lancashire BL6 5LE
Tel: 01257 480766

Specialities:

Unusual Perennials; Shrubs; Trees

Description:

This family-run nursery specialises in plants for traditional cottage gardens such as David Austin roses. It also stocks traditional stone troughs, basketware, planters, rockery plants, composts and water features. Helpful and knowledgeable staff are happy to give advice.

Directions:

The centre is located on the A6 just south of Adlington and 7 miles west of Bolton.

29 Daisy Nook Garden Centre

Daisy Nook, Failsworth, Manchester M35 9WJ
Tel: 01616 814245
Specialities: Garden Products, Shrubs

30 Darby's Garden Centre

Church Street, West Houghton, Bolton, Manchester BL5 3QW
Tel: 01942 818430
e-mail: darbysgarden@btconnect.com
Specialities: Plants, Shrubs,Handcrafted Ornamental Ironwork

31 Fir Tree Farm Christmas Tree Retailers

The Nook, Appley Bridge, Wigan, Manchester WN6 9JB
Tel: 01257 252607
Specialities: Xmas Trees

32 Garden Needs

1 Radford Street, Salford, Manchester M7 4NT
Tel: 01617 925448

33 Gardening World

Brunel Street, Horwich, Bolton, Manchester BL6 5NX
Tel: 01204 693644
Specialities: Ferns, Shrubs

34 Garsides Garden Centre

8 Oldham Road, Failsworth, Manchester M35 0JE
Tel: 01616 814789
Specialities: Greenhouses and Sheds, Turf, Hanging Baskets

35 Giants Seat Nursery

Radcliffe, Manchester M26 1EW
Tel: 01617 233462
Specialities: Annuals/Biennials/Bedding, Shrubs

36 Golden Days Garden Centre

Manchester Road, Cheadle, Manchester SK8 2NZ
Tel: 01614 283098
Specialities: Garden Furniture, Herbaceous Perennials, Shrubs

37 Golden Days Garden Centre

Back Lane, Appley Bridge, Standish, Wigan, Manchester WN6 8RS
Tel: 01257 423355
Specialities: Trees, Garden Furniture

38 Golden Springs Farm

Harrop Edge Road, Mottram, Hyde, Manchester SK14 6SJ
Tel: 01457 765575

39 Hawthorn Bank Nursery

4 Hawthorn Bank, Bolton, Manchester BL2 3JB
Tel: 01204 306895
Specialities: Seasonal Plants, Annuals/Biennials/Bedding

40 Heaton Fold Horticultural Centre

Overdale Drive, Bolton, Manchester BL1 5BU
Tel: 01204 336848

41 Higher Stanley Bank Farm

New Moss Rd, Cadishead, Manchester M44 5JT
Tel: 01617 753886

42 Holland Hydroponics

17 Rondin Road, Ardwick, Manchester M12 6BF
Tel: 01612 735995
Specialities: Hydroponics

43 Holland Nurseries

Darwen Road, Bromley Cross, Bolton,
Manchester BL7 9AB
Tel: 01204 301788
Specialities: Herbaceous Perennials, Trees

44 Kenyon Lane Nurseries Ltd

Kenyon Lane, Culcheth, Warrington,
Manchester WA3 4AU
Tel: 01925 763324

45 Kiwi Nurseries

Bradley Lane, Stretford, Manchester M32 8RH
Tel: 01619 765007
Specialities: Annuals/Biennials/Bedding, Shrubs

46 Kiwi Nurseries

Arbour Lane, Standish, Wigan,
Manchester WN6 0YJ
Tel: 01257 472511
Specialities: Herbaceous Perennials, Shrubs

47 L & T Stone & Landscape Supplies Ltd

Unit 15, Miry Lane Industrial Estate, Wigan,
Manchester WN6 7TG
Tel: 01942 497997
e-mail: info@ltstonesupplies.co.uk
website: www.ltstonesupplies.co.uk
Specialities: Stone, Landscape Materials

48 Laurel Nurseries & Garden Centre Ltd

Atherton Road, Hindley, Wigan,
Manchester WN2 3XD
Tel: 01942 526258
Specialities: Greenhouses and Sheds, Garden Products

49 Linnyshaw Garden Centre

Moss Lane, Worsley, Manchester M28 3NE
Tel: 01617 992050
Specialities: Garden Products, Shrubs

50 Lomax Nurseries

Adswood Road, Cheadle Hulme,
Manchester SK8 5QY
Tel: 01614 851824
Specialities: Herbaceous Perennials, Garden Furniture, Shrubs

51 Lymefield Garden Centre Ltd

Broadbottom, Hyde, Manchester SK14 6AG
Tel: 01457 764686
Specialities: Garden Products, Restaurant/Coffee Shop

52 Mason Nursery

Bradshaw Road, Tottington, Bury,
Manchester BL8 3PP
Tel: 01204 887618
Specialities: Annuals/Biennials/Bedding, Shrubs

53 Mckinsey Healthy Herbs

Orrell Rd, Orrell, Wigan, Manchester WN5 8QZ
Tel: 01695 632825
Specialities: Herbs, Organic Plants

54 Milton Nurseries Ltd

Cricket Street, Wigan, Manchester WN6 7TP
Tel: 01942 825445
Specialities: Annuals/Biennials/Bedding, Shrubs

55 Mockridge Nurseries

Rayner Lane, Ashton-under-Lyne,
Manchester OL7 0AB
Tel: 01613 717944
Specialities: Annuals/Biennials/Bedding, Shrubs

56 Moss Bank Nurseries Ltd

Hooten Lane, Leigh, Manchester WN7 3DG
Tel: 01942 671787
Specialities: Annuals/Biennials/Bedding, Herbaceous Perennials

57 Newbank Garden Centre

Bury Road, Radcliffe, Manchester M26 2XW
Tel: 01617 245050
Specialities: Restaurant/Coffee Shop, Water Products and Features

58 Newbank Garden Centre

Turf Lane, Royton, Oldham, Manchester OL2 6JH
Tel: 01706 844150

59 Newcroft Nurseries

40 Newcroft Road, Urmston, Manchester M41 9NN
Tel: 01617 486036
Specialities: Annuals/Biennials/Bedding, Fuchsia, Xmas Trees

60 Newgate Nurseries

1 Newgate Cottages, Plodder Lane, Bolton, Manchester BL5 1AH
Tel: 01204 655230
Specialities: Seasonal Plants, Annuals/Biennials/Bedding

61 Park Farm Shop & Garden Centre

Manchester Road, Walmersley, Bury, Manchester BL9 5NP
Tel: 01706 823577
Specialities: Farm Shop, Restaurant/Coffee Shop

62 Parkers Garden Centre

448-452 Chester Road, Manchester M16 9HL
Tel: 01618 774247
Specialities: Garden Products, Paving

63 Parkers Garden Centre

Carrington Road, Urmston, Manchester M41 6HT
Tel: 01618 481110

64 Phedar Nursery

Bunkers Hill, Romiley, Stockport, Manchester SK6 3DS
Tel: 01614 303772

65 Primrose Cottage Nursery & Garden Centre

Ringway Road, Manchester, Manchester M22 5WF
Tel: 01614 371557
Specialities: Annuals/Biennials/Bedding, Garden Products, Shrubs

66 Ridgeway Nurseries Ltd

6 Ridgeway Road, Timperley, Altrincham, Manchester WA15 7EY
Tel: 01619 047008

67 Sholvermoor Nurseries

Sholvermoor Farm, Ripponden Road, Oldham, Manchester OL4 2LR
Tel: 01616 245553
Specialities: Herbaceous Perennials, Homegrown

68 Stakehill Nurseries Ltd

Stakehill Lane, Middleton, Manchester M24 2RX
Tel: 01616 433075
Specialities: Annuals/Biennials/Bedding, Herbaceous Perennials

69 Stephen H Smith Garden & Leisure

Irwell Valley Branch, Radcliffe Moor Road, Bradley Fold, Bolton, Manchester BL2 6RE
Tel: 01204 529949
Specialities: Herbaceous Perennials, Garden Furniture, Giftware

70 Summerseat Garden Centre

Railway Street, Bury, Manchester BL9 5QD
Tel: 01204 883048
Specialities: Restaurant/Coffee Shop, Garden Products

71 The Vicarage Botanical Gardens

Manchester Road, Carrington, nr Sale, Manchester M31 4AG
Tel/Fax: 0161 775 2750
e-mail: info@vicaragebotanicalgardens.co.uk
website: www.vicaragebotanicalgardens.co.uk

Specialities:

Display Gardens; Perennials; Shrubs

Description:

This 5-acre site was the overgrown garden of a Vicarage. It is now beautifully laid with woodland walks and an excellent plant centre. There are guided tours of the gardens every Wednesday from April to September.

Directions:

From Junction 8 of the M60 take the A6144 towards Lymm. The Gardens are about 3 miles along this road, just past the village of Carrington.

72 Ward Fuchsias

5 Pollen Close, Sale, Manchester M33 3LS
Tel: 01612 827434
Specialities: Fuchsia

73 Woodend Nursery

Unit 11 Warmco Industrial Estate, Manchester Road, Mossley, Ashton-under-Lyne, Manchester OL5 9AY
Tel: 01457 837803
Specialities: Conifers, Shrubs

74 Woodford Park Garden Centre

Chester Road, Woodford, Stockport,
Manchester SK7 1QS
Tel: 01614 390745

75 Worsley Hall Nurseries & Garden Centre

Leigh Road, Worsley, Manchester M28 2LJ
Tel: 01617 908792
Specialities: Annuals/Biennials/Bedding, Garden Products

76 Wyevale Garden Centre (Altrincham)

Green Lane, Timperley, Altrincham,
Manchester WA15 8QP
Tel: 01619 806036

77 Wyevale Garden Centre (Marple)

Otterspool, Dooley Lane, Marple, Stockport,
Manchester SK6 7HE
Tel: 01614 277211

78 Wyevale Garden Centre (World of Pets & Leisure)

Thorley Lane, Timperley, Altrincham,
Manchester WA15 7PJ
Tel: 01619 039299

MERSEYSIDE

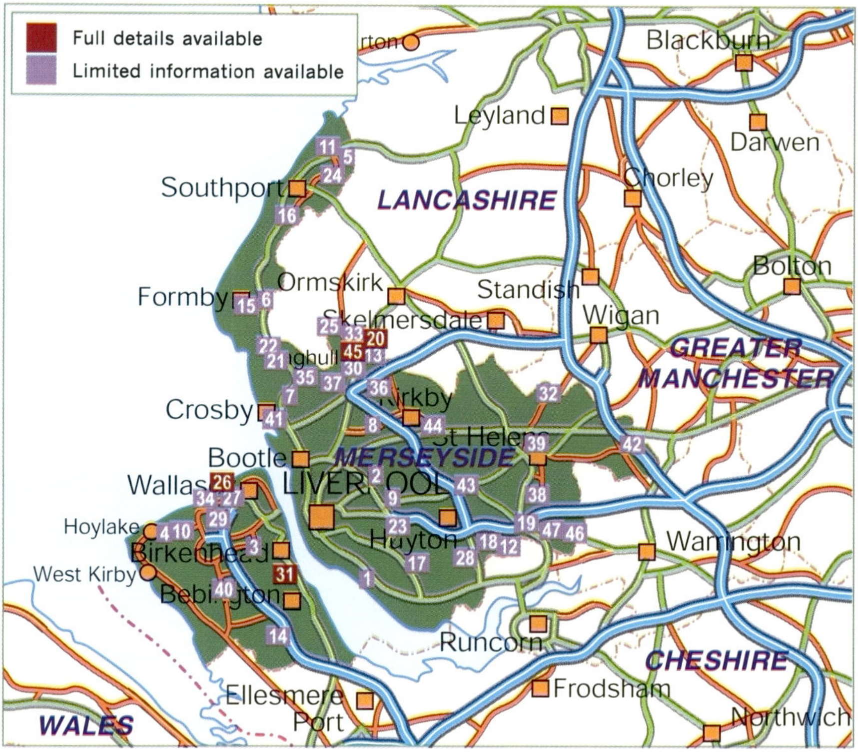

1 Aigburth Hall Nurseries

Aigburth Hall Avenue, Liverpool,
Merseyside L19 9EA
Tel: 01514 273120

2 Avant Gardens & Driveways

130 Almonds Green, Liverpool, Merseyside L12 5HT
Tel: 0800 458 0139
Specialities: Annuals/Biennials/Bedding, Shrubs, Herbaceous Perennials

3 Beresford Nurseries & Garden Centre

Storeton Road, Birkenhead, Merseyside CH42 9PG
Tel: 01516 081950
Specialities: Annuals/Biennials/Bedding, Greenhouses and Sheds

4 The Bigger Plant Co.Ltd

Orchard Nurseries, Fornalls Green Lane, Meols,
Wirral, Merseyside CH47 9RL
Tel: 01516 323532
Specialities: Specimen Trees, Shrubs

5 Box Tree Nursery

192 Cambridge Road, Southport,
Merseyside PR9 7LS
Tel: 01704 227465
Specialities: Annuals/Biennials/Bedding, Shrubs

6 Bridge Inn Nurseries

Moss Side, Formby, Liverpool, Merseyside L37 0AF
Tel: 01704 832355
Specialities: Shrubs, Herbaceous Perennials

7 Brownmoor Nursery

43 Brownmoor Lane, Liverpool,
Merseyside L23 0TD
Tel: 01512 841911
Specialities: Annuals/Biennials/Bedding, Shrubs

8 Buckels Nursery

Copplehouse Lane, Fazakerley, Liverpool,
Merseyside L10 0AG
Tel: 01515 252712
Specialities: Annuals/Biennials/Bedding, Shrubs, Herbaceous Perennials

9 C & D Garden & Pet Centre

297 East Prescot Road, Liverpool,
Merseyside L14 2DD
Tel: 01512 283143

10 Carr Farm Garden Centre Ltd

Birkenhead Road, Meols, Wirral,
Merseyside CH47 9RE
Tel: 01516 321457
Specialities: Restaurant/Coffee Shop, Aquatics

11 Cedars Garden Centre

86-88 Radnor Drive, Southport,
Merseyside PR9 9RS
Tel: 01704 212763
Specialities: Garden Products

12 Cronton Nurseries

Cronton Road, Widnes, Merseyside WA8 5QJ
Tel: 01514 242999

13 D.J & M Mason

15 The Smallholdings, Millbank lane, Liverpool,
Merseyside L31 9AT
Tel: 01515 261054

14 Fieldcrest Herbs & Flowers

Fieldcrest, Thornton Common Road, Wirral,
Merseyside CH63 0LT
Tel: 01513 348878
Specialities: Floristry, Herbs

15 Formby Garden Centre

1 Cable Street, Formby, Liverpool,
Merseyside L37 3LU
Tel: 01704 879198
Specialities: Garden Furniture, Garden Products

16 Four Seasons Garden Centre

224a Liverpool Road, Southport,
Merseyside PR8 4PB
Tel: 01704 560164
Specialities: Annuals/Biennials/Bedding, Herbaceous Perennials, Shrubs

17 Gateacre Garden Centre

Acrefield Road, Liverpool, Merseyside L25 5JW
Tel: 01514 286556

18 Gorse Lodge Nurseries

Cronton Road, Widnes, Merseyside WA8 5PX
Tel: 01514 233133

19 Harefield Cottage Water Gardens

Warrington Road, The Stoops, Rainhill, Prescot,
Merseyside L35 6PG
Tel: 01514 306647
Specialities: Aquatics, Water Products and Features

20 Hartley's Flowers & Plants

Robbin's Island, Northway, Lydiate, nr Ormskirk,
Merseyside L31 4EJ
Tel: 0800 78 313 96
e-mail: flowers.plants@virgin.net
website: www.hartleysnurseries.co.uk

Specialities:

Bedding Plants; Hanging Baskets; Florists

Description:

Hartley's are specialist growers of bedding plants and chrysanthemum blooms, producing more than ¼ **million plants** and over 2000 hanging baskets per year. From September through to Christmas, they produce chrysanthemum blooms for cut flower sales within their large flower shop.

Directions:

Hartley's is situated off Robbin's Island at Lydiate. This traffic island is the junction of the B5407 (Liverpool Road) and the A59 (Northway).

21 Ince Blundell Nurseries Ltd

Lady Green Lane, Liverpool, Merseyside L38 1QB
Tel: 01519 292915

22 Lady Green Nurseries & Garden Centre Ltd

Lady Green Lane, Liverpool, Merseyside L38 1QB
Tel: 01519 293635
Specialities: Restaurant/Coffee Shop, Garden Furniture, Giftware

23 Landlife Wildflowers Ltd

Court Hey Park, Liverpool, Merseyside L16 3NA
Tel: 01517 371819
Specialities: Wild Flowers

24 Larkins

94 Bispham Road, Southport, Merseyside PR9 7DF
Tel: 01704 222773
Specialities: Pet Products, Garden Products

25 Lydiate Barn Nurseries & Garden Centre

Southport Road, Lydiate, Liverpool, Merseyside L31 4EE
Tel: 01515 201238

26 Magill's Nursery

Leasowe Road, Wallasey, The Wirral, Cheshire CH45 8LW
Tel/Fax: 0151 639 6784
website: www.magillsnurseries.co.uk

Specialities:

Shrubs; Herbaceous; Herbs

Description:

This family owned nursery has been providing friendly personal service for more than 30 years. It offers high quality plants, shrubs and trees at reasonable prices. In addition to herbs, alpines, climbers and bedding plants, the nursery also stocks a good range of garden ornaments and hand-made pottery.

Directions:

Magill's Nursery is close to Junction 1 of the M53, near the Wallasey Village Flyover

27 Manor Garden Centre

Leasowe Road, Wallasey, Merseyside CH45 8LW
Tel: 01516 388227

28 Martins Garden Centre

Netherley Road, Tarbock Green, Prescot, Merseyside L35 1QE
Tel: 01514 878147
Specialities: Annuals/Biennials/Bedding, Garden Products

29 Mcgreal Garden Supplies

Leasowe Rd, Wirral, Merseyside CH46 2RE
Tel: 01516 390495

30 Northway Nurseries

Moss Lane, Lydiate, Liverpool, Merseyside L31 4DG
Tel: 01515 264191
Specialities: Shrubs, Herbaceous Perennials

31 Port Sunlight Garden Centre

The Causeway, Port Sunlight, Wirral CH62 5DY
Tel: 0151 645 6244 Fax: 0151 644 1492
e-mail: info@portsunlightgardens.co.uk
website: www.portsunlightgardens.co.uk

Specialities:

Bedding Plants; Trees & Shrubs; Coffee Shop

Description:

Set in the picturesque village of Port Sunlight, the Garden Centre offers a dazzling variety of plants, trees, shrubs and garden furniture. It also stocks a vast array of garden sundries and has an excellent coffee shop serving hot lunches daily.

Directions:

From junction 5 of the M53, take the A41 north towards Birkenhead. Follow the brown tourist signs for Port Sunlight. The Garden Centre is behind Hulme Hall, close to the Lady Lever Art Gallery.

32 Rectory Nurseries

97 Rectory Road, Ashton-in-Makerfield, Wigan, Merseyside WN4 0QD
Tel: 01942 715160

33 Rosemount Nurseries

Pygons Hill Lane, Liverpool, Merseyside L31 4JD
Tel: 01515 260583
Specialities: Annuals/Biennials/Bedding, Shrubs

34 Rostherne Nursery

Rostherne, Leasowe Road, Wirral,
Merseyside CH46 2RH
Tel: 01516 386240

35 Rushton's Nurseries & Garden Centre

Runnells Lane, Liverpool, Merseyside L23 1TR
Tel: 01519 242365
Specialities: Restaurant/Coffee Shop

36 Sandy Lane Nurseries

Sandy Lane, Melling, Liverpool, Merseyside L31 1EJ
Tel: 01515 263232
Specialities: Annuals/Biennials/Bedding, Shrubs

37 Sefton Meadows Garden & Home Centre

Sefton Meadows, Sefton Lane, Liverpool,
Merseyside L31 8BT
Tel: 01515 316688
Specialities: Restaurant/Coffee Shop

38 St. Helens Home & Garden Centre

Jubits Lane, Sutton Manor, St. Helens,
Merseyside WA9 4RT
Tel: 01744 811960
Specialities: Giftware, Garden Products

39 Suregrow Garden Centre Ltd

Merton Bank Road, St. Helens,
Merseyside WA9 1HY
Tel: 01744 27879

40 Thingwall Garden & Aquatic Centre

Lower Thingwall Lane, Wirral, Merseyside CH61 1AZ
Tel: 01516 480281
Specialities: Aquatics, Water Products and Features

41 Tony Almond DIY & Garden Centre

52-54 South Road, Waterloo, Liverpool,
Merseyside L22 5PQ
Tel: 01512 862460
Specialities: Garden Products

42 Trebaron Garden Centre

174 Southworth Road, Newton-le-Willows,
Merseyside WA12 0BS
Tel: 01925 224138
Specialities: Garden Products, Garden Furniture

43 Whitakers Garden Centre

Liverpool Road, Prescot, Merseyside L34 3LX
Tel: 01514 266455

44 White Moss Horticulture Ltd

Simonswood Moss Works, Perimeter Road,
Liverpool, Merseyside L33 3AN
Tel: 01515 472979

45 Windmill Nurseries

Liverpool Road, Lydiate, nr Ormskirk,
Lancashire L31 2LY
Tel: 0151 531 8555 Fax: 0151 526 0360

Specialities:

Shrubs & Trees; Roses; Composts

Description:

Windmill Nurseries takes its name from the adjacent 17th century windmill. Established 40 years ago, the nurseries sell a wide variety of plants, shrubs and trees along with rustic garden furniture, rockery stones, composts galore and an extensive selection of garden sundries.

Directions:

Windmill Nurseries are located off the A59 on the old Liverpool Road, 5 miles south of Ormskirk

46 Woodend Nursery

Warrington Road, Bold Heath, Widnes,
Merseyside WA8 3XL
Tel: 01514 244932
Specialities: Geranium, Annuals/Biennials/Bedding, Fuchsia

47 Wyevale Garden Centre (Bold Heath)

Mill Lane, Widnes, Merseyside WA8 3UU
Tel: 01514 246264

NORFOLK

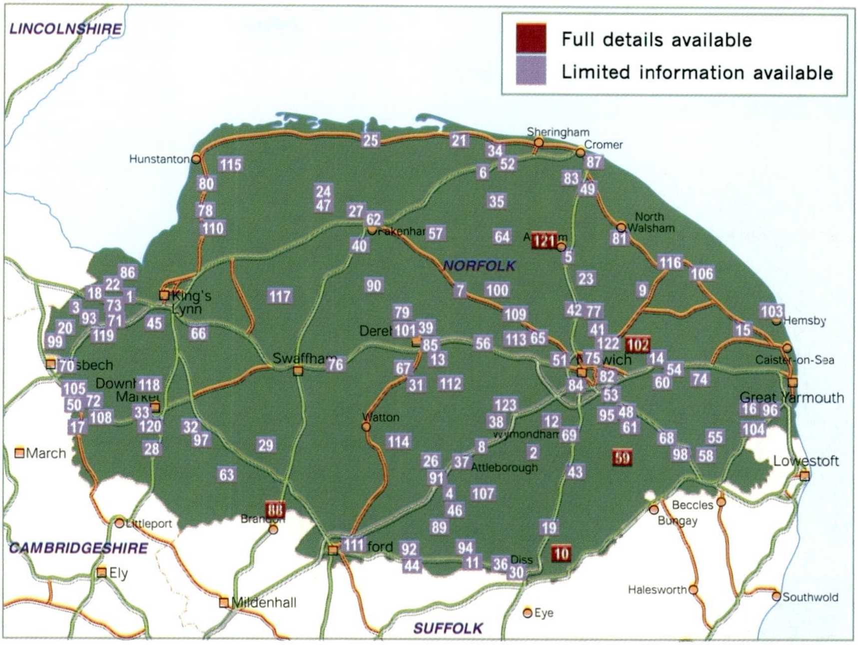

1 African Violets Centre

71 Station Road, Terrington St. Clement,
King's Lynn, Norfolk PE34 4PL
Tel: 01553 828374
Specialities: African Violets

2 Amulree Exotics

Norwich Road, Ashwellthorpe, Norfolk NR16 1EL
Tel: 01508 488101
Specialities: Exotics

3 Applegate Nurseries

Walnut Road, Walpole St. Peter, Wisbech,
Norfolk PE14 7PE
Tel: 01945 780224

4 Aromafolia

Barbers Farm, Leys Lane, Old Buckenham,
Attleborough, Norfolk NR17 1NT
Tel: 01953 887713
Specialities: Ornamentals, Herbaceous Perennials

5 Aylsham Garden Centre

Norwich Road, Aylsham, Norwich,
Norfolk NR11 6UD
Tel: 01263 735001

6 Bakers and Larners of Holt

8-12 Market Place, Holt, Norfolk NR25 6BW
Tel: 01263 712323
Specialities: Garden Machinery, Garden Furniture, Barbeques

7 Bawdeswell Garden Centre

Norwich Road, Bawdeswell, Dereham,
Norfolk NR20 4RZ
Tel: 01362 688387
Specialities: Garden Furniture, Garden Products

8 Best Thorpe Farm Produce & Plant Centre

Norwich Road, Besthorpe, Attleborough,
Norfolk NR17 2LA
Tel: 01953 456982
Specialities: Annuals/Biennials/Bedding, Shrubs

9 Bill LeGrice Roses

Wroxham Barns Ltd, Tunstead Road, Hoveton, Norwich, Norfolk NR12 8QU
Tel: 01603 784466
Specialities: Rose

10 Blacksmiths Cottage Nursery

Langmere Road, Langmere, nr Dickleburgh, Norfolk IP21 4QA
Tel: 01379 741136 / 740982
e-mail: blackcottnursery@aol.com
website: www.blackcottnursery.co.uk

SEE ADVERTISEMENT BELOW

Specialities:

Hardy Perennials; Geraniums; Drought Tolerant Plants; Woodland Plants

Description:

Small specialist nursery growing a wide range of hardy perennial plants with one of the largest collections of woodland perennials in the country. All of their plants are grown in peat-free compost and contain an additive to combat vine weevil. Also on site is the Potting Shed café which serves a range of refreshments and home-made cakes.

Directions:

Situated in the hamlet of Langmere, between Rushall and Dickleburgh, 5 miles northeast of Diss. From Dickleburgh village, take Harvey Lane past the School and Village Centre and follow the road for 1.5 miles, nursery is on your right.

11 Blooms of Bressingham

Low Road, Bressingham, Diss, Norfolk IP22 2AB
Tel: 01379 688585

12 Bobbin Bros

Main Road, Swardeston, Norfolk NR14 8DN
Tel: 01508 570286

13 Breckland Garden Centre

Dereham Road, Yaxham, Dereham, Norfolk NR19 1RF
Tel: 01362 696750
Specialities: Garden Products

14 British Wildflower Plants

31 Main Road, North Burlingham, Norwich, Norfolk NR13 4TA
Tel: 01603 716615
Specialities: Wild Flowers

15 Broadland Nurseries

Main Road, Ormesby St Michael, Great Yarmouth, Norfolk NR29 3LS
Tel: 01493 732694
Specialities: Herbaceous Perennials, Shrubs, Camellia

16 Burgh Castle Nurseries

Mill Road, Burgh Castle, Great Yarmouth, Norfolk NR31 9QS
Tel: 01493 780678
Specialities: Herbaceous Perennials, Shrubs

17 C & I D Martin

62 St. Peters Road, Upwell, Wisbech, Norfolk PE14 9EJ
Tel: 01945 772239

18 C D Bell and Sons

Westholme, Station Road North, Walpole Cross Keys, King's Lynn, Norfolk PE34 4HB
Tel: 01553 829058
Specialities: Annuals/Biennials/Bedding

19 Cherry Lane Nurseries

Ipswich Road, Pulham Market, Diss, Norfolk IP21 4XP
Tel: 01379 676418

20 Choice Landscapes

Priory Farm, Salts Road, West Walton, Wisbech, Norfolk PE14 7EF
Tel: 01945 585051
Specialities: Landscape Gardening

21 Cley Nurseries Ltd

Holt Road, Cley, Holt, Norfolk NR25 7TX
Tel: 01263 740892
Specialities: Lavender, Rose, Shrubs

22 Court Nurseries

83 Popes Lane, Terrington St. Clement, King's Lynn, Norfolk PE34 4NT
Tel: 01553 828409
Specialities: Annuals/Biennials/Bedding, Shrubs

23 Cranes Nursery

Coltishall Road, Buxton, Norwich, Norfolk NR10 5JD
Tel: 01603 279253
Specialities: Garden Products

24 Creake Plant Centre

Nursery View, South Creake, Fakenham, Norfolk NR21 9PW
Tel: 01328 823018
Specialities: Annuals/Biennials/Bedding, Shrubs

25 The Datura Nursery

Marsh Lane, Wells-next-The-Sea, Norfolk NR23 1EG
Tel: 01328 710330
Specialities: Annuals/Biennials/Bedding, Herbaceous Perennials

26 Deeply Rooted Garden Centre

Long Street, Attleborough, Norfolk NR17 1AW
Tel: 01953 453175
Specialities: Garden Products

27 Del's Nursery

Barsham Road, Barsham Road, Sculthorpe, Fakenham, Norfolk NR21 9NA
Tel: 01328 855556

28 Dents of Hilgay Farm Shop & Garden Centre

Steel's Drove, Hilgay, Downham Market, Norfolk PE38 0QH
Tel: 01366 385661
Specialities: Farm Shop, Garden Furniture

29 Didlington Nurseries

Field Lodge, Didlington, Thetford, Norfolk IP26 5AT
Tel: 01842 878288
Specialities: Annuals/Biennials/Bedding, Herbaceous Perennials, Shrubs

30 Diss Garden Centre

Victoria Road, Diss, Norfolk IP22 4JG
Tel: 01379 642873

31 Doubleday & Co

Hill Crest Cranworth, Cranworth, Thetford, Norfolk IP25 7SZ
Tel: 01362 820007

32 The Downham Country Garden Store

Bath Road, West Dereham, King's Lynn, Norfolk PE33 9TP
Tel: 01366 381190

33 Downham Garden Centre

Railway Road, Downham Market, Norfolk PE38 9EB
Tel: 01366 382384
Specialities: Garden Products, Greenhouses and Sheds

34 E M C Y Garden & Leisure

Weybourne Road, Kelling, Holt, Norfolk NR25 7ER
Tel: 01263 711574
Specialities: Garden Products, Giftware

35 Edgefield Nurseries

Norwich Road, Edgefield, Melton Constable, Norfolk NR24 2RL
Tel: 01263 587457
Specialities: Shrubs, Cut Flowers

36 F.W Myhill & Son Ltd

35 Mere Street, Diss, Norfolk IP22 4AD
Tel: 01379 642465

37 F.W Myhill & Son Ltd

Queens Square, Attleborough, Norfolk NR17 2AE
Tel: 01953 453104

38 F.W Myhill & Son Ltd

34-36 Fairland Street, Wymondham, Norfolk NR18 0AW
Tel: 01953 602272

39 F.W Myhill & Son Ltd

7 Church Street, Dereham, Norfolk NR19 1DJ
Tel: 01362 692975

40 Fakenham Garden Centre

Mill Road, Hempton, Fakenham, Norfolk NR21 7LH
Tel: 01328 863380
Specialities: Annuals/Biennials/Bedding, Pet Products, Aquatics

41 Farm Garden

Beeston Lane, Beeston, Norwich, Norfolk NR12 7BL
Tel: 01603 406660
Specialities: Composts, Herbaceous Perennials, Pots

42 The Fly Trap Plants

10 Newton Close, Newton St. Faith, Norwich, Norfolk NR10 3LH
Tel: 01603 710335
Specialities: Carnivorous Plants

43 Foundry Nurseries & Garden Centre

Morningthorpe, Norfolk NR15 2LJ
Tel: 01508 498263
Specialities: Annuals/Biennials/Bedding, Trees

44 Garboldisham Garden Centre

Oakdene, Hopton Road, Garboldisham, Diss, Norfolk IP22 2RQ
Tel: 01953 681326
Specialities: Herbaceous Perennials

45 Garden Classics

Bridge House, Saddlebow, King's Lynn, Norfolk PE34 3AR
Tel: 01553 617325

46 The Garden Enclosure

Banham Zoo, Kenninghall Road, Banham, Norfolk NR16 2HE
Tel: 01953 888043
Specialities: Garden Products, Giftware

47 Garden Force Garden Centre

Lester Rd, South Creake, Fakenham, Norfolk NR21 9PW
Tel: 01328 823734
Specialities: Hanging Baskets, Tropical, Herbaceous Perennials

48 Green Pastures Nursery

Mill Road, Bergh Apton, Norfolk NR15 1BQ
Tel: 01508 480734
Specialities: Farm Shop, Shrubs, Herbaceous Perennials

49 Groveland Garden Centre

Thorpe Market Rd, Roughton, Norwich, Norfolk NR11 8TB
Tel: 01263 833133
Specialities: Bonsai, Rare and Unusual Plants

50 Hazel Nurseries

Wisbech Road, Outwell, Wisbech, Norfolk PE14 8PG
Tel: 01945 772347
Specialities: Annuals/Biennials/Bedding, Herbaceous Perennials

51 Hellesdon Barns Garden Centre

Hellesdon Hall Road, Norwich, Norfolk NR6 5BB
Tel: 01603 485588
Specialities: Aquatics, Farm Shop

52 High Kelling Nursery Ltd

Selbrigg Road, High Kelling, Holt, Norfolk NR25 6RJ
Tel: 01263 712629
Specialities: Garden Products, Composts

53 Highway Garden and Leisure

Loddon Road, Framingham Pigot, Norfolk NR14 7PW
Tel: 01508 494665

54 Hillcrest Nurseries

South Burlingham Road, Lingwood, Norwich, Norfolk NR13 4ET
Tel: 01493 750014
Specialities: Annuals/Biennials/Bedding, Shrubs, Trees

55 Hillfield Nursery & Farm Shop

Mill Lane, Thorpe, Nr Haddiscoe, Norfolk NR14 6PA
Tel: 01508 548306
Specialities: Annuals/Biennials/Bedding, Shrubs, Farm Shop

56 Hockering Nursery

The Street, Hockering, Dereham, Norfolk NR20 3AJ
Tel: 01603 881290
Specialities: Annuals/Biennials/Bedding, Shrubs

57 Hoecroft Plants

Severals Grange, Holt Road, Wood Norton, Dereham, Norfolk NR20 5BL
Tel: 01362 684206
Specialities: Annuals/Biennials/Bedding, Herbaceous Perennials

58 The Home & Gardener Co

Unit 1 The Raveningham Centre, Beccles Road, Raveningham, Norwich, Norfolk NR14 6NU
Tel: 01508 480976
Specialities: Garden Products

59 Homestead Nurseries

Kirstead Green, Norwich NR15 1EB
Tel: 01508 550342 Fax: 01508 550444
e-mail: david@homesteadnurseries.co.uk
website: www.homesteadnurseries.co.uk

SEE ADVERTISEMENT BELOW

Specialities:

- Alpines
- Herbs
- Perennials
- Shrubs
- Trees
- Terracotta Pots
- Seeds
- Composts
- Seasonal Bedding

Description:

This is a family-run business where they pride themselves on their large range of high quality plants at very competitive prices. They stock more than 450 trees and over 200 rose varieties as well as a fantastic selection of shrubs, perennials and alpines, along with bedding and basket plants.

Directions:

Homestead Nurseries are located just off the B1332, about 8 miles south of Norwich

60 Homestead Nursery & Florist

Chapel Road, Strumpshaw, Norwich,
Norfolk NR13 4PA
Tel: 01603 715450
Specialities: Annuals/Biennials/Bedding, Floristry

61 Honeydale Nursery

The Street, Bergh Apton, Norfolk NR15 1BN
Tel: 01508 550387
Specialities: Annuals/Biennials/Bedding, Trees

62 Howell's Nurseries

The Nursery, Trap Lane, Northerb By-Pass,
Fakenham, Norfolk NR21 9RN
Tel: 01328 862681
Specialities: Annuals/Biennials/Bedding, Hanging Baskets

63 Hythe Alpines

Whitegates, Methwold Hythe, Thetford,
Norfolk IP26 4QW
Tel: 01366 728543
Specialities: Alpines/Rock Plants

64 The Knot Garden Nursery

Moat Farm, Wood Dalling, Norwich,
Norfolk NR11 6RL
Tel: 01263 584288
Specialities: Annuals/Biennials/Bedding, Shrubs

65 Lakeside Nursery

Taverham Lane, Costessey, Norwich,
Norfolk NR8 5BE
Tel: 01603 742865
Specialities: Annuals/Biennials/Bedding, Herbaceous Perennials

66 Laydons Garden Enhancements

East Winch Rd, Blackborough End, King's Lynn,
Norfolk PE32 1SF
Tel: 01553 840952
Specialities: Water Products and Features

67 Letton Road Nurseries

36 Letton Road, Shipdham, Thetford,
Norfolk IP25 7NE
Tel: 01362 820910
Specialities: Annuals/Biennials/Bedding, Shrubs

68 Loddon Garden & Seed Centre

Bridge Street, Loddon, Norwich, Norfolk NR14 6LZ
Tel: 01508 520632
Specialities: Garden Products, Seeds

69 Lodge Farm Nursery

Brick Kiln Lane, Swainsthorpe, Norfolk NR14 8PY
Tel: 01508 471104
Specialities: Annuals/Biennials/Bedding, Shrubs

70 Manor House Farm Nusery

Green Lane, Wisbech, Norfolk PE14 7BJ
Tel: 01945 463595

71 Matthews Nursery

Main Road, Terrington St. John, Wisbech,
Norfolk PE14 7RR
Tel: 01945 880825
Specialities: Annuals/Biennials/Bedding, Shrubs

72 Mike Harrison

Hall Road, Outwell, Wisbech, Norfolk PE14 8PE
Tel: 01945 772424

73 Moat Road Nursery

Moat Road, Terrington St. Clement, King's Lynn,
Norfolk PE34 4PN
Tel: 01553 828723
Specialities: Shrubs, Exotics, Cut Flowers

74 Moulton Nurseries Plant Centre

Reedham Road, Moulton St. Mary, Norwich,
Norfolk NR13 3AP
Tel: 01493 750458
Specialities: Hanging Baskets, Restaurant/Coffee Shop

75 Mousehold Garden Centre

63 Mousehold Lane, Norwich, Norfolk NR7 8HP
Tel: 01603 413272
Specialities: Garden Products

76 Necton Garden Centre

Tuns Road, Necton, Swaffham, Norfolk PE37 8EH
Tel: 01760 723612

77 New Farm Products

10 Buxton Road, Frettenham, Norwich,
Norfolk NR12 7NG
Tel: 01603 897253
Specialities: Annuals/Biennials/Bedding, Herbaceous Perennials, Farm Shop

78 Norfolk Bamboo Company

Vine Cottage, The Drift, Ingoldisthorpe, King's Lynn,
Norfolk PE31 6NW
Tel: 01485 543935
Specialities: Bamboos

79 Norfolk Herbs

Blackberry Farm, Dillington, Dereham,
Norfolk NR19 2QD
Tel: 01362 860812
Specialities: Herbs

80 Norfolk Lavender Ltd

Caley Mill, Lynn Road, Heacham, King's Lynn,
Norfolk PE31 7JE
Tel: 01485 570384
Specialities: Heathers

81 North Walsham Garden Centre

Norwich Road, North Walsham, Norfolk NR28 0DR
Tel: 01692 407524
Specialities: Rose, Fruit Trees, Restaurant/Coffee Shop

82 Norwich Heather & Conifer Centre

54a Yarmouth Road, Norwich, Norfolk NR7 0HE
Tel: 01603 439434
Specialities: Conifers, Heathers

83 Norwich Road Nursery

Norwich Road, Roughton, Norwich,
Norfolk NR11 8NA
Tel: 01263 761581
Specialities: Annuals/Biennials/Bedding, Shrubs, Trees

84 Notcutts Garden Centre

Daniels Road, Norwich, Norfolk NR4 6QP
Tel: 01603 453155

85 Old Hall Nurseries

Dumpling Green, Dereham, Norfolk NR19 1HP
Tel: 01362 693102
Specialities: Annuals/Biennials/Bedding, Shrubs

86 Ornamental Conifers

22 Chapel Road, Terrington St. Clement,
King's Lynn, Norfolk PE34 4ND
Tel: 01553 828874
Specialities: Conifers, Ornamentals

87 Overstand Cottage Garden Centre

6 Mundesley Road, Overstrand, Cromer,
Norfolk NR27 0PU
Tel: 01263 576654
Specialities: Annuals/Biennials/Bedding, Shrubs, Garden Furniture

88 P & R Garden Supplies

Fengate Drove, Brandon, Suffolk IP27 0PW
Tel: 01842 814800
e-mail: info@p-rgardensupplies.co.uk
website: www.p-rgardensupplies.co.uk

Specialities:

Plants arranged in "Plants for Places" order, making it easier to select a plant for every garden situation; Hanging Basket & Pot Filling Service; Wide Selection of Pots.

Products and Services:

- Aquatics
- Credit Cards Accepted
- Fruit Trees
- Garden Products
- Information Desk
- Large Car Park
- Pet Products
- Seeds and Bulbs

Directions:

Brandon is on the A1065, about 10 miles north of Mildenhall. P & R is located in a retail park on the Weeting (north) side of the railway crossing.

Description:

A family-run garden centre, P&R Garden Supplies was established in 2002 and stocks shrubs, conifers, perennials, herbs, alpines as well their very popular bedding plants and hanging baskets. Hardy plants are arranged in "Plants for Places" order, helping you to chose the perfect plant for every garden situation. The nursery also stocks a complete range of composts, soil improvers, bark, shingle, slate, cobbles, topsoil and manure – all at very competitive prices. In addition P & R also stocks a wide range of garden sundries and there's a good range of weed-killers, plant feeds and pest control products from the top market brands. P & R has a membership scheme which gives discounts and special offers. If a plant or product is not in stock, P & R offers an ordering service. The Aquatics Centre are Koi and Cichlid specialists. The Pet Centre caters for all pets, wild birds, poultry and non-domestic animals

89 P W Plants

Sunnyside, Heath Road, Kenninghall,
Norfolk NR16 2DS
Tel: 01953 888212
Specialities: Annuals/Biennials/Bedding, Herbaceous Perennials

90 Packwell Nurseries

Rudds Drift, Mileham, King's Lynn,
Norfolk PE32 2RD
Tel: 01328 701308
Specialities: Annuals/Biennials/Bedding, Herbaceous Perennials

91 Peter Beales Retail Ltd

London Road, Attleborough, Norfolk NR17 1AY
Tel: 01953 454707

92 The Plantsman's Preference

Lynwood, Hopton Road, Garboldisham, Diss,
Norfolk IP22 2QN
Tel: 01953 681439
Specialities: Grasses, Geranium, Herbaceous Perennials

93 Plumridge Nurseries

Ninwood House, Mill Road, Walpole Highway,
Wisbech, Norfolk PE14 7QP
Tel: 01945 780203

94 Poplar Farm Nursery

Fersfield Road, Bressingham, Diss,
Norfolk IP22 2AP
Tel: 01379 688235
Specialities: Annuals/Biennials/Bedding, Shrubs

95 Pots of Plants

2 Alberta Piece, Alpington, Norfolk NR14 7NS
Tel: 01508 494480
Specialities: Pot Plants

96 Progress Nurseries

Beccles Road, Bradwell, Great Yarmouth,
Norfolk NR31 8QD
Tel: 01493 662064
Specialities: Annuals/Biennials/Bedding, Shrubs

97 Quaymount Quality Plants

The Nursery, The Row, Wereham, King's Lynn,
Norfolk PE33 9AY
Tel: 01366 500691
Specialities: Annuals/Biennials/Bedding

98 Reads Nursery

Hales Hall, Hales Green, Loddon,
Norfolk NR14 6QW
Tel: 01508 548395
Specialities: Trees, Annuals/Biennials/Bedding

99 Rokewood Nurseries

Longacre, Walton Road, Wisbech, Norfolk PE14 7AG
Tel: 01945 583033

100 The Romantic Garden

The Street, Swannington, Norwich,
Norfolk NR9 5NW
Tel: 01603 261488
Specialities: Topiary, Ornamentals, Shrubs

101 Rooms & Blooms

Riverside Centre, Old Swaffham Road, Dereham,
Norfolk NR19 2DL
Tel: 01362 698722

102 Salhouse Garden Centre & Nurseries

Honeycombe Road, Salhouse, Norwich,
Norfolk NR13 6JP
Tel: 01603 722250 Fax: 01603 722246
website: www.salhousegardencentre.co.uk

SEE ADVERTISEMENT PAGE 195

Specialities:

Large A-Z of Shrubs Usual & Unusual; Italian Specimen Plants & Shrubs

Description:

This family-run business enjoys a beautiful location and offers a large selection of good priced shrubs, Italian specimens, roses, alpines, perennials, heathers, grasses, ferns, herbs, conifers, climbers, ornamental trees, camellias, rhododendrons, azaleas, hedging, fruit trees and bushes. Also in stock are water features, houseplants, seeds, bulbs, seed potatoes, onion sets, pots, greeting cards and gifts. Also on site is an excellent coffee shop.

Directions:

From the A1151, about 7 miles northeast of Norwich, turn right on the B1140 to Salhouse. The centre is just off this road in the village.

103 Scratby Garden Centre

Hall Farm, Beach Road, Scratby, Great Yarmouth,
Norfolk NR29 3AJ
Tel: 01493 733149
Specialities: Trees, Annuals/Biennials/Bedding

104 Southview Garden & Aquatic Centre

Beccles Road, Fritton, Great Yarmouth, Norfolk NR31 9EU
Tel: 01493 488212
Specialities: Aquatics, Garden Products, Herbaceous Perennials

105 Star Bulb Co

Outwell Road, Outwell, Wisbech, Norfolk PE14 8PQ
Tel: 01945 773594
Specialities: Bulbs

106 Sutton Garden Centre and Water Features

The Street, Sutton, Norwich, Norfolk NR12 9RA
Tel: 01692 580424
Specialities: Aquatics, Greenhouses and Sheds

107 Tall Trees Nusery

New Buckenham Road, Old Buckenham, Attleborough, Norfolk NR17 1PW
Tel: 01953 860412
Specialities: Trees, Shrubs

108 Tanrose Nurseries

Downham Road, Outwell, Wisbech, Norfolk PE14 8SL
Tel: 01945 772415
Specialities: Seasonal Plants, Annuals/Biennials/Bedding

109 Taverham Nursery Centre

Fir Covert Road, Taverham, Norwich, Norfolk NR8 6HT
Tel: 01603 860522
Specialities: Aquatics, Greenhouses and Sheds, Shrubs

110 Thaxters Garden Centre

49 Hunstanton Road, Dersingham, King's Lynn, Norfolk PE31 6NA
Tel: 01485 541514
Specialities: Greenhouses and Sheds, Garden Products

111 Thetford Garden Centre

Kilverstone, Thetford, Norfolk IP24 2RL
Tel: 01842 763267
Specialities: Restaurant/Coffee Shop, Garden Furniture, Giftware

112 Thorncroft Clematis Nursery

The Lings, Reymerston, Norwich, Norfolk NR9 4QG
Tel: 01953 850407
Specialities: Clematis

113 Urban Jungle

Ringland Lane, Costessey, Norwich,
Norfolk NR8 5BG
Tel: 01603 744997
Specialities: Exotics

114 Walnut Tree Garden Nursery

Flymoor Lane, Rocklands, Attleborough,
Norfolk NR17 1BP
Tel: 01953 488163
Specialities: Shrubs, Trees, Herbaceous Perennials

115 Wards Nurseries

Foundry Lane, Ringstead, Hunstanton,
Norfolk PE36 5LE
Tel: 01485 525242
Specialities: Annuals/Biennials/Bedding, Herbaceous Perennials

116 Wayford Nursery

Wayford, Norwich, Norfolk NR12 9LJ
Tel: 01692 580226
Specialities: Annuals/Biennials/Bedding, Shrubs

117 West Acre Gardens

West Acre, King's Lynn, Norfolk PE32 1UJ
Tel: 01760 755562
Specialities: Herbaceous Perennials, Grasses, Alpines/Rock Plants

118 Whispering Trees Nursery

West Way, Wimbotsham, King's Lynn,
Norfolk PE34 3QB
Tel: 01366 388752
Specialities: Fruit Trees

119 Whitegate Nurseries

Main Road, Terrington St. John, Wisbech,
Norfolk PE14 7RP
Tel: 01945 880241
Specialities: Annuals/Biennials/Bedding, Shrubs

120 Windmill Farm Produce & Garden Centre

Downham Market Bypass, Denver, Downham Market, Norfolk PE38 0DH
Tel: 01366 388414

121 Woodgate Nursery

Cawston Road, Aylsham, Norfolk NR11 6UH
Tel/Fax: 01263 731510

SEE ADVERTISEMENT PAGE 196

Specialities:

Perennials; Vegetable Plants; Friendly Customer Service

Description:

This family-owned nursery prides itself on its customer friendly service with a small team of helpful staff. The nursery offers a wide range of plants with between 800 and 1000 perennials, as well as a good selection of trees, shrubs, climbers, conifers and heathers.

Directions:

At the A140 roundabout south of Aylsham, take the B1145 west towards Cawston, The nursery is about 1 mile along this road on the left hand side.

122 Wyevale Garden Centre (Sprowston)

Blue Boar Lane, Norwich, Norfolk NR7 8RJ
Tel: 01603 412239

123 Wymondham Garden Centre

Tuttles Lane East, Wymondham, Norfolk NR18 0EP
Tel: 01953 600282
Specialities: Garden Products

NORTHAMPTONSHIRE

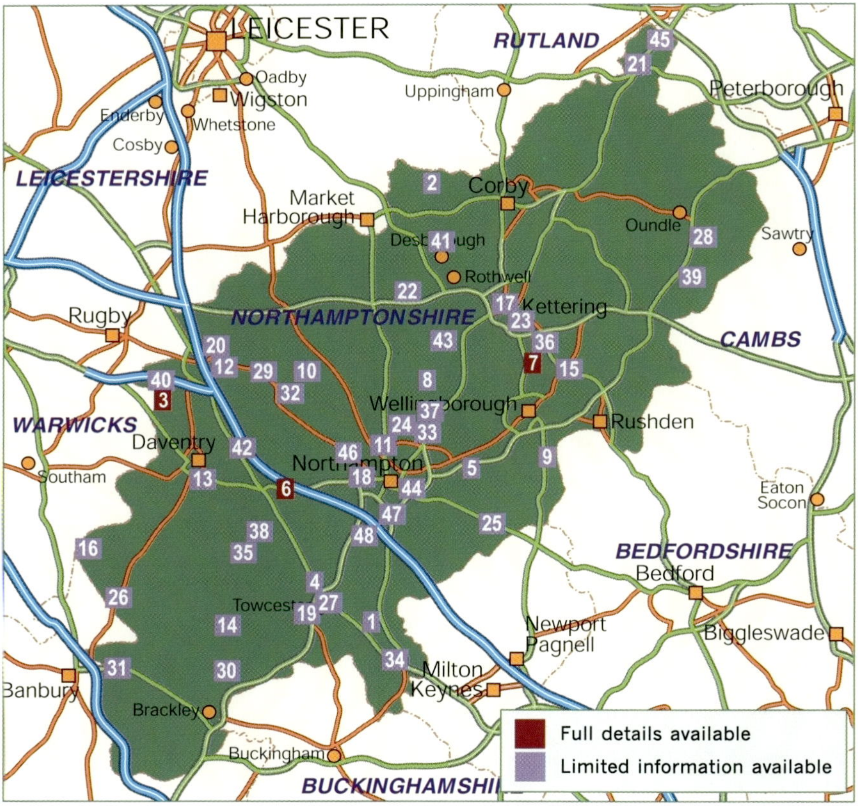

1 Alderton Plant Nursery

Spring Lane, Alderton, Towcester,
Northamptonshire NN12 7LW
Tel: 01327 811253
Specialities: Shrubs, Herbaceous Perennials

2 Ashley Herb Farm

45 Main Street, Ashley, Market Harborough,
Northamptonshire LE16 8HG
Tel: 01858 565722
Specialities: Herbs

3 Barby Nurseries

Daventry Road, Barby, nr Rugby,
Warwickshire CV23 8TS
Tel: 01788 890314 Fax: 01788 890715

Specialities:

Hanging Baskets; Acers; Aquatics

Products and Services:

- Aquatics
- Barbeques
- Seasonal Displays
- Credit Cards Accepted
- Fruit Trees
- Garden Design
- Garden Furniture
- Garden Products
- Houseplants
- Information Desk
- Large Car Park
- Plant Guarantee
- Plants
- Seeds and Bulbs
- Trees

Directions:

From Daventry, take the A361 until you come to the Barby/Welton crossroads. Turn left towards Barby. Keep going until you pass the Braunston turn on your left. The nurseries are 50 yards further, on the left

4 Bell Plantation Garden Centre

Watling Street, Towcester,
Northamptonshire NN12 6GX
Tel: 01327 354126
Specialities: Garden Furniture, Garden Products, Fencing

5 Billing Garden Store

The Causeway, Great Billing, Northampton,
Northamptonshire NN3 9EX
Tel: 01604 404550

6 Bliss Lane Nursery

Bliss Lane, Flore, Northampton NN7 4LJ
Tel: 01327 340918 Fax: 01327 342283

SEE ADVERTISEMENT ON PAGE 200

Specialities:

Nursery; Farm Shop; Tea Room

Products and Services:

- Seasonal Displays
- Christmas Trees
- Christmas Wreaths
- Credit Cards Accepted
- Cut Flowers
- Disabled Facilities
- Farm Shop
- Floristry
- Garden Design
- Information Desk
- Large Car Park
- Light Lunches/Tea Room
- Open Gardens
- Plant Guarantee
- Plants
- Seeds and Bulbs

Directions:

From the A45, about 7 miles west of Northampton, follow the brown and white signs for the Nursery.

7 Bosworth's Nurseries and Garden Centre

Finedon Road, Burton Latimer, Kettering,
Northamptonshire NN15 5QA
Tel: 01536 722635 Fax: 01536 420305
e-mail: info@bosworths-garden-centre.co.uk
website: www.bosworths-garden-centre.co.uk

SEE ADVERTISEMENT ON PAGE 200

Specialities:

Quality Plants; Garden Design & Construction; Coffee Shop

Products and Services:

- Barbecues
- Seasonal Displays
- Credit Cards Accepted
- Disabled Facilities
- Dried & Artificial Flowers
- Fencing
- Fireworks
- Garden Construction
- Garden Design
- Garden Furniture
- Garden Products
- Giftware
- Greenhouses & Sheds
- Hot Tubs
- Home-grown Plants
- Information Desk
- Large Car Park
- Plant Guarantee
- Plants
- Restaurant/Coffee Shop
- Seeds and Bulbs
- Swimming Pools

Directions:

Less than 1 mile from Junction 10 of the A14. From Junction 10, go into Burton Latimer, straight over two roundabouts and turn right at the third one.

8 Cedar Nurseries

Poplars Lane, Holcot, Northampton,
Northamptonshire NN6 9SW
Tel: 01604 781267

9 Cherry Blossom Bonsai

56 Eastfield Rd, Wollaston, Wellingborough,
Northamptonshire NN29 7RU
Tel: 01933 665383
Specialities: Bonsai

10 Coton Manor Plant Nursery

Coton, Northampton, Northamptonshire NN6 8RQ
Tel: 01604 740219
Specialities: Herbaceous Perennials, Shrubs, Bulbs

11 Cramden Nursery

Harborough Road North, Northampton,
Northamptonshire NN2 8LU
Tel: 01604 842365
Specialities: Geranium, Pelargonium, Lavender

12 Crick Garden Centre

West Haddon Road, Crick, Northampton,
Northamptonshire NN6 7SQ
Tel: 01788 824774
Specialities: Garden Products, Trees

13 Dennetts of Daventry

Wood Farm Nurseries, London Road, Daventry,
Northamptonshire NN11 4ND
Tel: 01327 702288
Specialities: Annuals/Biennials/Bedding, Shrubs

14 Elizabeth House Nursery

The Green, Lois Weedon, Towcester,
Northamptonshire NN12 8PN
Tel: 01327 860056
Specialities: Shrubs, Herbaceous Perennials

15 Finedon Garden Centre

35 Irthlingborough Road, Wellingborough,
Northamptonshire NN9 5EH
Tel: 01933 680212
Specialities: Herbaceous Perennials, Shrubs, Paving

16 Firtree Nursery

Lower Boddington, Daventry,
Northamptonshire NN11 6YP
Tel: 01327 262343
Specialities: Annuals/Biennials/Bedding, Trees

Bliss Lane Nursery

Bliss Lane, Flore, Northampton NN7 4LJ
Tel: 01327 340918 Fax: 01327 342283

Nursery - Farm Shop - Tea Room

Opening Times:

Nursery: Every Day 9:00 - 6:00
Tea Room: Thu - Sun 10:00 - 5:00
Farm Shop: Thu - Sun 10:00 - 5:00

Seasonal Displays
Credit Cards Accepted
Cut Flowers
Disabled Facilities
Display Gardens
Farm Shop
Floristry
Garden Construction
Garden Design
Information Desk
Large Car Park
Plant Guarantee
Plants
Restaurant/Tea Room
Seeds and Bulbs

Bosworth's Nurseries & Garden Centre

Finedon Road, Burton Latimer, Kettering, Northants NN15 5QA
Tel: 01536 722635 Fax: 01536 420305
e-mail: info@bosworths-garden-centre.co.uk
website: www.bosworths-garden-centre.co.uk

Opening Times

9am-6pm, Monday to Saturday
10am-5pm, Sunday
Closed Christmas Day, Boxing Day and New Year's Day

Everything for your garden....

Barbecues
Seasonal Displays
Credit Cards Accepted
Disabled Facilities
Dried & Artificial Flowers
Fencing
Fireworks
Garden Construction
Garden Construction
Garden Design
Garden Furniture
Garden Products
Giftware
Greenhouses & Sheds
Hot Tubs
Home-grown Plants
Information Desk
Large Car Park
Plant Guarantee
Plants
Restaurant/Coffee Shop
Seeds and Bulbs
Swimming Pools

17 The Garden Shop

194 London Road, Kettering,
Northamptonshire NN15 6DT
Tel: 01536 412448
Specialities: Garden Products

18 Golby's Garden Centre

Tollgate Way, Millway, Northampton,
Northamptonshire NN5 6ES
Tel: 01604 752155
Specialities: Garden Products

19 Green Lane Garden Centre

Brackley Road, Towcester,
Northamptonshire NN12 6JD
Tel: 01327 350593
Specialities: Garden Products

20 Greenfellows Garden Centre

Yelvertoft Road, Crick, Northampton,
Northamptonshire NN6 7TF
Tel: 01788 824324
Specialities: Garden Furniture, Herbaceous Perennials

21 Hillfield Nursery

1 Slate Drift, Collyweston, Stamford,
Northamptonshire PE9 3PG
Tel: 01780 444251
Specialities: Hanging Baskets, Annuals/Biennials/Bedding

22 Hiqh Bank Nursery

Highbank, Harrington, Northampton,
Northamptonshire NN6 9NQ
Tel: 01536 711802
Specialities: Annuals/Biennials/Bedding, Shrubs

23 Ise Garden Centre

Warkton Lane, Barton Seagrave, Kettering,
Northamptonshire NN15 5AA
Tel: 01536 519792
Specialities: Herbaceous Perennials

24 Moulton College Garden Centre

West Street, Moulton, Northampton,
Northamptonshire NN3 7RR
Tel: 01604 494206

25 The Nursery

Yardley, Hastings, Northampton,
Northamptonshire NN7 1HJ
Tel: 01604 696241

26 Otter Hollow Nursery

10 Appletree Road, Chipping Warden, Banbury,
Northamptonshire OX17 1LQ
Tel: 01295 660269

27 Payne's Nurseries

192 Watling Street East, Towcester,
Northamptonshire NN12 6DB
Tel: 01327 350377
Specialities: Trees, Shrubs, Herbaceous Perennials

28 Plantation

Manor House, Oundle Road, Polebrook,
Peterborough, Northamptonshire PE8 5LQ
Tel: 01832 274755
Specialities: Fruit Trees, Ornamentals, Hardy Plants

29 Plantsman

West Haddon Nurseries, Northampton Road, West Haddon, Northampton, Northamptonshire NN6 7AR
Tel: 01788 510206
Specialities: Garden Products

30 Plenti-Full-Pots

Falcutt Hall Farm, Brackley,
Northamptonshire NN13 5QY
Tel: 01280 850479
Specialities: Hanging Baskets, Pots, Herbaceous Perennials

31 Purely Plants

Main Road, Middleton Cheney, Banbury,
Northamptonshire OX17 2PW
Tel: 01295 760236

32 Ravensthorpe Nursery

6 East Haddon Road, Ravensthorpe, Northampton,
Northamptonshire NN6 8ES
Tel: 01604 770548
Specialities: Herbaceous Perennials, Shrubs, Trees

33 Rob's Nursery

19A Church Street, Moulton, Northampton,
Northamptonshire NN3 7SP
Tel: 01604 495073
Specialities: Shrubs, Herbaceous Perennials

34 Roman Way Garden Centre

8 Watling Street, Potterspury, Towcester,
Northamptonshire NN12 7QN
Tel: 01908 542427
Specialities: Specimen Plants, Herbaceous Perennials

35 Rose Farm Nurseries

Maidford Rd, Litchborough, Towcester,
Northamptonshire NN12 8JH
Tel: 01327 830346
Specialities: Shrubs, Iris

36 Seasons Garden Centre

Cranford Road, Burton Latimer, Kettering,
Northamptonshire NN15 5LZ
Tel: 01536 724009

37 Smart Plants Nursery

Sandy Hill Lane, Moulton, Northampton,
Northamptonshire NN3 7JB
Tel: 01604 454106
Specialities: Annuals/Biennials/Bedding, Herbaceous Perennials

38 Spring Farm Nursery

Northampton Road, Litchborough, Towcester,
Northamptonshire NN12 8JB
Tel: 01327 830799
Specialities: Annuals/Biennials/Bedding, Shrubs

39 Tandee Nursery

Barnwell Road, Barnwell, Peterborough,
Northamptonshire PE8 5RJ
Tel: 01832 293755
Specialities: Shrubs, Herbaceous Perennials

40 Village Farm Nurseries

Onley Lane, Barby, Rugby,
Northamptonshire CV23 8TB
Tel: 01788 891608
Specialities: Annuals/Biennials/Bedding, Hanging Baskets, Shrubs

41 West Lodge Plants and Gardens Ltd

West Lodge Farm, Back Lane, Desborough,
Kettering, Northamptonshire NN14 2SH
Tel: 01536 764255

42 Whilton Locks Garden Store

Whilton Locks, Whilton, Daventry,
Northamptonshire NN11 2NH
Tel: 01327 843100

43 Willows Nursery

Walgrave, Northampton,
Northamptonshire NN6 9QA
Tel: 01536 791371
Specialities: Herbaceous Perennials, Shrubs

44 Workbridge Centre

The Old Laundry, Bedford Road, Northampton,
Northamptonshire NN4 7AD
Tel: 01604 823476
Specialities: Garden Products, Restaurant/Coffee Shop

45 Wothorpe Nurseries

Kettering Rd, Wothorpe Hill, Stamford,
Northamptonshire PE9 3JG
Tel: 01780 763268
Specialities: Garden Products

46 Wyevale Garden Centre (Harlestone Heath)

Harlestone Road, Northampton,
Northamptonshire NN5 6UJ
Tel: 01604 751346

47 Wyevale Garden Centre (Northampton)

Newport Pagnell Road, Wootton, Northampton,
Northamptonshire NN4 6HP
Tel: 01604 765725

48 Youngs Nurseries

33 Northampton Road, Blisworth, Northampton,
Northamptonshire NN7 3DW
Tel: 01604 858574
Specialities: Alpines/Rock Plants, Floristry, Rose

NORTHUMBERLAND

1 Azure Garden Centre

Station Road, Cramlington,
Northumberland NE23 8BJ
Tel: 01670 733762

2 Belsay Hall

Belsay, Newcastle Upon Tyne,
Northumberland NE20 0DX
Tel: 01661 881069

Situated just west of Berwick-on-Tweed and clearly sign-posted off the A1, **Berwick Garden Centre** is well worth a visit. It has a comprehensive range of plants, trees, shrubs and gardening accessories and specialises in aquatics. The garden centre is open 7 days a week and has an excellent licensed restaurant and coffee shop.

Aquatics
Barbeques
Books & Stationery
Seasonal Displays
Clothing
Coffee Shop
Conservatories
Credit Cards Accepted
Disabled Facilities
Farm Shop
Fencing
Fireworks
Fruit Trees
Garden Furniture
Garden Machinery
Garden Products
Giftware
Greenhouses and Sheds
Houseplants
Information Desk
Large Car Park
Licensed Restaurant
Paving
Pet Products
Plant Guarantee
Plants
Seeds and Bulbs
Trees

FOR ALL YOUR GARDEN PLANTS.....

Berwick Garden Centre

East Ord, Berwick-upon-Tweed TD15 2NS
Tel: 01289 302383
e-mail: info@berwick-gardencentre.co.uk
website: www.berwick-gardencentre.co.uk

Opening Times
8.00am - 6.00pm Monday to Saturday
11.00am - 5.00pm Sundays

3 Berwick Garden Centre

East Ord, Berwick-upon-Tweed TD15 2NS
Tel: 01289 302383
e-mail: info@berwick-gardencentre.co.uk
website: www.berwick-gardencentre.co.uk

SEE ADVERTISEMENT PAGE 204

Specialities:

Aquatics, Farm Shop, Licensed Restaurant

Products and Services:

- Aquatics
- Barbeques
- Books & Stationery
- Seasonal Displays
- Clothing
- Coffee Shop
- Conservatories
- Credit Cards Accepted
- Disabled Facilities
- Farm Shop
- Fencing
- Fireworks
- Fruit Trees
- Garden Furniture
- Garden Machinery
- Garden Products
- Giftware
- Greenhouses and Sheds
- Houseplants
- Information Desk
- Large Car Park
- Licensed Restaurant
- Paving
- Pet Products
- Plant Guarantee
- Plants
- Seeds and Bulbs
- Trees

Directions:

Follow brown signs on A1 by-pass of Berwick-upon-Tweed to Berwick Garden Centre

4 Bide-A-Wee Cottage Gardens

Stanton, Stantonbury, Morpeth,
Northumberland NE65 8PR
Tel: 01670 772238

5 Cagro Horticulture

Catmire Plantation, Rothley, Morpeth,
Northumberland NE61 4JX
Tel: 01670 774407
e-mail: s.askew@tiscali.co.uk
website: www.members.lycos.co.uk/bigrothley

Specialities:

Acers (Maples); Ornamental Trees and Shrubs; Garden Plants (Container Grown)

Description:

Cargo Horticulture is located at Catmire Plantation which is part of a Capability Brown landscape. The 15-acre woodland nursery is located near Rothley Castle –a local visitor attraction – and Wallington Hall.

Directions:

From Morpeth, take the B6343 to about 1 mile west of Scots Gap, then the B6342 to Rothley/Rothbury. At Rothley crossroads, turn right on minor road signposted Longwitton/Morpeth. Catmire Plantation is 300 yards along this road

6 Chipchase Castle Nursery

Wark, Hexham, Northumberland NE48 3NT
Tel: 01434 230083
Specialities: Geranium

7 Clifton Plants

Kirkley Mill Farm Yard, Kirkley Mill, Berwick Mill, Ponteland, Newcastle Upon Tyne NE20 0BQ
Tel: 01661 820033 Fax: 01661 820034
e-mail: graeme@cliftonplants.com
website: www.cliftonplants.com
Specialities: Fuchsias

8 Craster Towers Stable Yard Farm Shop & Nursery

Craster Tower, Craster, Alnwick,
Northumberland NE66 3SS
Tel: 01665 576898
Specialities: Farm Shop

9 Dobbies Garden World

Street House Farm, Street Houses, Ponteland,
Newcastle Upon Tyne, Northumberland NE20 9BT
Tel: 01661 820202

10 Down To Earth Garden Centre

Tyne Mills Industrial Estate, Hexham,
Northumberland NE46 1XL
Tel: 01434 600920

11 Ford Nursery

Castle Gardens, Berwick-upon-Tweed,
Northumberland TD15 2PZ
Tel: 01890 820379
Specialities: Shrubs, Herbaceous Perennials, Trees

12 The Garden Centre

Wheatridge, Seaton Delaval, Whitley Bay,
Northumberland NE25 0QW
Tel: 01912 370246

13 Glendale Garden Centre

7-9 South Road, Wooler,
Northumberland NE71 6QE
Tel: 01668 281003
Specialities: Pet Products, Annuals/Biennials/Bedding, Herbaceous Perennials

14 Greenscene Garden Design

The Smithy, Old Great North Road, Blagdon,
Newcastle Upon Tyne NE13 6DQ
Tel: 01670 789688
e-mail: scenegreen@aol.com
website: www.greenscenegardendesign.co.uk
Specialities: Garden Centre, Garden Design and Construction

15 Halls of Heddon

West Heddon Nursery, Heddon-on-The-Wall, Newcastle Upon Tyne, Northumberland NE15 0JS
Tel: 01661 852445
Specialities: Annuals/Biennials/Bedding

16 Halls of Heddon

The Nursery, Ovington, Prudhoe, Northumberland NE42 6EE
Tel: 01661 832467
Specialities: Aster (Michaelmas Daisy)

17 Heighley Gate Garden Centre

Morpeth, Northumberland NE61 3DA
Tel: 01670 513416
Specialities: Farm Shop, Restaurant/Coffee Shop, Annuals/Biennials/Bedding

18 Herterton House Garden and Nursery

Hartington, Cambo, Morpeth, Northumberland NE61 4BN
Tel: 01670 774278

19 Hexam Herbs

Chesters Walled Garden, The Chesters Humshaugh, Hexham, Northumberland NE46 4BQ
Tel: 01434 681483
Specialities: Herbs

20 Northern Ark Nursery

West Road, Longhorsley, Morpeth, Northumberland NE65 8RA
Tel: 07967 317344
e-mail: northernark@tiscali.co.uk
Specialities: Truly Hardy Perennials and Shrubs in Peat Free compost; Mail Order Available

21 Ryal Nursery

East Farm Cottage, Ryal, Northumberland NE20 0SA
Tel: 01661 886562 Fax: 01661 886918
e-mail: alpines@ryal.freeserve.co.uk
Specialities: Alpines

22 Shades of Green Garden Centre Ltd

North Seaton Industrial Estate, Freeman Way, Ashington, Northumberland NE63 0YB
Tel: 01670 850818
Specialities: Garden Furniture, Pots

23 Shield Green Nurseries

Shield Green, Hexham, Northumberland NE46 2JU
Tel: 01434 600191

24 Stannington Nurseries Ltd

58 Station Road, Stannington, Morpeth, Northumberland NE61 6NH
Tel: 01670 789377
Specialities: Ornamentals, Shrubs, Alpines, Tomatoes

25 Tyne Valley Garden Centre

Mickley Square, Mickley, Stocksfield, Northumberland NE43 7BT
Tel: 01661 843263
Specialities: Aquatics, Farm Shop, Annuals/Biennials/Bedding

26 William Wilson

Lane Nurseries, Callerton Lane, Ponteland, Newcastle Upon Tyne, Northumberland NE20 9EG
Tel: 01661 822333
Specialities: Annuals/Biennials/Bedding, Herbaceous Perennials

27 The Willows Garden Centre Ltd

Unit 1 Willowburn Trading Est., Alnwick, Northumberland NE66 2PF
Tel: 01665 605533
Specialities: Garden Furniture

NOTTINGHAMSHIRE

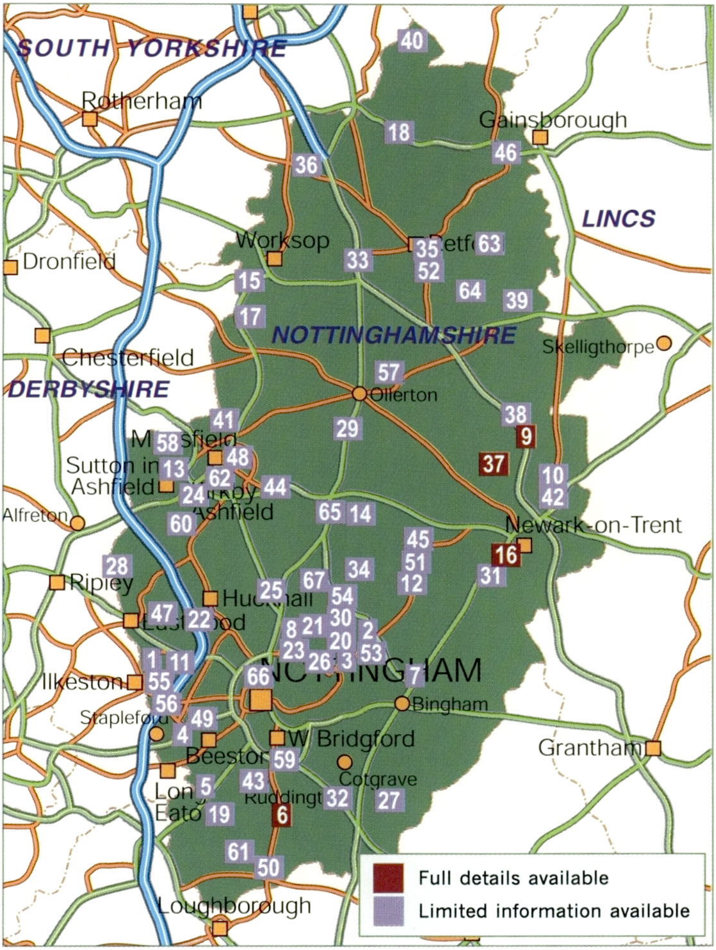

1 Andersen's Nurseries

Awsworth Lane, Cossall, Nottingham,
Nottinghamshire NG16 2SA
Tel: 01159 301884
Specialities: Exotic Fruit Trees, Grasses

2 Aquatic Centre

Lowdham Road, Bulcote, Nottingham,
Nottinghamshire NG14 5GT
Tel: 01159 312986
Specialities: Aquatics, Water Products and Features

3 Ashdale Cottage Garden Plants

204 Lambley Lane, Gedling, Nottingham,
Nottinghamshire NG4 4PB
Tel: 01159 666060
Specialities: Cottage Garden Plants

4 Bardhills Garden Centre

Toton Lane, Stapleford, Nottingham,
Nottinghamshire NG9 7JB
Tel: 01159 490019
Specialities: Garden Products

5 Barton Nurseries

Manor Road, Barton-in-Fabis, Nottingham, Nottinghamshire NG11 0AA
Tel: 01159 830233
Specialities: Annuals/Biennials/Bedding, Shrubs

6 Bradmore Garden Centre

Pendock Lane, Bradmore, Nottingham NG11 6PQ
Tel: 0115 984 7990 Fax: 0115 940 6175
e-mail: bradm3@yahoo.co.uk
website: www.bradmoregardencentre.co.uk

SEE ADVERTISEMENT ABOVE

Specialities:

Full Service Garden Centre; Coffee Shop

Description:

Family owned and operated, Bradmore Garden Centre has been in business since 1989 and provides full customer service from its modern retail outlet. In addition to a wide selection of plants and trees, the Centre also has a stylish coffee shop.

Directions:

Bradmore is on the A60, about 6 miles south of Nottingham. Coming from Nottingham, go through the village and then take the second turn on the left.

7 Bridgeford Garden Centre

Fosse Road, East Bridgford, Nottingham, Nottinghamshire NG13 8LA
Tel: 01949 20055
Specialities: Aquatics, Giftware, Herbaceous Perennials

8 Brookfields Garden Centre

431 Mapperley Plains, Nottingham, Nottinghamshire NG3 5RW
Tel: 01159 268200
Specialities: Annuals/Biennials/Bedding, Conservatories, Paving

9 Carlton Walled Garden Nursery

Church Lane, Carlton-on-Trent, Newark NG23 6LP
Tel: 01636 822282

SEE ADVERTISEMENT PAGE 209

Specialities:

2 Acres of Display Gardens; Rare & Unusual Perennials and Shrubs; Shade-loving Plants

Description:

Carlton Walled Garden Nursery offers 2 acres of inspirational planting, including many rare and unusual perennials. A large area is devoted to various shade-loving plants and there's also a herb

and vegetable maze set in a square of old Victorian fruit trees. A new white garden sits next to established old fig trees and bamboos and grasses grow beside a 'dry-river bed' path leading to an ornamental Victorian pond.

Directions:

Carlton-on-Trent is 1 minute from the A1, about 7 miles north of Newark. The Nursery is signposted from this junction

10 Cedar Lodge Nurseries

White Moor Lane, Collingham, Newark, Nottinghamshire NG23 7QH
Tel: 01636 893360

11 Cool Temperate Nursery

Trinity Farm, Awsworth Lane, Cossall, Nottingham, Nottinghamshire NG16 2RZ
Tel: 01159 162673
Specialities: Hedging, Fruit Trees

12 Crink Lane Nurseries

Crink Lane, Southwell, Nottinghamshire NG25 0QH
Tel: 01636 812706
Specialities: Annuals/Biennials/Bedding, Shrubs

13 Dalesforth Garden Centre Ltd

Skegby Lane, Sutton-in-Ashfield, Nottinghamshire NG17 3DH
Tel: 01623 557817
Specialities: Annuals/Biennials/Bedding, Turf, Greenhouses and Sheds

14 Dalesforth Nurseries

Cockett Lane, Farnsfield, Newark, Nottinghamshire NG22 8LF
Tel: 01623 883187
Specialities: Annuals/Biennials/Bedding, Herbaceous Perennials, Rose

15 Darfoulds Nursery

High Park, Darfoulds, Worksop, Nottinghamshire S80 3DU
Tel: 01909 472881
Specialities: Annuals/Biennials/Bedding

16 Devon Croft Nurseries

81 Farndon Road, Newark,
Nottinghamshire NG24 4SQ
Tel/Fax: 01636 704013
e-mail: james@devoncroftnursery.co.uk
website: www.devoncroftnursery.co.uk

SEE ADVERTISEMENT ABOVE

Specialities:

Summer Bedding, Hanging Baskets, Hostas, Bamboo

Products and Services:

- Plants
- Trees & Shrubs
- Seeds and Bulbs
- Fruit Trees
- Garden Products
- Credit Cards Accepted
- Large Car Park

Directions:

From the Newark by-pass (A46), take the B6166 (Farndon Road) into Newark. The nursery is sign-posted half a mile on the right hand side.

17 Dukeries Garden Centre

Welbeck, Welbeck, Worksop,
Nottinghamshire S80 3LT
Tel: 01909 476506

18 Everton Nurseries

Gainsborough Road, Everton, Doncaster,
Nottinghamshire DN10 5BW
Tel: 01777 817851
Specialities: Annuals/Biennials/Bedding, Herbaceous Perennials

19 Field House Nurseries

Leake Road, Gotham, Nottingham,
Nottinghamshire NG11 0JN
Tel: 01159 830278
Specialities: Primula, Seeds

20 Flamstead Nurseries

Park Lane, Lambley, Nottingham,
Nottinghamshire NG4 4PY
Tel: 01159 312547
Specialities: Herbaceous Perennials, Shrubs

21 Floralands Garden Centre

Catfoot Lane, Lambley, Nottingham,
Nottinghamshire NG4 4QL
Tel: 01159 268137

22 Garden Mania

Unit 1 The Aerodrome, Watnall Road, Hucknall, Nottingham, Nottinghamshire NG15 6EN
Tel: 01159 680780
Specialities: Annuals/Biennials/Bedding, Shrubs

23 Greenacres Nursery

322 Spring Lane, Mapperley, Nottingham, Nottinghamshire NG3 5RQ
Tel: 01159 262951
Specialities: Annuals/Biennials/Bedding, Herbaceous Perennials

24 Greenhills Garden Centre

Cauldwell Road, Sutton-in-Ashfield, Nottinghamshire NG17 5LB
Tel: 01623 554418
Specialities: Garden Products, Trees, Herbaceous Perennials

25 Greenwood Gardens

Ollerton Road, Arnold, Nottingham, Nottinghamshire NG5 8PR
Tel: 01159 205757
Specialities: Bonsai

26 Hilltop Nurseries

Lambley Lane, Gedling, Nottingham, Nottinghamshire NG4 4PB
Tel: 01159 612054
Specialities: Annuals/Biennials/Bedding, Shrubs

27 Hortors Nurseries and Farm Shop

Grove Farm, Hickling Lane, Kinoulton, Nottingham, Nottinghamshire NG12 3ED
Tel: 01949 81312
Specialities: Hanging Baskets

28 Jacksdale Garden Centre Ltd

Main Road, Jacksdale, Nottingham, Nottinghamshire NG16 5HR
Tel: 01773 605215
Specialities: Products, Restaurant/Coffee Shop, Pots

29 Lakeside Garden Shop

Rufford Country Park, Ollerton, Newark, Nottinghamshire NG22 9DG
Tel: 01623 824153
Specialities: Garden Furniture, Garden Products, Giftware

30 Lily Vale Nurseries

Park Lane, Lambley, Nottingham, Nottinghamshire NG4 4PY
Tel: 01159 313506
Specialities: Annuals/Biennials/Bedding, Shrubs

31 Liz Garden Centre

Hawton Lane, Farndon, Newark, Nottinghamshire NG24 3SD
Tel: 01636 610199
Specialities: Garden Products, Trees

32 Moores Nurseries & Garden Centre Ltd

156 Melton Road, Stanton-on-The-Wolds Keyworth, Nottingham, Nottinghamshire NG12 5BQ
Tel: 01159 373717
Specialities: Annuals/Biennials/Bedding, Garden Products

33 Morton Nurseries

Ranby, Retford, Nottinghamshire DN22 8HW
Tel: 01777 702530

34 New Hollybeck Nurseries

Hollybeck House, Oxton Road, Southwell, Nottinghamshire NG25 0RW
Tel: 01159 655450

35 Normans Nurseries

London Road, Retford, Nottinghamshire DN22 7JE
Tel: 01777 703301

36 Northern Garden Centre

Blyth Road, Oldcotes, Worksop, Nottinghamshire S81 8JE
Tel: 01909 731600
Specialities: Garden Furniture, Garden Products

37 Norwell Nurseries & Gardens

Woodhouse Road, Norwell, Nr Newark, Nottinghamshire NG23 6JX
Tel: 01636 636337
e-mail: wardha@aol.com
website: www.Norwellnurseries.co.uk

Specialities:

Rare & Unusual Herbaceous Perennials, Cottage Garden & Shade Plants

Description:

A nursery set within an acre of beautiful gardens containing over 2,000 different plants. Colour themed beds, woodland, scree and stunning

herbaceous borders hold propagating material for the inspirational nursery. Lectures nationally. Recommended in Gardeners Favourite Nurseries and the Good Gardens Guide.

Directions:

Six miles north of Newark and just 1.5 miles from A1

Open daily:

(10am-5pm) March until Oct 20th except Sat & Tues. (Also open Sat in May and June). (Closed August).

38 Old England Nurseries

Barrel Hill Road, Sutton-on-Trent, Newark, Nottinghamshire NG23 6PR
Tel: 01636 821329
Specialities: Annuals/Biennials/Bedding, Shrubs

39 Orchard Nursery Garden Centre

Main Street, Laneham, Retford, Nottinghamshire DN22 0ND
Tel: 01777 228882

40 Palms Springs

Prospect Farm, Misson Springs, Doncaster, Nottinghamshire DN10 6ET
Tel: 01302 772139
Specialities: Grasses, Bamboos, Hardy Garden Palms

41 Park Hall Plants

The Walled Garden, Park Hall Road, Mansfield Woodhouse, Mansfield, Nottinghamshire NG19 8QX
Tel: 01623 629071

42 Pearson's Nurseries

White Moor Lane, Collingham, Newark, Nottinghamshire NG23 7QH
Tel: 01636 893030
Specialities: Annuals/Biennials/Bedding, Shrubs

43 Quality Conifers

9a St. Marys Crescent, Ruddington, Nottingham, Nottinghamshire NG11 6FQ
Tel: 01158 461548
Specialities: Conifers

44 Rainworth Garden Centre Ltd

Southwell Road East, Rainworth, Mansfield, Nottinghamshire NG21 0EW
Tel: 01623 796316
Specialities: Hit Tubs

45 Reg Taylor Nurseries

Corkhill Lane, Normanton, Southwell, Nottinghamshire NG25 0PR
Tel: 01636 813184
Specialities: Customer Service

46 Retford & Gainsborough Garden Centre

Bar Road, Saundby, Retford, Nottinghamshire DN22 9EU
Tel: 01427 848784
Specialities: Garden Products

47 Reuben Shaw & Sons

Hollydene Nursery, Moorgreen, Newthorpe, Nottingham, Nottinghamshire NG16 2FF
Tel: 01773 714326
Specialities: Floristry, Annuals/Biennials/Bedding

48 Sherwoods

7-9 Sherwood Hall Road, Mansfield, Nottinghamshire NG18 2NH
Tel: 01623 624923
Specialities: Paving, Fencing

49 Showglads

105 Derby Road, Bramcote, Nottingham, Nottinghamshire NG9 3GZ
Tel: 01159 255498
Specialities: Gladiolus

50 Six Acre Nurseries

Loughborough Road, Costock, Loughborough, Nottinghamshire LE12 6XB
Tel: 01509 856079

51 Southwell Garden Centre

Fiskerton Road, Southwell, Nottinghamshire NG25 0TH
Tel: 01636 812886

52 T J Barker & Son

London Road, Retford, Nottinghamshire DN22 7JG
Tel: 01777 703165

53 Tall Trees Garden Centre

Main Road, Bulcote, Nottingham, Nottinghamshire NG14 5GT
Tel: 01159 312356
Specialities: Annuals/Biennials/Bedding, Aquatics, Restaurant/Coffee Shop

54 Timmermans Garden Centre

Lowdham Lane, Woodborough, Nottingham,
Nottinghamshire NG14 6DN
Tel: 01159 664033

55 Trinity Farm Organics

Awsworth Lane, Cossall, Nottingham,
Nottinghamshire NG16 2RZ
Tel: 01159 442545
Specialities: Organic

56 Trowell Garden Centre

Stapleford Road, Trowell, Nottingham,
Nottinghamshire NG9 3TG
Tel: 01159 326920
Specialities: Garden Products, Garden Furniture, Giftware

57 Walesby Garden Centre

Sandhurst, Brake Road, Walesby, Newark,
Nottinghamshire NG22 9NQ
Tel: 01623 860382
Specialities: Garden Products

58 Westfield Garden Centre

Beck Lane, Skegby, Sutton-in-Ashfield,
Nottinghamshire NG17 3AH
Tel: 01623 554515
Specialities: Greenhouses and Sheds, Garden Machinery, Shrubs

59 Wheatcroft Garden Centre

Landmere Lane, Edwalton, Nottingham,
Nottinghamshire NG12 4DE
Tel: 01159 216060

60 Wilbourn Garden Centre

60 Station Street, Kirkby-in-Ashfield, Nottingham,
Nottinghamshire NG17 7AU
Tel: 01623 753100

61 Wilford Bulb Co Ltd

69-71 Main Street, East Leake, Loughborough,
Nottinghamshire LE12 6PF
Tel: 01509 852905
Specialities: Lilium (Lily), Shrubs, Trees

62 Wilsons

13 Garth Rd, Mansfield,
Nottinghamshire NG18 5AQ
Tel: 01623 625827
Specialities: Hanging Baskets

63 Windmill View Garden Centre

Retford Road, South Leverton, Retford,
Nottinghamshire DN22 0HE
Tel: 01427 880965
Specialities: Garden Products, Hanging Baskets, Restaurant/Coffee Shop

64 Windrush Nurseries

Longacre, Upton, Retford,
Nottinghamshire DN22 0RA
Tel: 01777 248709
Specialities: Herbaceous Perennials, Shrubs

65 Wonderland Garden Centre

White Post, Farnsfield, Newark,
Nottinghamshire NG22 8HX
Tel: 01623 883395
Specialities: Garden Furniture, Annuals/Biennials/Bedding

66 The Worm That Turned

120-122 Derby Road, Nottingham,
Nottinghamshire NG1 5FB
Tel: 01159 482800

67 Yeomans' Exotics

2 Carrington Lane, Calverton, Nottingham,
Nottinghamshire NG14 6HQ
Tel: 01159 654350
Specialities: Exotics, Rare and Unusual Plants

OXFORDSHIRE

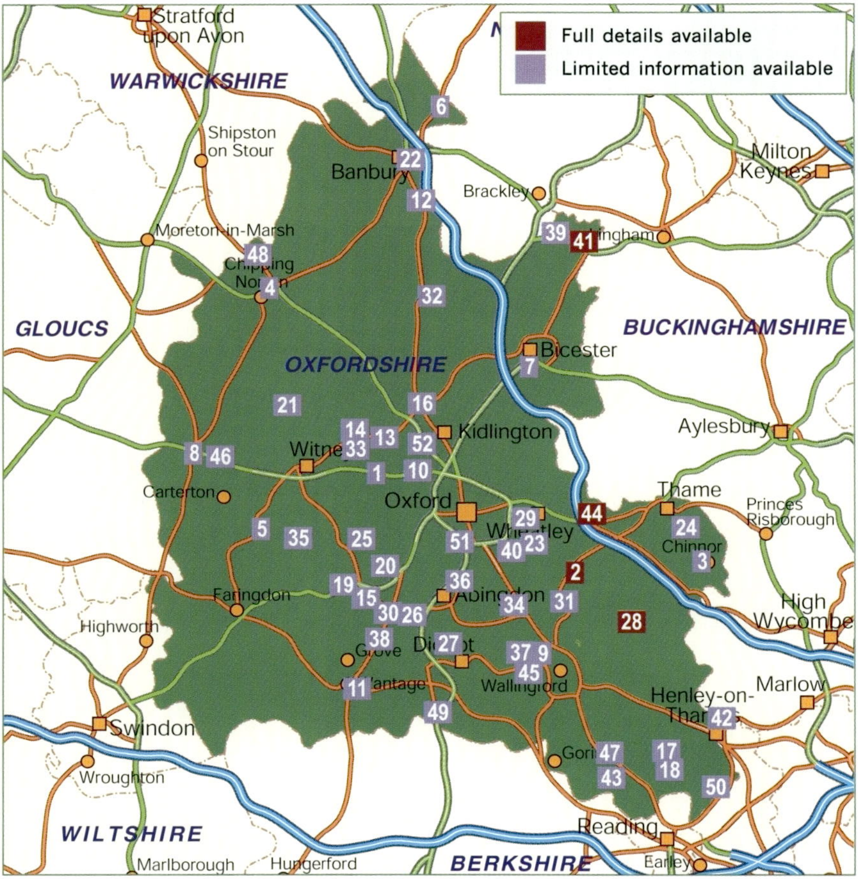

1 A40 Nursery & Garden Centre

67 Old Witney Road, Eynsham, Witney, Oxfordshire OX29 4PU
Tel: 01865 883394
Specialities: Garden Products

2 Acorn Nursery

Milton Road, Stadhampton, Oxford OX44 7XX
Tel: 01865 891663
e-mail: alan@acornplants.co.uk
website: www.acornplants.co.uk

Specialities:

Hardy Perennials; Shrubs & Grasses; Herbs

Description:

Acorn Nursery is a small high quality family-run plant nursery and small holding. It sells a huge variety of plants and trees, as well as free range eggs. The Nursery has built a reputation for stocking even the most rare and unusual plants.

Directions:

From Junction 7 of the M40, take the A329 towards Stadhampton. Continue along this road for 4 miles and you will come to Acorn Nursery.

3 The Andrews Garden Co

Grafton Orchard, Chinnor, Oxfordshire OX39 4DS
Tel: 01844 353540

4 Applegarth Nurseries

Hardy Plant Centre, Banbury Road,
Chipping Norton, Oxfordshire OX7 5SY
Tel: 01608 684990
Specialities: Annuals/Biennials/Bedding, Herbaceous Perennials

5 Bampton Garden Plants

The Nursery, Buckland Road, Bampton,
Oxfordshire OX18 2AA
Tel: 01993 852233

6 Barn Farm Plants

Wardington, Banbury, Oxfordshire OX17 1SN
Tel: 01295 758080
Specialities: Annuals/Biennials/Bedding, Herbaceous Perennials, Garden Furniture

7 Blooms of Bressingham

Oxford Road, Bicester, Oxfordshire OX25 2NY
Tel: 01869 242248

8 Burford Garden Company

Shilton Road, Burford, Oxfordshire OX18 4PA
Tel: 01993 823117

9 Busy Baskets

142 Wantage Road, Wallingford,
Oxfordshire OX10 0LU
Tel: 01491 833831

10 Cassington Nurseries

Yarnton Road, Cassington, Witney,
Oxfordshire OX29 4DY
Tel: 01865 882550
Specialities: Hanging Baskets, Annuals/Biennials/Bedding

11 Charlton Park Garden Centre

Charlton Road, Wantage, Oxfordshire OX12 8EP
Tel: 01235 772700

12 Cotefield Nurseries

2 Cotefield Cottages, Oxford Road, Bodicote,
Banbury, Oxfordshire OX15 4AQ
Tel: 01295 275222
Specialities: Annuals/Biennials/Bedding, Herbaceous Perennials

13 Fieldfare Nurseries

Church Road, Church Hanborough, Witney,
Oxfordshire OX29 8AA
Tel: 01993 703554

14 Freeland Nurseries

Wroslyn Road, Freeland, Witney,
Oxfordshire OX29 8AQ
Tel: 01993 881430

15 Frosts at Millets Farm

Kingston Rd, Frilford, Abingdon,
Oxfordshire OX13 5HB
Tel: 01865 391923

16 The Good Value Plant Nursery

2 Bunkers Hill, Kidlington, Oxfordshire OX5 3BA
Tel: 01865 372452
Specialities: Garden Products

17 Greenshoots

Peppard Common, Henley-on-Thames,
Oxfordshire RG9 5LB
Tel: 01491 628933
Specialities: Annuals/Biennials/Bedding, Herbaceous Perennials

18 The Herb Farm

Peppard Road, Sonning Common, Reading,
Oxfordshire RG4 9NJ
Tel: 01189 724220
Specialities: Herbs

19 The Herb Garden

Kingston House Estate, Race Farm Lane,
Kingston Bagpuize, Oxfordshire OX13 5AU
Tel: 01235 224869
Specialities: Herbs, Lavender

20 Hill's Nurseries

Appleton Common, Appleton, Abingdon,
Oxfordshire OX13 5QN
Tel: 01865 862081

21 Hilltop Garden Centre & Nursery

Witney Road, Ramsden, Chipping Norton,
Oxfordshire OX7 3AS
Tel: 01993 868403
Specialities: Annuals/Biennials/Bedding, Garden Products

22 Hurrans Garden Centre

Compton Road, Banbury, Oxfordshire OX16 2PR
Tel: 01295 266300
Specialities: Restaurant/Coffee Shop

23 Jennings Farm Shop

Wheatley Rd, Garsington, Oxford, Oxfordshire OX44 9DY
Tel: 01865 361311
Specialities: Garden Products

24 Lashlake Nurseries

Chinnor Road, Towersey, Thame, Oxfordshire OX9 3QZ
Tel: 01844 212392

25 Little Brook Nurseries

Standlake Road, Northmoor, Witney, Oxfordshire OX29 5AY
Tel: 01865 300737
Specialities: Trees, Shrubs

26 Marcham Plants

Hyde Farm Nurseries, Abingdon Road, Marcham, Abingdon, Oxfordshire OX13 6NX
Tel: 01865 391054

27 Mathews

7 Harwell Road, Sutton Courtenay, Abingdon, Oxfordshire OX14 4BN
Tel: 01235 520648

28 Michael Gee's Market Garden

Garden Cottage, Pyrton, Watlington, Oxfordshire OX49 5AN
Tel: 01491 612489 / 612479

Specialities:

Bedding Plants; Hanging Baskets; Patio Plants

Description:

A delightful plant centre where Michael Gee has been growing a wide variety of quality bedding plants, perennials and vegetable plants for more than 30 years. Open daily all year round, the centre also stocks an extensive selection of composts, large grow bags, bark, Fison's Tomorite, Bamboo canes and many other garden accessories.

Directions:

From Junction 6 of the M40, take the B4009 towards Watlington. About 3 miles along this road, turn right on minor road to Pyrton. Follow signs for the nursery which is close to the church.

29 Mill View Farm Plant Centre

Ladder Hill, Wheatley, Oxfordshire OX33 1HY
Tel: 01865 873488
Specialities: Annuals/Biennials/Bedding, Herbaceous Perennials, Shrubs

30 Murray Maclean

Collins Farm, Kingston Road, Frilford, Abingdon, Oxfordshire OX13 5NX
Tel: 01865 391242
Specialities: Trees, Shrubs, Hedging

31 Newington Nursery

Newington, Wallingford, Oxfordshire OX10 7AW
Tel: 01865 400533

32 Nicholson Nurseries Ltd

The Park, North Aston, Bicester, Oxfordshire OX25 6HL
Tel: 01869 340342
Specialities: Landscape Gardening, Annuals/Biennials/Bedding, Herbaceous Perennials

33 North Leigh Garden Centre

Park Road, North Leigh, Witney, Oxfordshire OX29 8AQ
Tel: 01993 881372
Specialities: Annuals/Biennials/Bedding, Herbaceous Perennials

34 Notcutts Garden Centre

Mattocks Roses, Nuneham Courtenay, Oxford, Oxfordshire OX44 9PY
Tel: 01865 343454

35 Ransons Nurseries

The Nursery, Cote, Bampton, Oxfordshire OX18 2EG
Tel: 01993 850521
Specialities: Rose, Annuals/Biennials/Bedding, Xmas Trees

36 Robert Mattock Roses

The Rose Nurseries, Lodge Hill, Abingdon, Oxfordshire OX14 2JD
Tel: 01865 391247
Specialities: Rose

37 Root One

High Road, Brightwell-cum-Sotwell, Wallingford, Oxfordshire OX10 0PT
Tel: 01491 836277

38 Steventon Road Nurseries

Steventon Road, East Hanney, Wantage,
Oxfordshire OX12 0HS
Tel: 01235 868828
Specialities: Herbaceous Perennials, Grasses

39 Swallows Nursery

Mixbury, Brackley, Oxfordshire NN13 5RR
Tel: 01280 847721
Specialities: Shrubs, Climbers, Trees

40 Thamesdown Farm Nurseries

155-157 Oxford Road, Garsington, Oxford,
Oxfordshire OX44 9AU
Tel: 01865 361265

41 Tile House Plants

Tile House Farm, Finmere, Buckingham MK18 4AS
Tel: 01280 848358
e-mail: tilehouseplants@fsmail.net
website: www.bestofbuckingham.com

SEE ADVERTISEMENT BELOW

Specialities:

Hardy Perennials; cottage garden plants

Description:

This family-run business specialising in hardy perennials and cottage garden plants also stocks a selection of more unusual plants. The range includes plants for the rockery, the gravel garden and raised beds, as well as the conventional herbaceous border. Cream teas.

Directions:

Finmere is on the A421, about 5 miles west of Buckingham

Follow the

42 Toad Hall Garden Centre

Marlow Road, Fawley, Henley-on-Thames,
Oxfordshire RG9 3AG
Tel: 01491 574615
Specialities: Annuals/Biennials/Bedding, Garden Furniture

43 Waterfield Nursery

Collins End, Goring Heath, Reading,
Oxfordshire RG8 7RH
Tel: 01491 681541
Specialities: Houseplants, Seasonal Plants

44 Waterperry Gardens

Waterperry, Nr Wheatley, Oxfordshire OX33 1JZ
Tel: 01844 339254/339226
e-mail: office@waterperrygardens.co.uk
website: www.waterperrygardens.co.uk

SEE ADVERTISEMENT BELOW

Specialities:

Quality Plant Centre, Large Ornamental Gardens, Teashop

Products and Services:

- Books and Stationery
- Seasonal Displays
- Clothing
- Credit Cards Accepted
- Disabled Facilities
- Display Gardens
- Fruit Trees
- Garden Furniture
- Garden Products
- Giftware
- Large Car Park
- Plant Guarantee
- Plants
- Restaurant/Coffee Shop
- Seeds and Bulbs
- Trees

Directions:

From London (M40) J8 and from the north J8a. Waterperry Gardens is 2.5 miles from the A40. Turn off at Wheatley and follow the brown signposts with the Tourist Board Rose symbol.

45 Wear's Nursery

High Road, Brightwell-cum-Sotwell, Wallingford, Oxfordshire OX10 0PS
Tel: 07790 425284
Specialities: Herbaceous Perennials, Geranium

46 Whitehill Farm Nursery

Whitehill Farm, Burford, Oxfordshire OX18 4DT
Tel: 01993 823218
Specialities: Ornamental Grasses, Bamboos

47 Woodcote Nursery & Garden Centre

Reading Road, Woodcote, Reading, Oxfordshire RG8 0QX
Tel: 01491 680335
Specialities: Seasonal Plants, Garden Products, Garden Furniture

48 Wyatts Countryside Centre

Hill Barn Farm, Great Rollright, Chipping Norton, Oxfordshire OX7 5SH
Tel: 01608 684835
Specialities: Garden Products, Farm Shop, Childrens Play Area

49 Wyevale Garden Centre (Chilton)

Newbury Road, Chilton, Didcot,
Oxfordshire OX11 0QN
Tel: 01235 833900

50 Wyevale Garden Centre (Henley)

Reading Road, Harpsden Woods, Harpsden,
Henley-on-Thames, Oxfordshire RG9 4AE
Tel: 01189 403078

51 Wyevale Garden Centre (Oxford)

Southern By-Pass, Oxford, Oxfordshire OX1 5AR
Tel: 01865 326066

52 Yarnton Nurseries Garden Centre

Sandy Lane, Yarnton, Kidlington,
Oxfordshire OX5 1PA
Tel: 01865 372124
Specialities: Garden Products, Restaurant/Coffee Shop, Giftware

RUTLAND

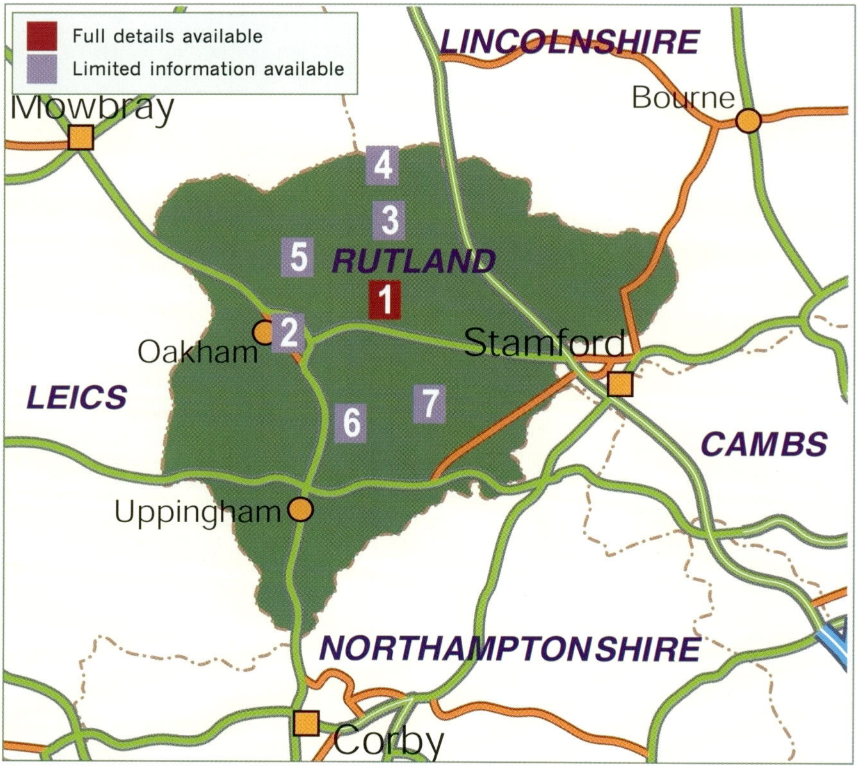

1 Barnsdale Gardens

The Avenue, Exton, Oakham, Rutland LE15 8AH
Tel: 01572 813200 Fax: 01572 813346
e-mail: office@barnsdalegardens.co.uk
website: www.barnsdalegardens.co.uk

SEE ADVERTISEMENT PAGE 221

Specialities:

Display Gardens; Specialist Plant Nursery.

Products and Services:

- Books, stationery & Gifts
- Seasonal Displays
- Credit Cards Accepted
- Disabled Facilities
- Plants by mail-order
- Information Desk
- Large Car Park
- Groups catered for
- Restaurant/Coffee Shop
- Year-round courses and events

Directions:

The Gardens are signposted with brown and white signs from the A606 about 3 miles east of Oakham

2 Grafters Home and Garden Centre

9 Melton Road, Oakham, Rutland LE15 6AX
Tel: 01572 759176

3 Greetham Garden Centre

Oakham Road, Greetham, Oakham, Rutland LE15 7NN
Tel: 01572 813100
Specialities: Clematis

4 The Herb Nursery

Thistleton, Main Street, Thistleton, Oakham, Rutland LE15 7RE
Tel: 01572 767658

5 Rutland Garden Centre

Ashwell Road, Ashwell, Nr Oakham, Rutland LE15 7QN
Tel: 01572 723583
Specialities: Annuals/Biennials/Bedding

6 Rutland Water Garden Nursery

Lyndon Road, Manton, Oakham, Rutland LE15 8RN
Tel: 01572 737711
Specialities: Hanging Baskets, Annuals/Biennials/Bedding

7 Tylers Garden Nursery

18 Well Cross, Edith Weston, Oakham, Rutland LE15 8HG
Tel: 01780 720072

SHROPSHIRE

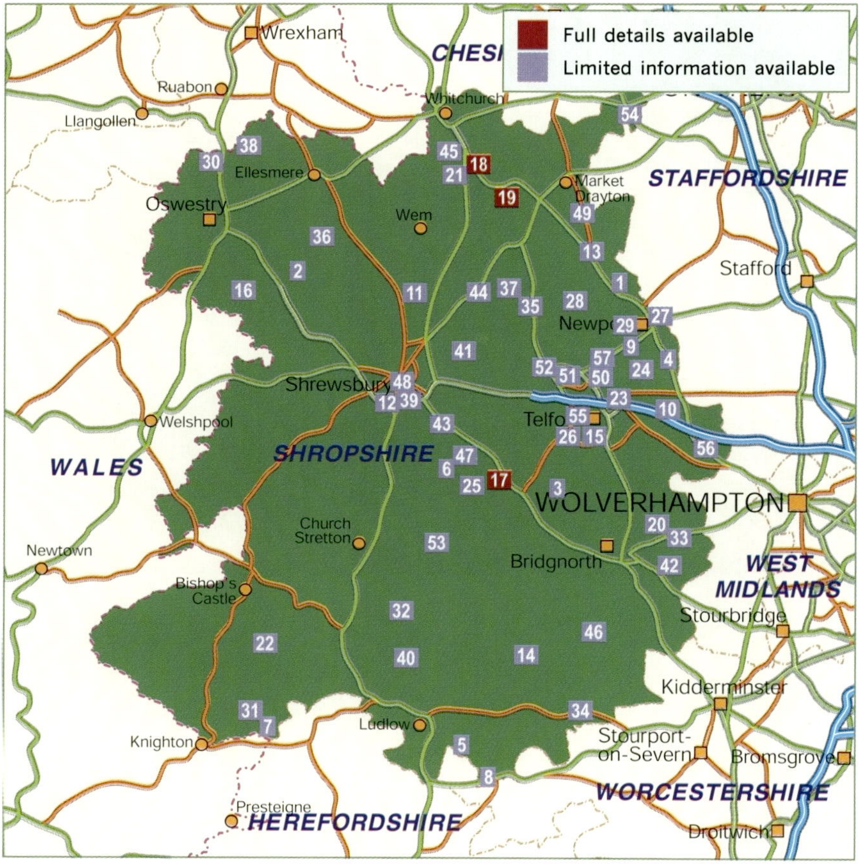

1 Barlow Nurseries

Aylesland House, Pickstock, Newport, Shropshire TF10 8AH
Tel: 01952 550933
Specialities: Annuals/Biennials/Bedding, Herbaceous Perennials

2 Barncroft Nurseries

Olden Lane, Ruyton XI Towns, Shrewsbury, Shropshire SY4 1JD
Tel: 01939 261619
Specialities: Garden Products

3 Bishtons Nursery

Posenhall, Broseley, Shropshire TF12 5BE
Tel: 01952 883766
Specialities: Herbaceous Perennials, Shrubs

4 Bloomsbury Nursery

Grange Acre, Bloomsbury, Shifnal, Shropshire TF11 8RT
Tel: 01952 691203
Specialities: Annuals/Biennials/Bedding, Shrubs

5 Bower Farm Nurseries

Bower Farm, Greete, Ludlow, Shropshire SY8 3BS
Tel: 01584 878081
Specialities: Shrubs, Herbaceous Perennials

6 Brooklands Nursery

Cound Moor, Cound, Shrewsbury,
Shropshire SY5 6BA
Tel: 01694 731667
Specialities: Climbers, Shrubs, Herbaceous Perennials

7 Bucknell Nurseries

Timber Yard, Bucknell, Shropshire SY7 0EL
Tel: 01547 530606
Specialities: Hedging, Shrubs, Trees

8 Burford Garden Company

Burford House, Burford, Tenbury Wells,
Shropshire WR15 8HQ
Tel: 01584 810777

9 Cheswell Nurseries

78 Wellington Road, Lilleshall, Newport,
Shropshire TF10 9AG
Tel: 01952 677324
Specialities: Shrubs

10 Claire Austin Hardy Plants

The Stone House, Cramp Pool, Shifnal,
Shropshire TF11 8PE
Tel: 01952 463700
Specialities: Herbaceous Perennials, Shrubs

11 Country Garden Plant Centre

Black Birches, Hadnall, Shrewsbury,
Shropshire SY4 3DH
Tel: 01939 210380
Specialities: Rose

12 Dobbies Garden World

Bayston Hill Nurseries, Hereford Road, Bayston Hill,
Shrewsbury, Shropshire SY3 0DA
Tel: 01743 874261

13 Garden Ornaments At Hinstock

Tettenhall House, Newport Road, Hinstock,
Market Drayton, Shropshire TF9 2TH
Tel: 01952 550870
Specialities: Garden Ornaments

14 The Gobbett Nursery

Farlow, Farlow, Kidderminster,
Shropshire DY14 8TD
Tel: 01746 718647
Specialities: Syringa (Lilac), Magnolia, Shrubs

15 Grange Farm Garden Centre

Stirchley Road, Stirchley, Telford,
Shropshire TF3 1EH
Tel: 01952 594346
Specialities: Annuals/Biennials/Bedding, Herbaceous Perennials

16 Hall Farm Nursery

Vicarage Lane, Kinnerley, Oswestry,
Shropshire SY10 8DH
Tel: 01691 682135
Specialities: Annuals/Biennials/Bedding, Shrubs

17 Harley Nursery & Plant Centre

Harley, nr Much Wenlock, Shropshire SY5 6LN
Tel: 01952 510241 Fax: 01952 510570
e-mail: njmurphy@hotmail.co.uk

Specialities:

Fruit trees; Ornamental Trees

Description:

Harley Nursery began in a small way as a Pick Your Own fruit farm in the 1970s. In 1990 it was bought by the Murphy family and the 6 acre site now offers a wide choice of fruit and ornamental trees, all types of plants, shrubs, and bedding plants. The range of fruit trees (including apples, pears, plums, damsons, peaches, nectarines and apricots) offers the best selection in Shropshire.

Directions:

Harley Nursery is on the A458 about 2 miles NW of Much Wenlock

18 Heathwood Nurseries

The Meadows, Higher Heath, Whitchurch,
Shropshire SY13 2JA
Tel: 01948 840120
e-mail: sales@heathwood.co.uk
website: www.heathwood.co.uk

SEE ADVERTISEMENT PAGE 224

Specialities:

Garden Shrubs; Hedging; Trees; Climbers & Fruit Trees

Description:

The nurseries stock an excellent range of home-grown shrubs, hedging, trees and climbers. They

grow plants from 20cms up to 20ft and everything in between. Heathwood Nurseries also stock pavers, slabs, gravel and tree planting sundries.

Directions:

Heathwood Nurseries is on the A41, about 5 miles south of Whitchurch.

19 Hillcrest Conifers & Garden Plants

Woodfield, Fauls Green, Market Drayton, Shropshire SY13 2BB
Tel: 01630 639851

Specialities:

Conifers; Trees & Shrubs; Herbaceous Plants

Description:

Hillcrest was established in 1992, specialising in semi-mature conifers, ornamental trees and unusual shrubs, many of them new introductions. Today, the nursery offers 750 varieties of plants and shrubs and 200 kinds of conifer, including drought-resistant species.

Directions:

From the A53/A41 junction about 3 miles south-west of Market Drayton, take the A53 for ½ mile. Then turn right towards Lostford, then first right to Fauls Green. After 2 miles the nursery is on the left hand side.

20 Hillview Hardy Plants

Worfield, Bridgnorth, Shropshire WV15 5NT
Tel: 01746 716454
Specialities: Herbaceous Perennials, Hardy Plants

21 Holly Farm Nurseries

Whitchurch Road, Prees, Whitchurch, Shropshire SY13 2DD
Tel: 01948 840630
Specialities: Annuals/Biennials/Bedding, Shrubs

22 Kempton Nurseries

Kempton, Lydbury North, Shropshire SY7 0JG
Tel: 01588 660250
Specialities: Annuals/Biennials/Bedding, Shrubs

23 Lakeside Plant Centre

Woodhouse Lane, Off Salisbury Avenue, Priorslee, Telford, Shropshire TF2 9SX
Tel: 01952 204466
Specialities: Herbaceous Perennials, Restaurant/ Coffee Shop, Farm Shop

24 Lilyhurst Plant Centre

Lilyhurst House, Lilyhurst, Shifnal,
Shropshire TF11 8RL
Tel: 01952 606546
Specialities: Annuals/Biennials/Bedding, Shrubs

25 Lower Spring Nursery

Kenley, Shrewsbury, Shropshire SY5 6PA
Tel: 01952 510589
Specialities: Annuals/Biennials/Bedding, Shrubs

26 Meadowdale Nurseries

Wellington Road, Horsehay, Telford,
Shropshire TF4 3BT
Tel: 01952 502470
Specialities: Shrubs, Herbaceous Perennials

27 Mere Park Garden Centre

Stafford Road, Newport, Shropshire TF10 9BY
Tel: 01952 813500

28 Mildene Nursery

Great Bolas, Telford, Shropshire TF6 6QD
Tel: 01952 541314
Specialities: Shrubs, Herbaceous Perennials

29 Millingtons Garden Centre

101 High Street, Newport, Shropshire TF10 7AY
Tel: 01952 811190
Specialities: Garden Products

30 Moreton Park Garden Centre

Gledrid, Chirk, Wrexham, Shropshire LL14 5DG
Tel: 01691 777722
Specialities: Garden Furniture

31 Mynd Hardy Plants

Heavens Above, Mynd, Bucknell,
Shropshire SY7 0BD
Tel: 01547 530459
Specialities: Herbaceous Perennials, Hardy Plants, Shrubs

32 Mynd Hardy Plants

Delbury Hall Estate, Diddlebury, Craven Arms,
Shropshire SY7 9DH
Tel: 01584 841222
Specialities: Herbaceous Perennials, Shrubs, Hardy Plants

33 New Barnes Nurseries

Bridgnorth Road, Hilton, Bridgnorth,
Shropshire WV15 5PB
Tel: 01746 710874
Specialities: Hedging, Herbaceous Perennials, Trees

34 Newbridge Nursery

Cleobury Mortimer, Cleobury Mortimer,
Kidderminster, Shropshire DY14 8HB
Tel: 01299 270701

35 Nobridge Nurseries

Cold Hatton, Telford, Shropshire TF6 6QB
Tel: 01952 541319
Specialities: Annuals/Biennials/Bedding

36 The Nursery

Weston Lullingfields, Shrewsbury,
Shropshire SY4 2AW
Tel: 01939 260307
Specialities: Annuals/Biennials/Bedding, Herbaceous Perennials

37 Oakgate Garden Centre and Nursery

Ellerdine, Telford, Shropshire TF6 6RL
Tel: 01939 250428
Specialities: Restaurant/Coffee Shop, Annuals/Biennials/Bedding, Garden Products

38 Pentre Nurseries

Pentre House, St. Martins, Oswestry,
Shropshire SY11 3EA
Tel: 01691 772475
Specialities: Garden Products, Hanging Baskets

39 Percy Throwers Garden & Leisure

Oteley Rd, Shrewsbury, Shropshire SY2 6QW
Tel: 01743 352311

40 Potters Farm Nursery

Lower Hayton, Ludlow, Shropshire SY8 2AY
Tel: 01584 823394
Specialities: Trees, Conifers

41 Roden Nurseries

Roden Lane, Roden, Telford, Shropshire TF6 6BP
Tel: 01952 770951

42 S.M.R Plant Nursery

Cleck Farm, Upper Farmcote, Bridgnorth, Shropshire WV15 5PS
Tel: 01746 710765
Specialities: Shrubs

43 Severndale Nurseries

Cross Houses, Shrewsbury, Shropshire SY5 6JR
Tel: 01743 761212
Specialities: Garden Products

44 Shawbury Garden Centre & Fruit Farm

Edgebolton, Shrewsbury, Shropshire SY4 4EL
Tel: 01939 251173

45 Springwood Nurseries

3 Steel Road, Tilstock, Whitchurch, Shropshire SY13 3LG
Tel: 01948 880397
Specialities: Herbaceous Perennials, Landscape Gardening

46 Stanley Farm Nurseries

Stanley Farm, Chorleywood, Bridgnorth, Shropshire WV16 6PS
Tel: 01746 718602
Specialities: Herbaceous Perennials, Shrubs

47 Stevenshill Nurseries

Harnage Cressage, Shrewsbury, Shropshire SY5 6EG
Tel: 01952 510255
Specialities: Annuals/Biennials/Bedding, Shrubs

48 Sutton Grange Nursery

Oteley Road, Shrewsbury, Shropshire SY2 6QL
Tel: 01743 355201

49 Sydnal Lane Nursery

Sydnal Lane, Woodseaves, Market Drayton, Shropshire TF9 2AS
Tel: 01630 653161
Specialities: Herbaceous Perennials, Shrubs

50 Tafs Garden Co

Gower Street, St. Georges, Telford, Shropshire TF2 9BQ
Tel: 01952 620184
Specialities: Garden Products, Greenhouses and Sheds, Fencing

51 Telford Garden Supplies

Hadley Park Road, Leegomery, Telford, Shropshire TF1 6PY
Tel: 01952 248955
Specialities: Garden Products

52 Watkins Nurseries

Apley Castle Gardens, Telford, Shropshire TF1 6DD
Tel: 01952 242393
Specialities: Annuals/Biennials/Bedding

53 Wenlock Edge Nurseries

Lushcott, Easthope, Much Wenlock, Shropshire TF13 6EE
Tel: 01694 771430
Specialities: Annuals/Biennials/Bedding, Herbaceous Perennials, Shrubs

54 Westholme Nurseries

London Road, Woore, Crewe, Shropshire CW3 9RQ
Tel: 01630 647289
Specialities: Hanging Baskets, Annuals/Biennials/Bedding, Herbaceous Perennials

55 Wildlife Flora

Princes End, Dawley Bank, Telford, Shropshire TF4 2JW
Tel: 01952 506206
Specialities: Wild Flowers

56 Wyevale Garden Centre (Albrighton)

Newport Road, Albrighton, Wolverhampton, Shropshire WV7 3EE
Tel: 01902 374200

57 Wyevale Garden Centre (Telford)

Station Road, Donnington, Telford, Shropshire TF2 8JY
Tel: 01952 677733

SOMERSET

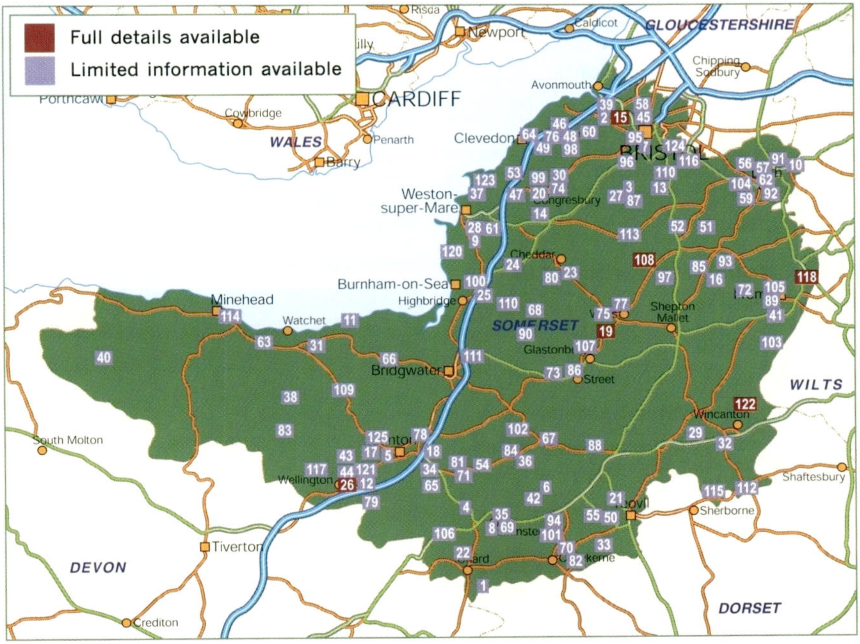

1 The Abbey Nursery

Forde Abbey, Forde Abbey, Chard,
Somerset TA20 4LU
Tel: 01460 220088
Specialities: Herbaceous Perennials

2 Anita's Nursery

The Tynings, Martcombe Road, Easton-in-Gordano,
Bristol, Somerset BS20 0QE
Tel: 01275 376379
Specialities: Garden Products, Restaurant/Coffee Shop

3 Arne Herbs

Limeburn Nurseries, Limeburn Hill, Chew Magna,
Bristol, Somerset BS40 8QW
Tel: 01275 333399
Specialities: Herbs, Wild Flowers

4 Ashill Nursery

Park Barn Lane, Park Barn, Ashill, Ilminster,
Somerset TA19 9LX
Tel: 01823 480869
Specialities: Shrubs, Herbaceous Perennials, Trees

5 Avery Nursery & Plant Centre

Silk Mills Road, Roughmoor, Taunton,
Somerset TA1 5AA
Tel: 01823 288324

6 Avon Bulbs

Burnt House Farm, Middle Lambrook,
South Petherton, Somerset TA13 5HE
Tel: 01460 242177
Specialities: Bulbs

7 Avon Pet & Garden Centre

47 Bedminster Down Road, Bristol,
Somerset BS13 7AB
Tel: 01179 663761

8 B D Garden & Pet Supplies

6 Ditton Street, Ilminster, Somerset TA19 0BQ
Tel: 01460 52967

9 Banwell Garden Centre

Castle Hill Road, Castle Hill, Banwell,
Somerset BS24 9NX
Tel: 01934 822246

35 Dillington Nursery

Dillington House, Ilminster, Somerset TA19 9DT
Tel: 01460 57908

36 Drayton Nurseries

School Street, Drayton, Langport,
Somerset TA10 0LN
Tel: 01458 251926
Specialities: Garden Design, Landscape Gardening

37 Elmsley Nursery

Lower Norton Lane, Kewstoke, Weston-super-Mare,
Somerset BS22 9YT
Tel: 01934 415333

38 Elworthy Cottage Plants

Elworthy, Lydeard St. Lawrence, Taunton,
Somerset TA4 3PX
Tel: 01984 656427
Specialities: Herbaceous Perennials, Cottage Garden Plants

39 Evergreen Garden

6 Baltic Place, Pill, Bristol, Somerset BS20 0EJ
Tel: 01275 371771

40 Exmoor Forest Nurseries

Gallon House, Simonsbath, Minehead,
Somerset TA24 7JY
Tel: 01643 831585
Specialities: Shrubs, Trees

41 Florabunda Nurseries

Battlemoor, Little Keyford, Frome,
Somerset BA11 5BE
Tel: 01373 461473
Specialities: Garden Products

42 Floral Fireworks

Burnt House Farm, Middle Lambrook,
South Petherton, Somerset TA13 5HE
Tel: 01460 249060

43 Ford Nursery

Bloom Lane, Oake, Taunton, Somerset TA4 1BE
Tel: 01823 461961
Specialities: Climbers

44 Foxmoore

Greenfields Nurseries, Chelston, Wellington,
Somerset TA21 9HY
Tel: 01823 662462
Specialities: Fuchsia, Herbaceous Perennials

45 Garaways Garden Centre

Chantry Road, Bristol, Somerset BS8 2QE
Tel: 01179 733402

46 Garden Park

100 Clevedon Road, Tickenham, Clevedon,
Somerset BS21 6RE
Tel: 01275 856675

47 Garden Park

Puxton Lane, Hewish, Weston-super-Mare,
Somerset BS24 6TA
Tel: 01934 876313

48 Glenbrook Bonsai Nursery

Stone-Edge Batch, Tickenham, Clevedon,
Somerset BS21 6SF
Tel: 01275 858596
Specialities: Bonsai

49 Golden Acres Nursery

Alvington Lane, Brympton, Yeovil,
Somerset BA22 8TH
Tel: 01935 475613
Specialities: Garden Products

50 Golden Acres Nursery

35 Clevedon Road, Tickenham, Clevedon,
Somerset BS21 6RA
Tel: 01275 853863
Specialities: Farm Shop, Restaurant/Coffee Shop

51 Graham's Hardy Plants

Southcroft, North Road, Timsbury, Bath,
Somerset BA2 0JN
Tel: 01761 472187
Specialities: Garden Products, Hardy Plants

52 The Great Western Gladiolus Nursery

17 Valley View, Clutton, Bristol,
Somerset BS39 5SN
Tel: 01761 452036
Specialities: Gladiolus

53 Greenholm Nurseries Ltd

Lampley Road, Kingston Seymour, Clevedon,
Somerset BS21 6XS
Tel: 01934 833350

54 Greenshutters Nurseries

Fivehead, Taunton, Somerset TA3 6PT
Tel: 01460 281265

55 Greensleeves Nurseries

Lower Odcombe, Lower Odcombe, Yeovil, Somerset BA22 8TW
Tel: 01935 864166
Specialities: Annuals/Biennials/Bedding, Farm Shop, Herbaceous Perennials

56 Heather Farm Nurseries

Lansdown Lane, Bath, Somerset BA1 4NA
Tel: 01225 334100

57 Hedges Direct Ltd

The Brow, Charlcombe Lane, Lansdown, Bath, Somerset BA1 5TP
Tel: 01225 318003
Specialities: Hedging

58 Henleaze Garden Shop

146 - 148 Henleaze Road, Bristol, Somerset BS9 4NB
Tel: 01179 620418
Specialities: Pet Products, Annuals/Biennials/Bedding

59 Hillier Garden Centre Bath

Whiteway Road, Bath, Somerset BA2 2RG
Tel: 01225 421162

60 Hillside Nurseries

West Hill, Wraxall, Bristol, Somerset BS48 1PH
Tel: 01275 855657

61 Hutton Garden Centre

Banwell Road, Hutton, Weston-super-Mare, Somerset BS24 9UB
Tel: 01934 822795
Specialities: Shrubs, Viburnum, Rose

62 Impact Plants

37 Eastbourne Avenue, Bath, Somerset BA1 6EN
Tel: 01225 447962
Specialities: Trees, Shrubs, Hedging

63 J D Plants

Washford, Watchet, Somerset TA23 0NT
Tel: 01984 640051
Specialities: Herbaceous Perennials, Shrubs

64 Jasmine Cottage Gardens

26 Channel Road, Clevedon, Somerset BS21 7BY
Tel: 01275 871850

65 Junker's Nursery

Lower Mead, West Hatch, Taunton, Somerset TA3 5RN
Tel: 01823 480774
Specialities: Hardy Plants, Acer, Magnolia

66 Keenthorne Nurseries

Keenthorne, Nether Stowey, Bridgwater, Somerset TA5 1HZ
Tel: 01278 732338

67 Kelways

Barrymore Farm, Picts Hill, Langport, Somerset TA10 9EZ
Tel: 01458 250521

68 Lands End Nurseries

Lands End Farm, Wedmore, Somerset BS28 4UQ
Tel: 01934 710287

69 Lanes Garden Shop

17 Silver Street, Ilminster, Somerset TA19 0DH
Tel: 01460 57703

70 Lower Severalls Garden and Nursery

Crewkerne, Somerset TA18 7NX
Tel: 01460 73234
Specialities: Herbaceous Perennials, Herbs

71 Mallet Court Nursery

Higher Street, Curry Mallet, Taunton, Somerset TA3 6SY
Tel: 01823 481493
Specialities: Magnolia, Quercus (Oak)

72 Meadows Nursery

5 Rectory Cottages, Selwood Street, Mells, Frome, Somerset BA11 3PN
Tel: 01373 812268
Specialities: Phlox

73 Michael Kings Ltd

Autumn Leaves, Whitley Lane, Walton, Nr Street, Somerset BA16 9RW
Tel: 01458 443066
Specialities: Composts

74 Middlecombe Nursery

Wrington Road, Congresbury, Bristol, Somerset BS49 5AN
Tel: 01934 876337
Specialities: Exotics, Shrubs, Herbaceous Perennials

75 Mill Cottage Plants

The Mill, Henley Lane, Wookey, Wells,
Somerset BA5 1AW
Tel: 01749 676966
Specialities: Hydrangea, Ferns

76 Miss J Lindsay

Clevedon Road, Tickenham, Clevedon,
Somerset BS21 6RE
Tel: 01275 852439
Specialities: Passiflora (Passion Flower), Garden Products

77 Mogg's of Wells Ltd

Cherry Orchard, Ash Lane, Wells,
Somerset BA5 2RU
Tel: 01749 672394
Specialities: Garden Furniture, Garden Machinery

78 Monkton Elm Garden Centre

Monkton Heathfield, Taunton, Somerset TA2 8QN
Tel: 01823 412381
Specialities: Garden Products, Aquatics, Pet Products

79 Mousehole Nurseries

Budgetts Cross, West Buckland, Wellington,
Somerset TA21 9LG
Tel: 01823 661112

80 Nethercoa Nurseries

Cheddar Road, Clewer, Wedmore,
Somerset BS28 4JE
Tel: 01934 742881

81 Newport Mills Nursery

Newport Mills Farm, Wrantage, Taunton,
Somerset TA3 6DJ
Tel: 01823 490231
Specialities: Herbaceous Perennials

82 North Perrott Garden Centre

Townsend, North Perrott, Crewkerne,
Somerset TA18 7SR
Tel: 01460 77090

83 North Street Garden Shop

6 North Street, Wiveliscombe, Taunton,
Somerset TA4 2JY
Tel: 01984 623256

84 North Wing Nurseries

Water Street, Curry Rivel, Langport,
Somerset TA10 0HH
Tel: 01458 253247
Specialities: Geranium, Hanging Baskets, Fuchsia

85 Norton Green Farm Nursery

Wells Road, Stratton-on-The-Fosse, Radstock,
Somerset BA3 4RR
Tel: 01761 232137

86 Oaklands Nursery

The Roundabout, Street, Somerset BA16 0EP
Tel: 01458 841585

87 Pagans Hill Nursery

Pagans Hill, Chew Magna, Bristol,
Somerset BS40 8QJ
Tel: 01275 331260
Specialities: Hanging Baskets, Annuals/Biennials/Bedding, Herbaceous Perennials

88 Patricia Marrow

Kingsdon Nursery, Kingsdon, Somerton,
Somerset TA11 7LE
Tel: 01935 840232
Specialities: Shrubs, Rockery Plants, Herbaceous Perennials

89 Plus Garden Centre

Manor Road, Frome, Somerset BA11 4BS
Tel: 01373 453094

90 Poplar Farm

Burtle Road, Burtle, Bridgwater, Somerset TA7 8NB
Tel: 01278 723170

91 Poplar Nurseries

The Batch, Batheaston, Bath, Somerset BA1 7DR
Tel: 01225 858658
Specialities: Paving

92 Prior Park Garden Centre

Prior Park Road, Bath, Somerset BA2 4NF
Tel: 01225 427175
Specialities: Garden Products, Giftware, Aquatics

93 R T Herbs and Garden

Kilmersdon, Radstock, Somerset BA3 5TD
Tel: 01761 435470
Specialities: Herbs, Herbaceous Perennials, Wild Flowers

94 Richard Roper

Little Thatch Wiqborough, Wigborough,
South Petherton, Somerset TA13 5LP
Tel: 01460 240517

95 Riverside Garden Centre

Clift House Road, Southville, Bristol,
Somerset BS3 1RX
Tel: 08000 375796
Specialities: Garden Products, Restaurant/Coffee Shop

96 Rockery Bonsai Gardens

Highridge Road, Dundry, Bristol,
Somerset BS41 8JT
Tel: 01179 640394
Specialities: Bonsai, Rockery Plants

97 Rocky Mountain Nurseries

Maesbury, Wells, Somerset BA5 3HA
Tel: 01749 841014
Specialities: Garden Products, Giftware, Restaurant/Coffee Shop

98 Roundtrees at Tickenham

Church Lane, Tickenham, Clevedon,
Somerset BS21 6SD
Tel: 01275 858015

99 Roundtrees Garden Centre

Smallway, Congresbury, Bristol,
Somerset BS49 5AA
Tel: 01934 838237
Specialities: Garden Products, Garden Furniture, Restaurant/Coffee Shop

100 Sanders Garden World

Bristol Road, Brent Knoll, Highbridge,
Somerset TA9 4HJ
Tel: 01278 761111

101 Scotts Nurseries

Merriott, Somerset TA16 5PL
Tel: 01460 72306

102 Second Nature Plant Nursery

Croft House, Aller, Langport, Somerset TA10 0RA
Tel: 01458 259190

103 Sheila's Plants

68 Bunns Lane, West Woodlands, Frome,
Somerset BA11 5ES
Tel: 01373 836837
Specialities: Garden Products, Grasses

104 Slate & Stone

Turnpike Yard, Lower Bristol Road, Bath,
Somerset BA2 9EU
Tel: 01225 332661
Specialities: Paving, Water Products and Features

105 Slipps Garden Centre

Butts Hill, Frome, Somerset BA11 1HR
Tel: 01373 467013

106 Street Ash Nursery

Street Ash, Combe St. Nicholas, Chard,
Somerset TA20 3HZ
Tel: 01460 234736

107 Sweet Acre Nursery

Godney Road, Glastonbury, Somerset BA6 9AF
Tel: 01458 834602
Specialities: Herbaceous Perennials, Hanging Baskets, Shrubs

108 Tor Hole Plants

Hill Cottage, Tor Hole, Chewton Mendip, Radstock
BA3 4LS
Tel: 01761 241539
e-mail: Lesley.irving@mbzonline.net

Specialities:

Hardy Perennials; Moisture Lovers; Hellebores

Products and Services:

- Aquatics
- Children's Play Area
- Disabled Facilities
- Display Gardens
- Garden Design
- Plant Guarantee
- Plants

Directions:

From the A39 about 2 miles southwest of Chewton Mendip, take the B3135 towards Cheddar. One mile along this road, Tor Hole Plants is the second building on the left. Look out for the yellow signs.

109 Triscombe Nurseries

Triscombe, Bishops Lydeard, Taunton,
Somerset TA4 3HG
Tel: 01984 618267
Specialities: Herbaceous Perennials, Trees, Rose

110 Waggon Nursery

The Causeway, Mark, Highbridge,
Somerset TA9 4QJ
Tel: 01278 641597

111 The Walled Garden At Knowle

Bath Road, Knowle, Bridgwater, Somerset TA7 8PJ
Tel: 01278 686321

112 Wendy Alpines

Grass Roots, Stalbridge Road, Henstridge, Templecombe, Somerset BA8 0SA
Tel: 01963 363701

113 West Harptree Nursery

Bristol Road, West Harptree, Bristol, Somerset BS40 6HG
Tel: 01761 221370

114 West Somerset Garden Centre

Mart Road, Minehead, Somerset TA24 5BJ
Tel: 01643 703612
Specialities: Shrubs, Restaurant/Coffee Shop, Pots

115 Wheathill Lane Garden Centre

Wheathill Lane, Milborne Port, Sherborne, Somerset DT9 5EX
Tel: 01963 251500

116 White Gate Nurseries

Stockwood Hill, Keynsham, Bristol, Somerset BS31 2AN
Tel: 01179 862653

117 White Post Nursery

Langford Budville, Wellington, Somerset TA21 0RW
Tel: 01823 400234
Specialities: Herbaceous Perennials, Shrubs, Alpines/Rock Plants

118 White Row Garden & Gift Shop

White Row Farm, Beckington, Frome, Somerset BA11 6TN
Tel: 01373 830024

Specialities:

Hanging Baskets; Shrubs; Plants; Gifts.

Products and Services:

- Herbaceous & Bedding
- Herbs & Veg Plants
- Planters & pots
- Wide range of Gifts
- Cafe/Tearoom
- Farm Shop

Directions:

White Row Farm is off the main roundabout on the A36 at Beckington about 4 miles north of Frome, at the junction with the A361.

Description:

White Row Plant Centre won a Gold Medal at the Hampton Court Flower Show in 2006 for its display in the "Inspiring Spaces" category. The display reflected the link between Agriculture and Horticulture.

Hanging baskets are made up on the site and a large proportion of the basket plants are home grown. A much larger range of plants for the garden are stocked including perennials, shrubs and climbers plus composts and garden sundries.

The Gift Shop stocks an excellent range of unusual gifts and cards, pretty bouquets and individual gifts can be made to customer requirements.

119 Whitehall at Whitchurch

Norton Lane, Whitchurch, Bristol, Somerset BS14 0BT
Tel: 01275 832296

120 Wick Farm Nursery

Brean Rd, Lympsham, Weston-super-Mare, Somerset BS24 0HD
Tel: 01278 751275
Specialities: Herbaceous Perennials

121 Willowbrook Nursery & Garden Centre

West Buckland, Wellington, Somerset TA21 9HX
Tel: 01823 461324
Specialities: Aquatics, Pet Products

122 Wincanton Garden Centre

Bruton Road, Wincanton, Somerset BA9 8HA
Tel: 01963 33134

SEE ADVERTISEMENT PAGE 235

Specialities:

Bedding Plants, Coffee Shop, Paving, Fencing, Sheds

Products and Services:

- Aquatics
- Barbeques
- Books & Stationary
- Seasonal Displays
- Credit Cards Accepted
- Disabled Facilities

- Fencing
- Fruit Trees
- Garden Furniture
- Garden Products
- Greenhouses and Sheds
- Houseplants
- Information Desk
- Large Car Park
- Paving
- Pet Products
- Plant Guarantee
- Plants
- Restaurant/Coffee Shop
- Seeds and Bulbs
- Trees

Directions:

From the A303 take the Wincanton turn off, follow the signs to Wincanton racecourse and Wincanton Garden Centre is situated on the B3081 Wincanton to Bruton Road.

123 Woodspring Nursery

Woodspring Gardens, Collum Lane, Kewstoke, Weston-super-Mare, Somerset BS22 9YX
Tel: 01934 512481

124 Wyevale Garden Centre (Keynsham)

Hicks Gate, Keynsham, Bristol, Somerset BS31 2AD
Tel: 01179 778945

125 Wyevale Garden Centre (Taunton)

Pen Elm Hill, Taunton, Somerset TA2 6PE
Tel: 01823 323777

STAFFORDSHIRE

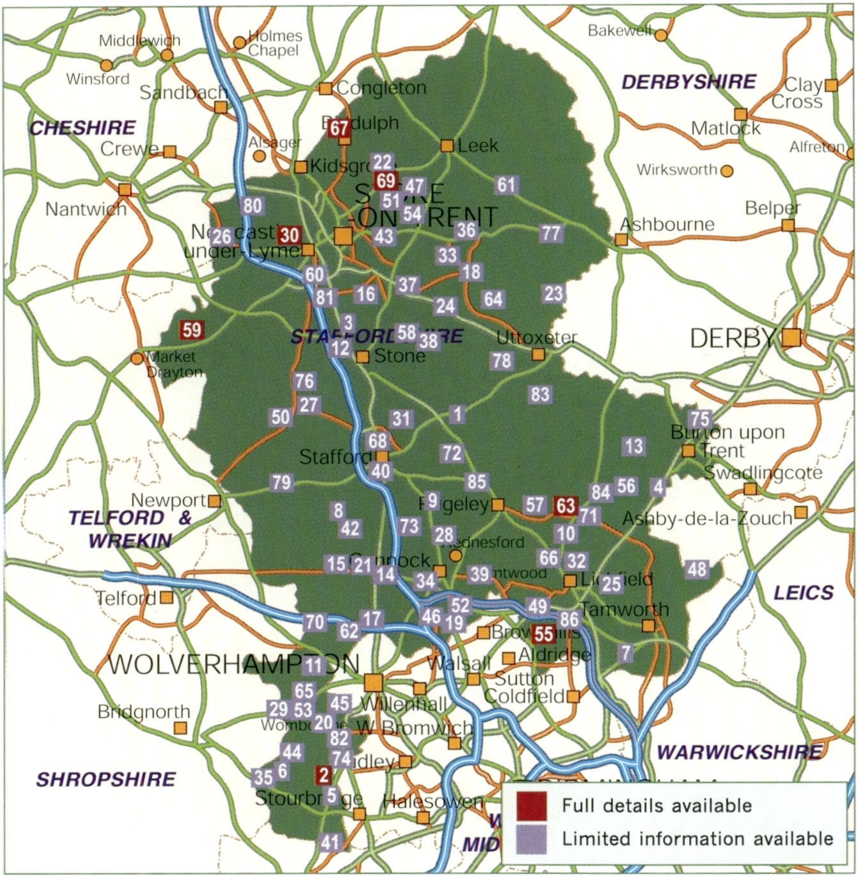

1 Amerton Garden Centre

within Amerton Farm & Craft Centre, Amerton, Stafford, Staffordshire ST18 0LA
Tel: 01889 271998
Specialities: Herbaceous Perennials, Exotics, Trees

2 Ashwood Nurseries

Ashwood Lower Lane, nr Kingswinford, Staffordshire DY6 0AE
Tel: 01384 401996 Fax: 01384 401108

Specialities:
Shrubs, perennials, conifers and alpines.

Products and Services:

- Books and Stationery
- Childrens Play Area
- Seasonal Displays
- Credit Cards Accepted
- Disabled Facilities
- Display Gardens
- Dried and Artificial Flowers
- Fruit Trees
- Garden Furniture
- Garden Products
- Giftware
- Houseplants
- Information Desk
- Large Car Park
- Plants
- Restaurant/Coffee Shop
- Seeds and Bulbs
- Trees

Directions:
From the A449 north of Kidderminster before Wall Heath, turn left into Ashwood Lower Lane. The nursery is ½ mile along this road on the left.

Description:
Ashwood is a traditional working nursery situated in the lovely South Staffordshire countryside, adjacent to the Staffordshire and Worcestershire canal. **The nursery produces** a large proportion of the plants you will find on sale in the beautifully maintained retail areas. They include an impressive selection of shrubs, trees, fruit, roses, climbing plants, herbaceous perennials, alpines and heathers. You will probably also be tempted by the extensive range of indoor plants, seasonal bedding and patio plants.

The nurseries are internationally famous for their speciality crops which are raised on the nursery. The hellebores are regarded as being some of the finest strains ever produced, attracting gardeners and growers from all over the world. Hardy cyclamen, lewisias, hepaticas, primula auricula and hydrangeas are important crops, both for seed and plant production. Also on site are a well-stocked garden shop, a quality gift shop and a delightful tea room. Everything in fact, to make a perfect day out.

3 Barlaston Old Road Nurseries

Old Road, Barlaston, Stoke-on-Trent, Staffordshire ST12 9EN
Tel: 01782 373960
Specialities: Herbaceous Perennials, Hanging Baskets, Aquatics

4 Barton Fields Garden Centre

Lichfield Road, Barton Under Needwood, Burton-on-Trent, Staffordshire DE13 8ED
Tel: 01283 711288

5 Beechwood Bonsai Nursery

Wolverhampton Road, Prestwood, Stourbridge, Staffordshire DY7 5AN
Tel: 01384 877847
Specialities: Bonsai

6 Bobbington

Manor Lane, Bobbington, Stourbridge, Staffordshire DY7 5EG
Tel: 01384 221020

7 Bournebrook Nursery

Sutton Road, Mile Oak, Tamworth, Staffordshire B78 3DX
Tel: 07866 437809

8 Bradley Garden Centre

Oak Lane, Bradley, Stafford, Staffordshire ST18 9EA
Tel: 01785 780208
Specialities: Annuals/Biennials/Bedding, Shrubs, Herbaceous Perennials

9 Brockton Nursery

Chase Road, Brocton, Stafford, Staffordshire ST17 0TL
Tel: 01785 661148

10 Bromley Hayes Garden Centre

Riley Hill Farm, Shaw Lane, Riley Hill, Lichfield, Staffordshire WS13 8HW
Tel: 01543 472230

11 Brownies Nursery

Stone Cottage, Wrottesley Park Road, Wrottesley Park, Wolverhampton, Staffordshire WV8 2HS
Tel: 01902 754987

12 Bury Bank Plant Centre

Bury Bank, Meaford, Stone, Staffordshire ST15 0QA
Tel: 01782 796353
Specialities: Shrubs, Trees, Conifers

13 Byrkley Garden Centre

Byrkley Drive, Rangemore, Burton-on-Trent, Staffordshire DE13 9RN
Tel: 01283 716467

14 C Piper & Sons Ltd

The Nurseries, Watling Street, Gailey, Stafford, Staffordshire ST19 5PR
Tel: 01902 790233
Specialities: Annuals/Biennials/Bedding, Shrubs, Ponsettias

15 Cottage Garden Roses

Woodlands House, Garden Lane, Stretton, Stafford, Staffordshire ST19 9LG
Tel: 01785 840217
Specialities: Rose

16 Country Cabin

10 Holding, Cocknage, Stoke-on-Trent, Staffordshire ST3 4AL
Tel: 01782 327429
Specialities: Herbaceous Perennials, Shrubs, Climbers

17 Coven Heath Nursery

Shaw Hall Lane, Coven Heath, Wolverhampton, Staffordshire WV10 7HE
Tel: 01902 783670
Specialities: Ornamentals, Conifers

18 D K Plants

19 Harbourne Road, Cheadle, Stoke-on-Trent, Staffordshire ST10 1JU
Tel: 01538 754460
Specialities: Alpines/Rock Plants, Rare and Unusual Plants, Herbaceous Perennials

19 De Ridder

35 Bentons Lane, Walsall, Staffordshire WS6 6EE
Tel: 01922 413542

20 Dimmingsdale Garden Centre & Nursery

Ebsdree Road, Dimmingsdale, Wolverhampton, Staffordshire WV4 4XQ
Tel: 01902 765040

21 Dobbies Garden World

Saxon Cross House, Watling Street, Gailey, Stafford, Staffordshire ST19 5PP
Tel: 01902 791555

22 Double Gates Nursery

Sytch Road, Brown Edge, Stoke-on-Trent, Staffordshire ST6 8QX
Tel: 01782 502120
Specialities: Annuals/Biennials/Bedding, Clematis

23 Doveleys Garden Centre

Doveleys Manor Park, Denstone, Uttoxeter, Staffordshire ST14 5BZ
Tel: 01889 591999
Specialities: Shrubs, Herbaceous Perennials, Conifers

24 Draycott Nurseries

Cheadle Road, Draycott, Stoke-on-Trent, Staffordshire ST11 9RG
Tel: 01782 392461
Specialities: Shrubs, Climbers, Annuals/Biennials/Bedding

25 Elifield Nurseries

Darnford Lane, Lichfield, Staffordshire WS14 9LB
Tel: 01543 432113

26 Elmside Plant Centre

Main Rd, Bowsey Wood, Crewe, Staffordshire CW3 9DH
Tel: 01270 820352

27 Fletcher's Country Garden Centre

Bridge Farm, Stone Road, Eccleshall, Stafford, Staffordshire ST21 6JY
Tel: 01785 851057
Specialities: Shrubs, Climbers, Rose

28 Fuchsia World

521 Pye Green Road, Hednesford, Cannock, Staffordshire WS12 4LP
Tel: 01543 422394
Specialities: Fuchsia

29 Gardenlands Garden & Leisure

Bridgnorth Rd, Pattingham, Wolverhampton, Staffordshire WV6 7EJ
Tel: 01902 700588

30 Garners Garden Centre

Cemetery Road, Silverdale, Newcastle-under-Lyme, Staffordshire ST5 6SH
Tel: 01782 636428 Fax: 01782 712674
e-mail: sales@garnersgardencentre.co.uk
website: garnersgardencentre.co.uk

Specialities:

Bedding Plants; Craft Centre; Restaurant

Products and Services:

- Aquatics
- Barbecues
- Books and Stationery
- Seasonal Displays
- Clothing
- Credit Cards Accepted
- Cut Flowers
- Disabled Facilities
- Dried and Artificial Flowers
- Farm Shop
- Fruit Trees
- Garden Furniture
- Garden Machinery
- Garden Products
- Giftware
- Houseplants
- Information Desk
- Large Car Park
- Pet Products
- Plant Guarantee
- Plants
- Restaurant/Coffee Shop
- Seeds and Bulbs
- Trees

Directions:

Garners is located just off the A525 near Keele University

Description:

This family-run business, established in 1987, always stocks a varied selection of indoor plants and a full range of appropriate plant foods, composts, pots and pot covers. The Outdoor Department offers a large and varied range of outdoor plants, shrubs, roses and trees, along with bedding plants, herbs

and conifers. It also stocks all types of pots and rockery stone. In the Barbecue Department, Garners only stocks the best quality brands, including such famous names as "Weber" for whom Garners are the main agents. Do take time to browse in the gift department which offers greeting cards, silk flowers, scented candles, flower arrangements books and CDs. Also on site is an extensive display of quality garden furniture and a Pet Department catering for all pet requirements. A major attraction is the Craft Department which provides a huge range of craft and DIY items. Round off your visit in the Barn Restaurant which serves freshly prepared home-made hot meals, coffee, tea and fresh Cream Cakes.

31 Greenheart Plants

Hopton Hall Lane, Hopton, Stafford, Staffordshire ST18 0AZ
Tel: 01785 257975
Specialities: Annuals/Biennials/Bedding, Hanging Baskets, Shrubs

32 G-Scapes Nursery at Curborough

Curborough Craft Centre, Watery Lane, Lichfield, Staffordshire WS13 8ES
Tel: 01543 256401
Specialities: Herbaceous Perennials, Shrubs, Annuals/Biennials/Bedding

33 Harewood Nurseries

Leek Road, Cheadle, Stoke-on-Trent, Staffordshire ST10 2EF
Tel: 01538 755281

34 Hatherton Aquatics and Nurseries

Watling Street, Cannock, Staffordshire WS11 1SH
Tel: 01543 503627
Specialities: Water Products and Features, Acer, Annuals/Biennials/Bedding

35 Haven Nurseries

Crab Lane, Bobbington, Stourbridge, Staffordshire DY7 5DZ
Tel: 01384 221543
Specialities: Geranium, Fuchsia

36 Hazles Cross Farm Nurseries

Hollins Lane, Kingsley, Stoke-on-Trent, Staffordshire ST10 2EP
Tel: 01538 752669
Specialities: Herbaceous Perennials

37 Heath House Farm

Heath House Farm, Caverswall Old Road, Forsbrook, Stoke-on-Trent, Staffordshire ST11 9BQ
Tel: 01782 396027
Specialities: Hanging Baskets, Shrubs, Annuals/Biennials/Bedding

38 Heathers Roses

Roseland Nursery, Moss Lane, Hilderstone, Stone, Staffordshire ST15 8RQ
Tel: 01889 505345

39 Heathhayes Nursery

Newlands Lane, Cannock, Staffordshire WS12 3HH
Tel: 01543 275360
Specialities: Annuals/Biennials/Bedding, Shrubs, Trees

40 Herbert Road Nurseries

Herbert Road, Stafford, Staffordshire ST17 9BH
Tel: 01785 257268

41 High Acres Nursery

High Acres, Sugar Loaf Lane, Iverley, Kidderminster, Staffordshire DY10 3PB
Tel: 01562 700405

42 Hill's Water Gardens

Bradley Garden Centre, Bradley, Stafford, Staffordshire ST18 9EA
Tel: 01785 780000
Specialities: Water Products and Features

43 Hillside Nurseries

67 Brookhouse Lane, Stoke-on-Trent, Staffordshire ST2 8NE
Tel: 01782 302460

44 Historic Herbs

Halfpenny Green Vinyards, Tom Lane, Bobbington, Stourbridge, Staffordshire DY7 5EP
Tel: 01384 221022
Specialities: Herbs

45 Hodges Nurseries

Radford Lane, Lower Penn, Wolverhampton, Staffordshire WV4 4XW
Tel: 01902 763760
Specialities: Annuals/Biennials/Bedding

46 Hollybush Nurseries

Warstone Road, Shareshill, Wolverhampton, Staffordshire WV10 7LX
Tel: 01922 418050
Specialities: Water Products and Features

47 Jacksons Nurseries

Thorney Edge Road, Bagnall, Stoke-on-Trent, Staffordshire ST9 9LD
Tel: 01782 502078
Specialities: Shrubs, Conifers, Herbaceous Perennials

48 Jackson's Nurseries

Main Street, Clifton Campville, Tamworth, Staffordshire B79 0AP
Tel: 01827 373307
Specialities: Fuchsia

49 John Brown Nurseries Ltd

The Gables, Walsall Road, Muckley Corner, Lichfield, Staffordshire WS14 0BG
Tel: 01543 452826
Specialities: Shrubs, Landscape Gardening, Trees

50 Johnson Hall Nurseries (P & M Edwards)

Newport Road, Eccleshall, Stafford, Staffordshire ST21 6JA
Tel: 01785 850400
Specialities: Annuals/Biennials/Bedding, Herbaceous Perennials, Shrubs

51 Ken Thursfield

Nursery End, Light Oaks Avenue, Light Oaks, Stoke-on-Trent, Staffordshire ST2 7NF
Tel: 01782 534943
Specialities: Annuals/Biennials/Bedding, Shrubs

52 Kingswood Nurseries Ltd

105 Norton Lane, Great Wyrley, Walsall, Staffordshire WS6 6NU
Tel: 01922 418291
Specialities: Ornamentals, Water Products and Features

53 Lealans Nursery and Garden Centre

Bridgnorth Road, Shipley, Wolverhampton, Staffordshire WV6 7EZ
Tel: 01902 700209

54 Luzlow Nursery

Luzlow, Bagnall, Stoke-on-Trent, Staffordshire ST9 9JZ
Tel: 01782 305001
Specialities: Annuals/Biennials/Bedding, Shrubs, Herbaceous Perennials

55 Lynn Nurseries

Lynn Lane, Lynn, Lichfield, Staffs. WS14 0EP
Tel: 01543 481918
e-mail: family@cartwright@g.mail.com
website: www.GardenPlants-SouthStaffs.com

Specialities:

Hanging baskets; Perennials; Stone ornaments

Description:

The owner of Lynn Nurseries, Jenny Cartwright, takes pride in the quality of the plants grown on her 4-acre site. The unusual herbaceous perennials give long lasting colour year after year. All plants are British sourced, container grown and most grown from cuttings on site, so the customer can be assured that

their plants will grow and be suitable for their growing conditions. Summer and winter hanging **baskets** are made on the premises.

Directions:

From the M6 Toll Road take the T4 exit then left on the A5127 south to Shenstone village. Follow the signs to Lynn (1.5m)

56 Manor Park Nurseries

Barton Gate, Barton Under Needwood, Burton-on-Trent, Staffordshire DE13 8AQ
Tel: 01283 716457

57 Mavesyn Nurseries

Mavesyn Ridware, Rugeley, Staffordshire WS15 3QB
Tel: 01543 493313
Specialities: Hanging Baskets, Shrubs, Annuals/Biennials/Bedding

58 Mount Nurseries

Spotacre, Moddershall, Stone, Staffordshire ST15 8RN
Tel: 01889 505627
Specialities: Annuals/Biennials/Bedding, Herbaceous Perennials, Shrubs

59 Mucklestone Nurseries

Rock Lane, Mucklestone, nr Market Drayton, Staffordshire TF9 4DN
Tel: 01630 674343 Mobile: 07985 425829
website: www.botanyplants.com

Specialities:

Perennials; Ornamental Trees – Acers & Sorbus

Description:

Located on the Shropshire/Staffordshire border, Mucklestone Nurseries are run by Brian Watkins who was originally a farmer before teaming up with William Friend who is passionate about plants – his father was a leading gardener. The Nurseries specialise in perennials and plants that prefer wetter and acidic soils, as well as a wide selection of ornamental trees with most types available. Open Thursday to Sunday – please ring first for an appointment on other days.

Directions:

The Nurseries are on the A51, two miles north of Loggerheads. Turn right at the crossroads.

60 Northwood Garden Centre

Clayton Road, Newcastle under Lyme,
Staffordshire ST5 4DH
Tel: 01782 635081
Specialities: Annuals/Biennials/Bedding, Herbaceous Perennials, Shrubs

61 Old Hall Nursery

Winkhill, Leek, Staffordshire ST13 7PN
Tel: 01538 308257
Specialities: Herbaceous Perennials, Herbs, Shrubs

62 Old Tree Nursery

Pendeford Hall Lane, Wolverhampton,
Staffordshire WV9 5ET
Tel: 01902 786042

63 P.D. Nurseries

Yoxall Road, Kings Bromley, nr Lichfield,
Staffordshire DE13 7JJ
Tel: 01543 472868

SEE ADJACENT ADVERTISEMENT

Specialities:

Hedging Plants; Large Trees & Conifers; Shrubs

Description:

Open every day from 10am to 5pm, P.D. Nurseries stocks a wide range of bedding plants, acers, bamboos, bay trees, conifers, heathers and perennials. Also available is an extensive range of hedging plants such as beech, blackthorn, conifers, hazel and privet.

Directions:

The Nurseries are signposted on the A515 just north of Kings Bromley

64 Paradise Nurseries and Water Garden Centre

Winnothdale, Tean, Stoke-on-Trent,
Staffordshire ST10 4HB
Tel: 01538 722566

65 Pavilion Garden Centre

Bridgnorth Road, Shipley, Wolverhampton, Staffordshire WV6 7EZ
Tel: 01902 701001
Specialities: Restaurant/Coffee Shop

66 The Plant Plot

Western Bypass, Stafford Road, Lichfield, Staffordshire WS13 8JA
Tel: 01543 262805
Specialities: Annuals/Biennials/Bedding, Herbaceous Perennials, Restaurant/Coffee Shop

67 Pointon's Plant Centre

Well Lane, Gillow Heath, Biddulph, Staffordshire ST8 6QS
Tel: 01782 512109 Fax: 01782 522354
e-mail: sales@pointonsplant centre
website: www.pointonsplantcentre.co.uk

Specialities:

Specimen conifers & Hollies; Aquatic plants; Rhododendrons

Description:

For more than 100 years this privately run nursery has been providing gardeners with healthy flourishing trees, shrubs and conifers. The Centre also stocks aquatic plants, annuals, seeds and compost. Refreshments are on sale and the site is wheelchair friendly and has disabled toilets. Ample car parking.

Directions:

From the A527 just north of Biddulph turn left on minor road to Gillow Heath.

68 Priory Garden Centre

18-22 Corporation Street, Corporation Street, Stafford, Staffordshire ST16 3LZ
Tel: 01785 246363

69 Proctor's Nursery

High Lane, Brown Edge, nr Stoke-on-Trent ST6 8RT
Tel: 07803 940759

Specialities:

Hedging plants; Japanese azaleas; Specimen trees

Description:

Barry Proctor opened his nursery in 2000 and within its 14 acres of grounds you'll find a huge range of plants of every kind although he specialises in hedging plants, Japanese azaleas and specimen trees. The nursery is open from 9am-5pm, every day of the week.

Directions:

From the A53, about 4 miles SE of Leek, take the B5051 to Brown Edge. The nursery is on this road.

70 R Spencer & Son

Yew Tree Cottage, Whitehouse Lane, Codsall Wood, Wolverhampton, Staffordshire WV8 1QS
Tel: 01902 842346

71 Rare and Exotic Plants

Woodshoot Nurseries, Cockshut Lane, Kings Bromley, Burton-on-Trent, Staffordshire DE13 7HN
Tel: 01543 472233
Specialities: Exotics

72 Roseacre Nursery

Great Haywood, Stafford, Staffordshire ST18 0RT
Tel: 01889 882288
Specialities: Rose, Shrubs, Trees

73 Rosedale Trees & Shrubs

Pillaton, Penkridge, Stafford, Staffordshire ST19 5RZ
Tel: 01785 713444

74 Sandiacre Farm Shop & Nurseries

Himley Lane, Himley, Dudley, Staffordshire DY3 4LN
Tel: 01902 893542
Specialities: Vegetables, Seasonal Plants, Fruit Bushes

75 Slaters Gardens

Dovecliff Road, Stretton on Dunsmore, Burton-on-Trent, Staffordshire DE13 0DJ
Tel: 01283 565650

76 Sleepy Hollow Nurseries

Badenhall, Eccleshall, Stafford, Staffordshire ST21 6LG
Tel: 01785 850403

77 Stanton Dale Nurseries

Cottage Farm, Stanton Dale, Ashbourne, Staffordshire DE6 2BX
Tel: 01538 308384
Specialities: Annuals/Biennials/Bedding, Herbaceous Perennials, Shrubs

78 Strawberry Garden Centre

Bramshall Road, Bramshall, Uttoxeter,
Staffordshire ST14 5BE
Tel: 01889 562292
Specialities: Shrubs, Annuals/Biennials/Bedding, Herbaceous Perennials

79 Swan Pit Nursery

Gnosall, Swan Pit, Gnosall, Stafford,
Staffordshire ST20 0EE
Tel: 01785 822128
Specialities: Annuals/Biennials/Bedding, Hanging Baskets, Shrubs

80 T Mountford

Quarry New Farm, Nantwich Road, Audley,
Stoke-on-Trent, Staffordshire ST7 8DW
Tel: 01782 720330

81 Trentham Garden Centre

Stone Road, Stoke-on-Trent, Staffordshire ST4 8AX
Tel: 01782 646644
Specialities: Annuals/Biennials/Bedding, Shrubs, Herbaceous Perennials

82 Whitecross Garden Centre

Orton Lane, Lower Penn, Wolverhampton,
Staffordshire WV4 4XA
Tel: 01902 895353

83 Willslock Cottage

Quee Lane, Willslock, Uttoxeter,
Staffordshire ST14 8RB
Tel: 01889 563405
Specialities: Hanging Baskets, Annuals/Biennials/Bedding

84 Woodhouse Green Nurseries

Woodhouses, Yoxall, Burton-on-Trent,
Staffordshire DE13 8NR
Tel: 01543 473183
Specialities: Annuals/Biennials/Bedding, Shrubs

85 Wyevale Garden Centre (Rugeley)

Wolseley Bridge, Wolseley Bridge, Stafford,
Staffordshire ST17 0YA
Tel: 01889 574884

86 Wyevale Garden Centre (Shenstone)

Birmingham Road, Shenstone, Lichfield,
Staffordshire WS14 0QQ
Tel: 01543 480095

SUFFOLK

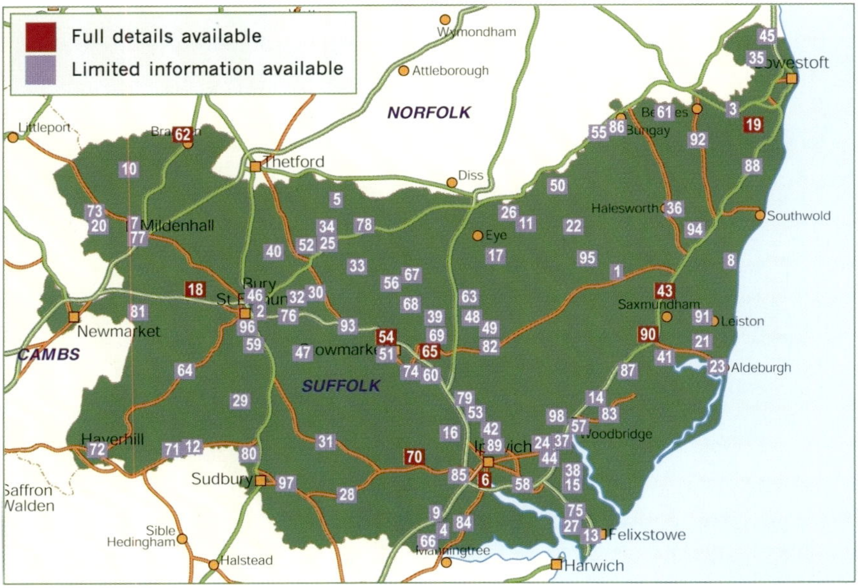

1 Ace of Spades Nursery

2 Mill Road, Bedingham, Woodbridge, Suffolk IP13 8LG
Tel: 07904 084503

2 Anglia Bonsai

1 Raedwald Drive, Bury St. Edmunds, Suffolk IP32 7DA
Tel: 01284 750710
Specialities: Bonsai

3 Barnby Nursery Centre

Swan Lane, Barnby, Beccles, Suffolk NR34 7QF
Tel: 01502 476284
Specialities: Annuals/Biennials/Bedding, Trees

4 Bentley Plants Ltd

Bergholt Rd, Bentley, Ipswich, Suffolk IP9 2DQ
Tel: 01473 312255
Specialities: Herbaceous Perennials, Trees

5 Blooms of Bressingham

High Street, Hopton, Diss, Suffolk IP22 2QX
Tel: 01953 681484

6 Bourne Garden Centre

Peppers Lane, Wherstead, Ipswich, Suffolk IP9 2AB
Tel: 01473 786617 Fax: 01473 686461
e-mail: clairebgc@hotmail.com
website: www.suffolkfoodhall.com

SEE ADVERTISEMENT PAGE 245

Specialities:

Unusual Plants, Planters, Garden Ornaments

Description:

Located close to the River Orwell and next to the Suffolk Food Hall, the Bourne Garden Centre stocks a vast array of plants, trees and shrubs, as well as garden tools, fertilisers and composts.

Directions:

From the A14 about 3 miles south of Ipswich, take the A137 south to Wherstead. Go through the village and then turn right on minor road towards Shotley. The centre is on the right about 100 yards along this road.

The Bourne Garden Centre is a traditional family owned business which forms part of the Suffolk Food Hall complex. With ten years experience and HTA membership, the garden centre offers a vast array of high quality and unusual plants, trees and shrubs. HTA gift vouchers are accepted.

Convieniently located just two miles south of Ipswich and within minutes of the A14/A12. The garden centre itself is housed in a 100year old barn conversion offering visitors fantastic views of the river Orwell in an area of outstanding natural beauty.

Once there you will be met by the wonderful staff which will make your visit a very personal experience. With a huge choice of gardening accessories, such as bird baths, ornaments, contempory pots and sundials, you will find everything to meet your gardening needs.

The Bourne Garden Centre will also cater for your seasonal needs. With trees, wreaths, mistletoe and holly all locally sourced from Suffolk itself. Also for the novice gardener there are seasonal classes and talks.

- Houseplants
- Shrubs
- Perennials
- Climbers
- Seeds
- Summer Bedding
- Play Area
- Cut Flowers
- Garden Design
- Information Desk
- Giftware
- Disabled Facilities

Adjacent is Suffolk Food Hall .....

Bourne Garden Centre

Peppers Lane, Wherstead, Ipswich, Suffolk IP9 2AB
Tel: 01473 786617
E-mail: claireBGC@hotmail.com
Website: www.suffolkfoodhall.com

Opening Hours
Monday to Saturday 9.00am - 6.00pm
Sunday 10.30am - 4.30pm

7 Brian & Pearl Sulman

54 Kingsway, Mildenhall, Bury St. Edmunds,
Suffolk IP28 7HR
Tel: 01638 712297
Specialities: Geranium, Pelargonium

8 Bridge Nurseries

Bridge House, Dunwich, Saxmundham,
Suffolk IP17 3DZ
Tel: 01728 648850
Specialities: Clematis

9 By Pass Nurseries

London Road, Capel St. Mary, Ipswich,
Suffolk IP9 2JR
Tel: 01473 310604
Specialities: Garden Furniture, Water Products and Features

10 C Matthews

18 High Street, Lakenheath, Brandon,
Suffolk IP27 9JS
Tel: 01842 860284

11 Castleden Trees

Park Farm, Depper Dip Lane, Hoxne, Eye,
Suffolk IP21 5BS
Tel: 01379 668197
Specialities: Trees, Hedgerow Shrubs

12 Clare Dutch & English Bulb Co

Riverside House, Stoke Road, Clare, Sudbury,
Suffolk CO10 8NS
Tel: 01787 277441
Specialities: Narcissus (Daffodil), Crocus, Tulipa (Tulip)

13 Conkers Pet & Garden Centre

110-112 Hamilton Road, Felixstowe,
Suffolk IP11 7AB
Tel: 01394 286124
Specialities: Pet Products, Garden Products

14 Crown Nursery

High Street, Ufford, Woodbridge, Suffolk IP13 6EL
Tel: 01394 460755
Specialities: Trees, Herbaceous Perennials

15 Crystal Garden Nursery

12 Ipswich Road, Newbourne, Suffolk IP12 4NS
Tel: 01473 736651

16 Dairy Farm Rose and Plant Centre

Lower Dairy Farm, Bramford, Ipswich,
Suffolk IP8 4JT
Tel: 01473 833359
Specialities: Rose, Clematis

17 Different Plants

The Melis Stud, Gate Farm, Cranley Green, Eye,
Suffolk IP23 7NX
Tel: 01379 870291
Specialities: Rare and Unusual Plants

18 Dogwood Garden Centre

Risby Barn, South Street, Risby,
Bury St Edmunds, Suffolk IP28 6QU
Tel/Fax: 01284 811055
website: www.risbybarn.co.uk

SEE ADVERTISEMENT PAGE 246

Specialities:

Perennials (English Crown), Nordmann Christmas Trees, Large range of Garden Orientated Gift Ideas

Products and Services:

- Books & Stationary
- Seasonal Displays
- Credit Cards Accepted
- Disabled Facilities
- Dried & Artificial Flowers
- Fencing
- Floristry
- Fruit Trees
- Garden Furniture
- Garden Products
- Giftware
- Information Desk
- Large Car Park
- Plant Guarantee
- Plants
- Restaurant/Coffee Shop
- Seeds and Bulbs
- Trees

Directions:

Risby Barn is 3 miles west of Bury St Edmunds, just off the A14.

19 Early Dawn Nurseries & Garden Centre

Rushmere Road, Carlton Colville,
Suffolk NR33 8HA
Tel: 01502 475150

SEE ADVERTISEMENT PAGE 248

Specialities:

Shrubs, Garden Equipment, Aquatics

Description:

Early Dawn has a magnificent collection of plants, shrubs and gardening equipment on display. Whether you are just looking for a packet of seeds, a pair of wellies or a new gas barbecue, you are sure to find what you are looking for.

Directions:

From the A146, approx. 5 miles southwest of Lowestoft, take the turning for Carlton Colville. Follow the signs for the Transport Museum, located in Chapel Road. Turn right at the church along Church Lane. Turn left at the school, along Hall Road. At the junction, go straight across then bear right at the mini roundabout. Follow the road along, passing Carlton Motors, and the nurseries are 500 yards down Rushmere Road on the right.

20 Elisabeth Goodwin Nurseries

Elm Tree Farm, Beeches Road, West Row,
Bury St. Edmunds, Suffolk IP28 8NP
Tel: 01638 713050
Specialities: Drought Resistant Plants

21 Elmwood Nurseries

Leiston Road, Knodishall, Saxmundham,
Suffolk IP17 1UQ
Tel: 01728 830797
Specialities: Farm Shop, Annuals/Biennials/Bedding

22 Etheridge Nurseries

Stradbroke, Pixey Green, Stradbroke, Eye,
Suffolk IP21 5NJ
Tel: 01379 384204
Specialities: Fruit Bushes, Floristry

23 The Exotic Garden Company

Hall Farm Shop, Saxmundham Road, Aldeburgh,
Suffolk IP15 5JD
Tel: 01728 454456

24 Firecrest Tree & Shrub Nursery

Hall Road, Little Bealings, Woodbridge,
Suffolk IP13 6LU
Tel: 01473 625937
Specialities: Acer, Specimen Trees, Trees

25 George Hill Nursery

Barningham Road, Stanton, Bury St. Edmunds,
Suffolk IP31 2AD
Tel: 01359 252085

26 Goldbrook Plants

Hoxne, Goldbrook, Hoxne, Eye, Suffolk IP21 5AN
Tel: 01379 668770
Specialities: Aster (Michaelmas Daisy), Hemerocallis (Daylily)

27 Goslings Farm

Longford House, High Road, Trimley St. Martin, Felixstowe, Suffolk IP11 0RJ
Tel: 01394 273361

28 Green Lawns Bonsai

Hadleigh Road, Boxford, Sudbury, Suffolk CO10 5JH
Tel: 01787 210501
Specialities: Bonsai

29 Hartest Plants

Gatefields Meadow, Shimpling, Bury St. Edmunds, Suffolk IP29 4EX
Tel: 01284 830171

30 Harveys Garden Plants

Great Green, Thurston, Bury St. Edmunds, Suffolk IP31 3SJ
Tel: 01359 233363
Specialities: Woodland Plants, Grasses, Hardy Plants

31 Hedgerows Farm Shop & Nursery

Hedgerows Farm, The Street, Brent Eleigh, Sudbury, Suffolk CO10 9NU
Tel: 01787 247772

32 Highmead Nurseries

Greensleeves, Meadow Lane, Thurston, Bury St. Edmunds, Suffolk IP31 3QG
Tel: 01359 231465

33 Hill Cottage Nursery

Four Ashes, Walsham-le-Willows, Bury St. Edmunds, Suffolk IP31 3BS
Tel: 01359 259358

34 Hillcrest Nurseries Ltd

Barningham Rd, Stanton, Bury St. Edmunds, Suffolk IP31 2DU
Tel: 01359 250327
Specialities: Annuals/Biennials/Bedding, Farm Shop

35 Holly Garden Nurseries

Flixton Road, Blundeston, Lowestoft, Suffolk NR32 5PL
Tel: 01502 730648

36 Holton Road Garden Centre

36 Holton Road, Halesworth, Suffolk IP19 8HG
Tel: 01986 872761

37 Home Meadows Nursery Ltd

Top St, Martlesham, Woodbridge, Suffolk IP12 4RD
Tel: 01394 382419

38 Katie's Garden

6 Ipswich Road, Newbourne, Suffolk IP12 4NS
Tel: 01473 736717
Specialities: Annuals/Biennials/Bedding

39 Kersey's Farm

Mendlesham Green, Stowmarket, Suffolk IP14 5RB
Tel: 01449 766104
Specialities: Equisetum

40 The Kitchen Garden Shop

Church Cottage, Church Lane, Troston, Bury St. Edmunds, Suffolk IP31 1EX
Tel: 01359 268322
Specialities: Herbs, Vegetables, Herbaceous Perennials

41 Ladybird Nurseries

Gromford Lane, Snape, Saxmundham, Suffolk IP17 1RD
Tel: 01728 688289
Specialities: Fuchsia, Geranium

42 Laurel Farm Garden Centre

Laurel Farm, Henley Road, Ipswich, Suffolk IP1 6TE
Tel: 01473 215984
Specialities: Shrubs, Annuals/Biennials/Bedding

43 Laurel Farm Herbs

Main Road (A12), Kelsale, Saxmundham, Suffolk, IP17 2RG
Tel: 01728 668223
e-mail: laurelfarmherbs@aol.com
website: laurelfarmherbs.co.uk

Specialities:

Hardy Usual and Unusual Herbs

Description:

Established in 1985 by Chris Seagon, Laurel Farm is now one of the leading Herb nurseries in East Anglia. The herb catalogue is free to download and gives a comprehensive listing of all the herbs. Credit cards are taken for internet and personal callers.

Directions:

Laurel Farm is on the A12, 2 miles north of Saxmundham and 3 miles south of Yoxford

44 Ling's Garden Store

Beardmore Park, Martlesham Heath, Ipswich, Suffolk IP5 3RX
Tel: 01473 610926
Specialities: Garden Furniture, Garden Products

45 Lound Plant Centre

Jay Lane, Lound, Lowestoft, Suffolk NR32 5LH
Tel: 01502 731431

46 Marlows DIY & Garden Centre

Marlow House, Hollow Road, Bury St. Edmunds, Suffolk IP32 7AP
Tel: 01284 731408

47 Martins Nurseries

Smallwood Green, Bradfield St. George, Bury St. Edmunds, Suffolk IP30 0AJ
Tel: 01449 737698

48 Mickfield Hostas

The Poplars, Wetheringsett Road, Mickfield, Suffolk IP14 5LH
Tel: 01449 711576

49 Mickfield Watergarden Centre Ltd

Debenham Road, Mickfield, Suffolk IP14 5LP
Tel: 01449 711336
Specialities: Aquatics, Water Products and Features

50 Middleton Farm Nurseries

The Street, Weybread, Diss, Suffolk IP21 5TQ
Tel: 01379 852294
Specialities: Annuals/Biennials/Bedding, Shrubs

51 Milpets

71 Ipswich Street, Stowmarket, Suffolk IP14 1BA
Tel: 01449 675947

52 Mrs S Wooster

Harebell Cottage, Low Street, Bardwell,
Bury St. Edmunds, Suffolk IP31 1AS
Tel: 01359 250145
Specialities: Campanula (Bellflower), Campanula

53 N.J Scott

Glencoe, Church Lane, Barham, Ipswich,
Suffolk IP6 0PU
Tel: 01473 830331

54 Narey's Garden Centre

Eden Nurseries, A14 Spikes Lane, Stowmarket,
Suffolk IP14 3QD
Tel: 01449 612559 Fax: 01449 612744

SEE ADVERTISEMENT BELOW

Specialities:

Own Bedding Plants; Water Features; Swimming Pool Supplies

Description:

This family-owned business has been established since 1954 – the oldest garden centre in the Stowmarket area. It offers a wonderful selection of plants, notably its magnificent bedding plants and huge range of shrubs and trees.

Directions:

Located in Spikes Lane, off the eastbound A14 just before the Stowmarket slip road.

55 New Atlantic Country Superstore

The Old Mill, Church Road, Earsham, Bungay,
Suffolk NR35 2TQ
Tel: 01986 891069

56 Norman's Farm

Mill Road, Wyverstone, Stowmarket,
Suffolk IP14 4SF
Tel: 01449 781081
Specialities: Syringa (Lilac)

57 Notcutts Garden Centre

Ipswich Road, Woodbridge, Suffolk IP12 4AF
Tel: 01394 445400

58 Notcutts Nurseries Ltd

40 Bluestem Road, Ransomes Industrial Estate,
Ipswich, Suffolk IP3 9RR
Tel: 01473 272721

59 Nowton Park Nursery

Nowton Park, Bury St. Edmunds, Suffolk IP29 5LU
Tel: 01284 749030
Specialities: Hanging Baskets, Annuals/Biennials/Bedding

60 Oakhill Plant Nursery

Jacks Green Road, Creeting St. Mary, Ipswich, Suffolk IP6 8NB
Tel: 01449 721092
Specialities: Annuals/Biennials/Bedding, Pot Plants

61 Old Hall Plants

1 The Old Hall, Barsham, Beccles, Suffolk NR34 8HB
Tel: 01502 717475
Specialities: Herbs, Houseplants

62 P & R Garden Supplies

Fengate Drove, Brandon, Suffolk IP27 0PW
Tel: 01842 814800
e-mail: info@p-rgardensupplies.co.uk
website: www.p-rgardensupplies.co.uk

Specialities:

Plants arranged in "Plants for Places" order, making it easier to select a plant for every garden situation; Hanging Basket & Pot Filling Service; Wide Selection of Pots.

Products and Services:

- Aquatics
- Credit Cards Accepted
- Fruit Trees
- Garden Products
- Information Desk
- Large Car Park
- Pet Products
- Seeds and Bulbs

Directions:

Brandon is on the A1065, about 10 miles north of Mildenhall. P & R is located in a retail park on the Weeting (north) side of the railway crossing.

Description:

A family-run garden centre, P&R Garden Supplies was established in 2002 and stocks shrubs, conifers, perennials, herbs, alpines as well their very popular bedding plants and hanging baskets. Hardy plants are arranged in "Plants for Places" order, helping you to chose the perfect plant for every garden situation. The nursery also stocks a complete range of composts, soil improvers, bark, shingle, slate, cobbles, topsoil and manure – all at very competitive prices. In addition P & R also stocks a wide range of garden sundries and there's a good range of weed-killers, plant feeds and pest control products from the top market brands. P & R has a membership scheme which gives discounts and special offers. If a plant or product is not in stock, P & R offers an ordering service. The Aquatics Centre are Koi and Cichlid specialists. The Pet Centre caters for all pets, wild birds, poultry and non-domestic animals

63 Park Green Nurseries

Wetheringsett, Park Green, Wetheringsett, Stowmarket, Suffolk IP14 5QH
Tel: 01728 860139
Specialities: Hosta, Ornamental Grasses

64 Paugers Plants

Bury Road, Depden, Bury St. Edmunds, Suffolk IP29 4BU
Tel: 01284 850527

65 Perfect Perennials

Earl Stonham, Stowmarket, Suffolk IP14 5HY
Tel: 01449 711320
e-mail: perper@hotmail.co.uk
website: www.perfectperennials.co.uk
Specialities: Herbaceous Perennials

66 The Place For Plants

Mill Road, East Bergholt, Colchester, Suffolk CO7 6UP
Tel: 01206 299224
Specialities: Annuals/Biennials/Bedding, Herbaceous Perennials

67 Plant Centre

Plant Centre, Station Road, Cotton, Stowmarket, Suffolk IP14 4NQ
Tel: 01449 781661

68 Potash Nursery

Cow Green, Bacton, Stowmarket, Suffolk IP14 4HJ
Tel: 01449 781671
Specialities: Fuchsia

69 Prestige Plants

Blacksmiths Lane, Middlewood Green, Stowmarket, Suffolk IP14 5EU
Tel: 01449 711144
Specialities: Topiary, Pots

70 Priory Plants

1 Covey Cottages, Hintlesham, Ipswich,
Suffolk IP8 3NY
Tel: 01473 652656
e-mail: sue.mann3@btinternet.com
website: www.prioryplants.co.uk
Specialities: Geranium, Agapanthus, Penstemon

71 R Bareham

The Nurseries, Stoke Road, Clare, Sudbury,
Suffolk CO10 8HH
Tel: 01787 277202

72 R Dare

The Nurseries, Wratting Road, Haverhill,
Suffolk CB9 0DA
Tel: 01440 706776

73 Raymond Rolfe

Bungalow, The Green, West Row, Bury St. Edmunds,
Suffolk IP28 8QH
Tel: 01638 715925

74 Roots & Shoots

Stowmarket Road, Badley, Ipswich, Suffolk IP6 8RR
Tel: 01449 721965
Specialities: Annuals/Biennials/Bedding, Herbs, Bulbs

75 Roselea Nursery

Kirton Rd, Trimley St. Martin, Felixstowe,
Suffolk IP11 0QL
Tel: 01394 448675
Specialities: Vegetable, Farm Shop

76 Rougham Hall Nurseries

A14 Rougham, Ipswich Road, Rougham,
Bury St. Edmunds, Suffolk IP30 9LZ
Tel: 01359 270577
Specialities: Herbaceous Perennials

77 Sandford Garden Centre

Barton Mills, Bury St. Edmunds, Suffolk IP28 6AH
Tel: 01638 713207

78 Sandy Lane Nursery

Townhouse Lane, Wattisfield, Diss,
Suffolk IP22 1NE
Tel: 01359 251678
Specialities: Trees, Ornamentals, Hedging

79 Shrubland Park Nurseries

Coddenham, Shrubland Park, Coddenham, Ipswich,
Suffolk IP6 9QJ
Tel: 01473 833187
Specialities: Conifers, Cacti And Succulents

80 Southwest Nurseries

6 St Catherines Road, Long Melford, Sudbury,
Suffolk CO10 9JU
Tel: 01787 374286

81 Sperrinks

Cherry Tree Nursery The Street Gazeley, The Street
Roundabout, Gazeley, Newmarket, Suffolk CB8 8RB
Tel: 01638 750768
Specialities: Annuals/Biennials/Bedding

82 Stonham Barnes Garden Centre

Stonham Barns, Pettaugh Road, Stonham Aspal,
Stowmarket, Suffolk IP14 6AT
Tel: 01449 711103

83 Swann Nursery

Eyke Road, Bromeswell, Woodbridge,
Suffolk IP12 2PN
Tel: 01394 460127
Specialities: Shrubs, Conifers, Hedging

84 Tattingstone Garden Centre

The Nursery, The Heath, Tattingstone, Ipswich,
Suffolk IP9 2LX
Tel: 01473 327777

85 Thompson & Morgan (UK) Ltd

Poplar Lane, Copdock, Ipswich, Suffolk IP8 3BU
Tel: 01473 688821
Specialities: Exotics, Flower, Vegetable

86 Three Willows Garden Centre

St Johns Road, Bungay, Suffolk NR35 1BW
Tel: 01986 893834

87 TLC Botanica

Chantry Farm, Eyke Road, Campsea Ashe,
Woodbridge, Suffolk IP13 0PZ
Tel: 01728 747113

88 Unusual Herbs and Edibles

23 Mill Lane, Wrentham, Beccles, Suffolk NR34 7JQ
Tel: 01502 675364
Specialities: Herbs, Vegetables, Organic Edible Plantrs

89 Victoria Nurseries

1 Kettlebaston Way, Ipswich, Suffolk IP4 2XX
Tel: 01473 253980

90 The Walled Garden

Park Road, Benhall, Saxmundham, Suffolk IP17 1JB
Tel/Fax: 01728 602510
e-mail: jim@thewalledgarden.co.uk
website: www.thewalledgarden.co.uk

SEE ADVERTISEMENT BELOW

Specialities:

Hardy Perennials; Terracotta Pots

Description:

The Walled Garden offers choice plants for sale in a beautiful setting. Complementing the nursery is a 1-acre walled garden of mixed perennial and shrub display borders which visitors can walk around.

Directions:

On the A12, follow brown tourist signs sited 200m west of the A1094 turn to Aldeburgh. Turn into Park Road. You will find the nursery in ½ mile on the right.

91 West End Nurseries

24 Westward House, Leiston, Suffolk IP16 4HX
Tel: 01728 830616

92 Winter Flora Plant Centre

Hall Farm, London Road, Weston, Beccles, Suffolk NR34 8TT
Tel: 01502 716810

93 Woolpit Nurseries

Old Stowmarket Road, Woolpit, Bury St. Edmunds, Suffolk IP30 9QS
Tel: 01359 240370

94 Woottens Plants

Blackheath Road, Wenhaston, Halesworth, Suffolk IP19 9HD
Tel: 01502 478258
Specialities: Herbaceous Perennials

95 Wrights Roses

Clements Nursery, The Street, Brundish, Nr Woodbridge, Suffolk IP13 8BL
Tel: 01379 388162
Specialities: Containerised Roses, Vegetables

96 Wyevale Garden Centre (Bury St Edmunds)

Rougham Road, Bury St. Edmunds, Suffolk IP33 2RN
Tel: 01284 755818

97 Wyevale Garden Centre (Sudbury)

Newton Road, Sudbury, Suffolk CO10 0PZ
Tel: 01787 373628

98 Wyevale Garden Centre (Woodbridge)

Grundisburgh Road, Hasketon, Woodbridge, Suffolk IP13 6HX
Tel: 01394 380022

SURREY

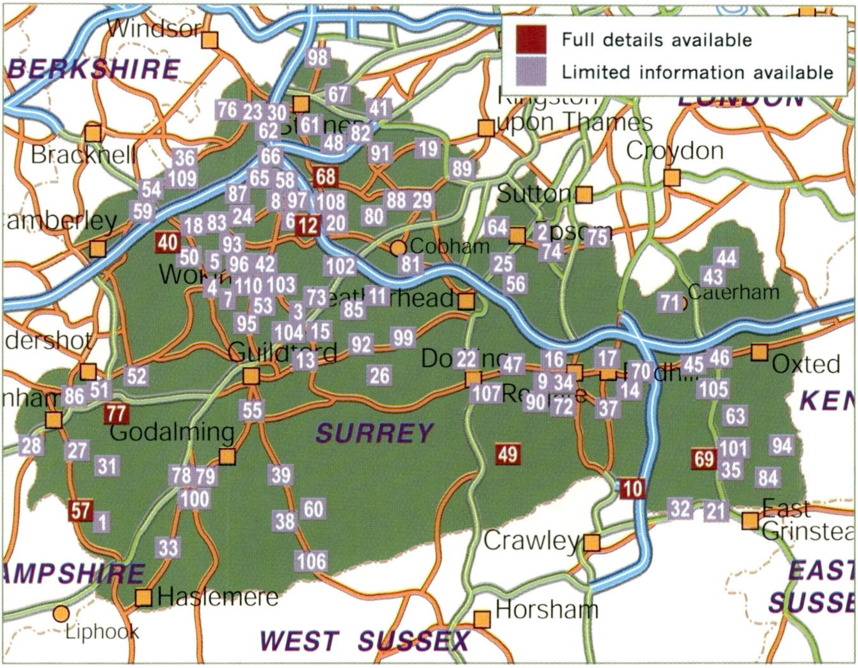

1 Avalon Garden Centre

Tilford Road, Churt, Farnham, Surrey GU10 2LL
Tel: 01428 604842
Specialities: Annuals/Biennials/Bedding, Shrubs

2 Beechcroft Nursery

127 Reigate Road, Epsom, Surrey KT17 3DE
Tel: 020 8393 4265
Specialities: Alpines/Rock Plants, Conifers, Annuals/Biennials/Bedding

3 Bentinck Nursery

Portsmouth Road, Ripley, Woking, Surrey GU23 6JA
Tel: 01483 225132

4 Blackhorse Nursery

Blackhorse Road, Woking, Surrey GU22 0QT
Tel: 01483 472486

5 Botany Barns (Knaphill Garden Centre)

Barrs Lane, Knaphill, Woking, Surrey GU21 2JW
Tel: 01483 481212
Specialities: Herbaceous Perennials, Shrubs

6 Bourne Valley Garden Centre

Woodham Park Road, Woodham, Weybridge, Surrey KT15 3TH
Tel: 01932 342123

7 Briarwood Nurseries

Saunders Lane, Woking, Surrey GU22 0NT
Tel: 01483 722353

8 Brox Road Nurseries

Brox Rd, Ottershaw, Chertsey, Surrey KT16 0LQ
Tel: 01932 872351

9 Buckland Nurseries

Buckland, Reigate Road, Reigate, Surrey RH2 9RE
Tel: 01737 242990
Specialities: Shrubs, Paving, Composts

10 Burstow Nurseries

Shipley Bridge, Horley, Surrey RH6 9SR
Tel/Fax: 01293 771942
e-mail: enquiries@burstownurseries.co.uk
website: www.burstownurseries.co.uk

Specialities:

Bedding Plants; Trees & Shrubs; Pots & Stoneware

Products and Services:

- Bird Baths & Tables
- Calor Gas
- Credit Cards Accepted
- Garden Shop
- Growing Media
- Hanging Baskets
- Landscaping Products
- Local Produce
- Gift Vouchers
- Ornamental/Fruit Trees
- Plants
- Qualified, friendly staff
- Seasonal Displays
- Vegetable Plants
- Seeds and Bulbs
- Wild Bird & Wildlife Products

Directions:

The Nursery is located under the Gatwick flight path, on the B2037 between Horley and Pound Hill, Crawley

Description:

Burstow Nurseries was founded in 1949 and was originally the walled garden for Burstow Hall which supplied produce to local shops and businesses. Over the past few years, the Nursery has continued to grow and has developed into a Garden Centre offering a "back to basics" approach for plant lovers.

In addition to growing a number of plants on site, the Nursery use a number of other local nurseries which enables them to offer a service for ordering plants not readily available from stock. The Nursery is passionate about plants and tries to select unusual and rare species in addition to the regular varieties.

As a member of the Horticultural Trades Association the Nursery participates in the National Garden Gift Voucher scheme.

11 Cedar Nursery

Horsley Road, Cobham, Surrey KT11 3JX
Tel: 01932 862473

12 Clay Lane Nursery

3 Clay Lane, South Nutfield, Nr Redhill, Surrey. RH1 4EG
Tel: 01737 823307
website: www.claylane-fuchsias.co.uk

Specialities:

Fuchsias, Original & Unusual Varieties.

Description:

A small nursery specialising in an interesting and unusual collection of Fuchsia varieties not normally available. All plants are grown at the nursery so help and advice is always at hand.

Directions:

From A25 at Nutfield follow sign to South Nutfield and then to Redhill Aerodrome. The nursery is 100m from the cross roads in Clay Lane.

13 Clandon Park Garden Centre

Clandon Park, West Clandon, Guildford, Surrey GU4 7RQ
Tel: 01483 222925
Specialities: Garden Furniture, Garden Products, Restaurant/Coffee Shop

14 Clay Lane Nursery

3 Clay Lane, South Nutfield, Redhill, Surrey RH1 4EG
Tel: 01737 823307

15 The Clematis Centre

Ripley Road, Send, Woking, Surrey GU23 7LT
Tel: 01483 223623
Specialities: Clematis

16 Clifton Nursery

Cliftons Lane, Reigate, Surrey RH2 9RA
Tel: 07778 659525

17 Colebrook Garden Centre

St. Annes Drive, Redhill, Surrey RH1 1AU
Tel: 01737 761949

18 Conifers Nursery

Bagshot Road, West End, Woking, Surrey GU24 9QR
Tel: 01276 858470
Specialities: Conifers

19 Creek Aquatic Garden Centre

427 Walton Road, West Molesey, Surrey KT8 2EJ
Tel: 020 8941 8758
Specialities: Aquatics, Water Products and Features

20 Crockford Park Nurseries

New Haw Road, Addlestone, Surrey KT15 2BU
Tel: 01932 853886
Specialities: Garden Products, Aquatics, Restaurant/Coffee Shop

21 Danecourt Nurseries

81 Copthorne Road, Felbridge, East Grinstead, Surrey RH19 2PB
Tel: 01342 326542

22 Denbies Kitchen Garden

Denbies Vine Estate, London Road, Dorking, Surrey RH5 6AA
Tel: 01306 880720
Specialities: Vines

23 Dunkirk Nursery

41 Clarence Street, Egham, Surrey TW20 9QY
Tel: 01784 432805
Specialities: Annuals/Biennials/Bedding

24 Evergreen Plant Supplies

Little Heath Nursery, Chobham, Woking, Surrey GU24 8RJ
Tel: 01276 857347
Specialities: Garden Products, Climbers, Shrubs

25 Farm Lane Nursery

Farm Lane, Ashtead, Surrey KT21 1LY
Tel: 01372 274400
Specialities: Shrubs, Herbaceous Perennials

26 Foliage Scented & Herb Plants

Walton Poor, Ranmore Common, Dorking, Surrey RH5 6SX
Tel: 01483 282273
Specialities: Herbs

27 Frensham Garden Centre

The Reeds Road, Frensham, Farnham, Surrey GU10 3BP
Tel: 01252 792545

28 Garden Style

Wrecclesham Hill, Wrecclesham, Farnham, Surrey GU10 4JX
Tel: 01252 735331
Specialities: Topiary, Shrubs, Conifers

29 Garsons Garden Centre

Garson Farm, Winterdown Road, Esher, Surrey KT10 8LS
Tel: 01372 460181

30 Ginkgo Garden Centre

Vicarage Road, Egham, Surrey TW20 8NT
Tel: 01784 473287
Specialities: Garden Products

31 Greenhills Nursery

Grange Road, Tilford, Farnham, Surrey GU10 2DY
Tel: 01252 795111
Specialities: Trees, Fruit Trees, Rose

32 Haskins Garden Centre

Snowhill Lane, Copthorne, Crawley, Surrey RH10 3EY
Tel: 01342 712545

33 Haslemere Garden Centre

Haslemere Rd, Brook, Godalming, Surrey GU8 5LB
Tel: 01428 685517

34 Heathfield Nurseries

Flanchford Road, Reigate, Surrey RH2 8AA
Tel: 01737 247641
Specialities: Bamboos, Annuals/Biennials/Bedding, Shrubs

35 Herons Bonsai Ltd

Wire Mill Lane, Newchapel, Lingfield, Surrey RH7 6HJ
Tel: 01342 832657
Specialities: Bonsai

36 Hillier Garden Centre Sunningdale

Sunningdale Nurseries, London Road, Windlesham, Surrey GU20 6LQ
Tel: 01344 623166

37 Hillside Plants Nursery

109 Horley Road, Earlswood, Redhill, Surrey RH1 5AS
Tel: 01737 765645

38 Hollyhocks

Unit 3b, Smithbrook Kilns, Cranleigh,
Surrey GU6 8JJ
Tel: 01483 268688

39 J & M Garden Nursery

High Banks, Birtley Road, Bramley, Guildford,
Surrey GU5 0LA
Tel: 01483 894855
Specialities: Garden Products

40 JW Elliott & Sons Ltd

Fenns Lane Nurseries, Fenns Lane, West End,
Woking, Surrey GU24 9QE
Tel: 01483 472210 Fax: 01483 797387
e-mail: sales@fennslanenursery.co.uk
website: www.fennslanenursery.co.uk

Specialities:

Japanese Maples; Field Grown Conifers; Ornamental & Fruit Trees

Description:

This family-run business was established in 1945 by the present owner's grandfather. It covers 31 acres with 7 acres dedicated to containerised plants, 95% of which are grown on site. Shrubs, trees and conifers are potted in spring from the field and the nursery also stocks an extensive selection of Japanese maples. The nursery also offers a garden maintenance and landscaping service.

Directions:

From the A322 turn left onto Brentmoor Road, then take first left onto Fenns Lane. The nursery is 100 yards on the right.

41 Jacquelines Garden Ltd

75 Vicarage Road, Sunbury-on-Thames,
Surrey TW16 7QD
Tel: 01932 781990

42 King Acre Landscapes

Wyevale Garden Centre, Egley Road, Woking,
Surrey GU22 0NH
Tel: 01483 750844
Specialities: Landscape Gardening

43 Knights Garden Centres

Rosedene Nursery, Woldingham Road, Woldingham,
Caterham, Surrey CR3 7LA
Tel: 01883 653142

44 Knights Garden Centres

Chelsham Place, Limpsfield Road, Chelsham,
Surrey CR6 9DZ
Tel: 01883 622340

45 Knights Ivy Mill Nursery

Bletchingley Road, Godstone, Surrey RH9 8NB
Tel: 01883 742665

46 Knights Nag's Hall Nursery

Oxted Road, Godstone, Surrey RH9 8DB
Tel: 01883 742275

47 Knights Plant Centre

Station Road, Betchworth, Surrey RH3 7DF
Tel: 01737 842099

48 Laleham Nurseries

Laleham Road, Shepperton, Surrey TW17 0JW
Tel: 01932 563322

49 Lilies Water Gardens

Broad Lane, Newdigate, Surrey RH5 5AT
Tel: 01306 631064 Fax: 01306 631693
e-mail: mail@lilieswatergardens.co.uk
website: www.lilieswatergardens.co.uk

Specialities:

Water Garden Plants; Herbaceous incl Dry-loving Plants

Products and Services:

- Credit Cards Accepted
- Mail orders/Internet orders
- Car Park
- Planting service
- Repotting service
- Trade discounts

Directions:

The nursery is in a pleasant, rural setting about 6 miles south of Dorking.

Opening hours:

Tuesday-Saturday 9am-5pm, March until the end of August.

Description:

Simon Harman, the owner of **Lilies**, has more than 17 years experience of growing **all types of plants for water gardens and surrounding areas** - including interesting, unusual and rare plants. He has won 17 awards at the R.H.S. Hampton Court Flower Show for water show gardens including 7 Gold Awards and, on 4 occasions, the much-coveted **Tudor Rose** award.

At Lilies you will find more than **80** varieties of **water lilies** and over **100 marginal and oxygenating plants**. There are also collections of **irises,ferns, grasses and herbaceous plants (dry and moisture loving)**. All plants are grown in a natural, duckweed-free environment.

On the nursery, there is also a range of **planting accessories** including floating baskets (ideal for tubs and deeper ponds); aqua planters, aquatic soil, planting baskets and other useful products. Simon offers a **consultancy service** covering plant removal,

preparation and implementation of planting schemes, pond maintenance and repotting. He is always happy to provide practical advice and guidance on all aspects of water gardening.

50 Lincluden Nursery

Shaftesbury Road, Bisley Green, Bisley, Woking, Surrey GU24 9EN
Tel: 01483 797005
Specialities: Climbers

51 Little Acre Nursery

St Georges Road, Badshot Lea, Farnham, Surrey GU9 9NT
Tel: 01252 782942

52 Littlebrook Fuchsias

Little Brook, Ash Green Lane West, Ash Green, Aldershot, Surrey GU12 6HL
Tel: 01252 329731
Specialities: Fuchsia

53 Loampits Farms

99 Westfield Road, Woking, Surrey GU22 9QR
Tel: 01483 725646

54 Longacres Nursery

London Road, Bagshot, Surrey GU19 5JB
Tel: 01276 476778
Specialities: Floristry, Shrubs, Trees

55 Loxhill Nursery Ltd

Old Portsmouth Road, Artington, Guildford, Surrey GU3 1LN
Tel: 01483 457979
Specialities: Trees, Shrubs, Herbaceous Perennials

56 Marsden Garden Centre

Pleasure Pit Road, Ashtead, Surrey KT21 1HU
Tel: 01372 273891
Specialities: Annuals/Biennials/Bedding, Garden Furniture

57 Millais Nurseries

Crosswater Farm, Crosswater Lane, Churt, Farnham, Surrey GU10 2JN
Tel: 01252 792698 Fax: 01252 792526
e-mail: sales@rhododendrons.co.uk
website: www.rhododendrons.co.uk

Specialities:

Rhododendrons; Azaleas; Magnolias; Japanese Maples

Products and Services:

- Display Gardens
- Information Desk
- Plant Guarantee
- Plants

Directions:

From the A287 turn into Jumps Road half a mile north of Churt village centre (signposted Tilford). After ¼ mile turn sharp left into Crosswater Lane.

Description:

This specialist rhododendron nursery is the result of the Millais family passion for rhododendrons. Family interest started with
J G Millais, a great naturalist who published the acclaimed two volume series 'Rhododendrons' (1917 and 1924) which described many rhododendrons for the first time. During the 1980's and 1990's, Ted and Romy Millais organised many plant hunting trips to the Himalayas, and introduced several wonderful new species. Under the management of their son David, the nursery has developed into one of the country's leading rhododendron growers.

The 5-acre nursery grows around 35,000 rhododendrons each year, one of the largest ranges in the country. It has won RHS Silver Gilt medals and the Rothschild Cup for best display of rhododendrons at Chelsea in 2005 and 2006. The nursery has everything from tiny Alpine gems to large woodland big-leaves specimens. Adjoining the plant centre is the trials garden where new varieties from around the world are planted out. The 6-acre woodland garden at Crosswater Farm is also open daily in May and features ponds, a stream and companion planting.

58 The Murray Road Nursery

Murray Road, Ottershaw, Chertsey, Surrey KT16 0HT
Tel: 01932 873812

59 Notcutts Garden Centre

Waterers Nurseries, London Road, Bagshot, Surrey GU19 5DG
Tel: 01276 472288

60 Notcutts Garden Centre

Guildford Road, Cranleigh, Surrey GU6 8LT
Tel: 01483 274222

61 Notcutts Garden Centre

Staines Road, Staines, Surrey TW18 2SF
Tel: 01784 460832

62 The Nurseries

Clockhouse Lane, Thorpe Lea, Egham, Surrey TW20 8PG
Tel: 01784 432351

63 Occasionally Yours

Lingfield Common Rd, Lingfield, Surrey RH7 6BZ
Tel: 01342 833937
Specialities: Farm Shop

64 Old Moat Garden Centre

Horton Lane, Epsom, Surrey KT19 8PQ
Tel: 01372 204251
Specialities: Herbaceous Perennials

65 The Otter Plant Centre

Murray Rd, Ottershaw, Chertsey, Surrey KT16 0HT
Tel: 01932 874875

66 Pantiles Nurseries

Almners Road, Lyne, Chertsey, Surrey KT16 0BJ
Tel: 01932 872195
Specialities: Shrubs, Trees

67 PJ's Palms and Exotics

41 Salcombe Road, Ashford, Surrey TW15 3BS
Tel: 01784 250181
Specialities: Exotics, Palms

68 The Plantation Nursery

Woburn Hill, Addlestone, Surrey KT15 2QG
Tel: 01932 833866 Fax: 01932 833872
e-mail: info@theplantationnursery.co.uk
website: www.theplantationnursery.co.uk

SEE ADVERTISEMENT BELOW

Specialities:

Specimen trees/shrubs, ornamentals, exotic plants

Description:

The nursery carries an extensive range of Bamboos, Conifers, Topiary, Palms, Citrus, Olives, Grasses, Rhododendron and many other unusual plants in various shapes and sizes. There is an excellent range of contemporary stylish architectural plants and more exotic Mediterranean trees and shrubs in stock.

Directions:

The Plantation is just 1 mile off the M25, junction 11, on the A317 towards Weybridge.

69 Plants 'n' Gardens

Eastbourne Road, Newchapel, Surrey RH7 6LF
Tel: 01342 833911
e-mail: paul@plantsandgardens.freeserve.co.uk
website: www.plants-n-gardens.co.uk

Specialities:
Japanese Maples; Hostas; Wisterias

Products and Services:
- Credit Cards Accepted
- Display Gardens
- Fruit Trees
- Garden Design
- Information Desk
- Large Car Park
- Plants
- Trees

Directions:
The Nursery is 150yds from the junction of the A22/B2028 in Newchapel. Take the A22 towards the M25 and the Nursery is on the left hand side.

Description:
Owner Paul Grimmer studied Horticulture at Wye College and is particularly passionate about unusual shrubs and perennials that are straightforward to grow. He has an expert knowledge of plants and design, and is happy to answer any queries. The Nursery was established in 1988 and has steadily expanded with projects such as the new display garden. The large glasshouse contains many Japanese Maples as well as seasonal bedding and basket plants. In the spring, a special promotion offers excellent discounts on trays. The growing tunnels are home to 16 varieties of wisteria as well as a range of hostas, rhododendrons and grasses. A unique feature of the Nursery is its display of stone statues from Zimbabwe.

70 Priory Farm

Sandy Lane, South Nutfield, Redhill,
Surrey RH1 4EJ
Tel: 01737 823500

71 Queens Park Community Garden Centre

Queens Park Rd, Caterham, Surrey CR3 5RB
Tel: 01883 383829

72 Reigate Garden Centre

143 Sandcross Lane, Reigate, Surrey RH2 8HH
Tel: 01737 248188

73 Ripley Nurseries

Portsmouth Road, Ripley, Woking, Surrey GU23 6EY
Tel: 01483 225090

74 Rochkam Nursery

139 Reigate Road, Epsom, Surrey KT17 3DW
Tel: 020 8394 2186
Specialities: Garden Products

75 S G Clarke Nursery

23 Croydon Road, Banstead, Surrey SM7 3BE
Tel: 020 8643 7979

76 Savill Garden Plant Centre

Wick Lane, Englefield Green, Egham,
Surrey TW20 0UU
Tel: 01784 435544

77 Seale Nurseries Specialist Rose Grower

Seale Nurseries, Seale Lane, Seale,
Farnham, Surrey GU10 1LD
Tel: 01252 782410
RHS medallist for roses

Seale
SPECIALIST ROSE GROWER
established 1948

Specialities:
Seale Super Roses; Climbing Roses; Scented Roses

Description:
The May family have been growing roses, and lots else besides, at Seale Nurseries since 1948. They hand raise everything they produce and use a minimum of protection and chemicals. Their roses are grown using their patented container growing system and may be planted at any time and will give superb results from the start.

Directions:

Seale Nurseries are in the village of Seale, just off the A31 Hog's Back, between Farnham and Guildford, 20 minutes from RHS Wisley.

78 Secretts Garden Centre

Old Portsmouth Road, Milford, Godalming, Surrey GU8 5HL
Tel: 01483 520550

79 Secretts of Milford Garden Centre

Hurst Farm, Chapel Lane, Milford, Godalming, Surrey GU8 5HU
Tel: 01483 520500

80 Seven Hills Garden Centre

Hillview Nursery, Seven Hills Road, Walton-on-Thames, Surrey KT12 4DD
Tel: 01932 844575

81 Seymours Garden & Leisure Centre

Stoke Road, Stoke D'Abernon, Cobham, Surrey KT11 3PU
Tel: 01932 862530

82 Shepperton Garden Centre

Nutty Lane, Shepperton, Surrey TW17 0RQ
Tel: 01932 781474

83 Sienna Flora

Bagshot Road, Chobham, Woking, Surrey GU24 8SJ
Tel: 01276 855408
Specialities: Garden Furniture, Garden Products, Giftware

84 Southon Plants

Mutton Hill, Dormansland, Lingfield, Surrey RH7 6NP
Tel: 01342 870150

85 Spring Reach Nursery

Long Reach, Ockham, Surrey GU23 6PG
Tel: 01483 284769
Specialities: Herbaceous Perennials, Clematis

86 Squires Garden Centre (Badshot Lea)

Badshot Lea Road, Badshot Lea, Farnham, Surrey GU9 9JX
Tel: 01252 333666

87 Squires Garden Centre (Chertsey)

Holloway Hill, Chertsey, Surrey KT16 0EA
Tel: 01932 563727

88 Squires Garden Centre (Hersham)

Burwood Road, Hersham, Walton-on-Thames, Surrey KT12 4AR
Tel: 01932 247579

89 Squires Garden Centre (Long Ditton)

Woodstock Lane North, Long Ditton, Surbiton, Surrey KT6 5HN
Tel: 020 8398 7170

90 Squires Garden Centre (Reigate)

Reigate Road, Buckland (A25), Reigate, Surrey RH2 9RE
Tel: 01737 247217

91 Squires Garden Centre (Shepperton)

Halliford Road, Shepperton, Surrey TW17 8RU
Tel: 01932 784121

92 Squires Garden Centre (West Horsley)

Epsom Road, West Horsley, Leatherhead, Surrey KT24 6AR
Tel: 01483 282911

93 Squires Garden Centre (Woking)

Littlewick Road, Woking, Surrey GU21 4XR
Tel: 01276 858446

94 Starborough Nursery

Starborough Road, Marsh Green, Edenbridge, Surrey TN8 5RB
Tel: 01732 865614
Specialities: Garden Products, Trees, Shrubs

95 Sutton Green Garden Nursery

Guildford Road, Sutton Green, Guildford,
Surrey GU4 7QA
Tel: 01483 232366

96 Toobees Exotics

20 Inglewood, Woking, Surrey GU21 3HX
Tel: 01483 722600

97 Transplant Nurseries

Otterhill Farm, Rowtown, Addlestone,
Surrey KT15 1HD
Tel: 01932 855918

98 Vermeulens Garden Centre

Horton Road, Staines, Surrey TW19 6AE
Tel: 01784 451737
Specialities: Garden Products, Garden Furniture

99 Vineries Garden Centre

Lower Road, Effingham, Leatherhead,
Surrey KT24 5JP
Tel: 01372 458934
Specialities: Bulbs, Pots

100 Wheeler Street Nursery

Wheeler Lane, Witley, Godalming, Surrey GU8 5QP
Tel: 01428 682638
Specialities: Annuals/Biennials/Bedding, Vegetables

101 Willow Trees Nursery

Newchapel Road, Lingfield, Surrey RH7 6BL
Tel: 01342 834961

102 The Wisley Plant Centre

RHS Garden, Wisley Garden, Wisley, Woking,
Surrey GU23 6QB
Tel: 01483 211113
Specialities: Herbaceous Perennials, Trees

103 Woking Nursery

Egley Road, Woking, Surrey GU22 0NH
Tel: 01483 723089
Specialities: Specimen Trees, Shrubs

104 Woodcote Park Nursery Ltd

Ripley Rd, Send, Woking, Surrey GU23 7LT
Tel: 01483 223504
Specialities: Climbers, Shrubs, Grasses

105 Woodham Nursery

Eastbourne Road, South Godstone, Godstone,
Surrey RH9 8EZ
Tel: 01342 892331

106 Wyevale Garden Centre (Alfold)

Horsham Road, Alfold, Cranleigh, Surrey GU6 8JE
Tel: 01403 752359

107 Wyevale Garden Centre (Dorking)

Reigate Road, Dorking, Surrey RH4 1NP
Tel: 01306 884845

108 Wyevale Garden Centre (Weybridge)

New Haw Road, Addlestone, Surrey KT15 2BU
Tel: 01932 845084

109 Wyevale Garden Centre (Windlesham Garden & Furniture Centre)

London Road, Windlesham, Surrey GU20 6LL
Tel: 01344 621411

110 Wyevale Garden Centre (Woking)

Egley Road, Woking, Surrey GU22 0NH
Tel: 01483 714861

SUSSEX - EAST

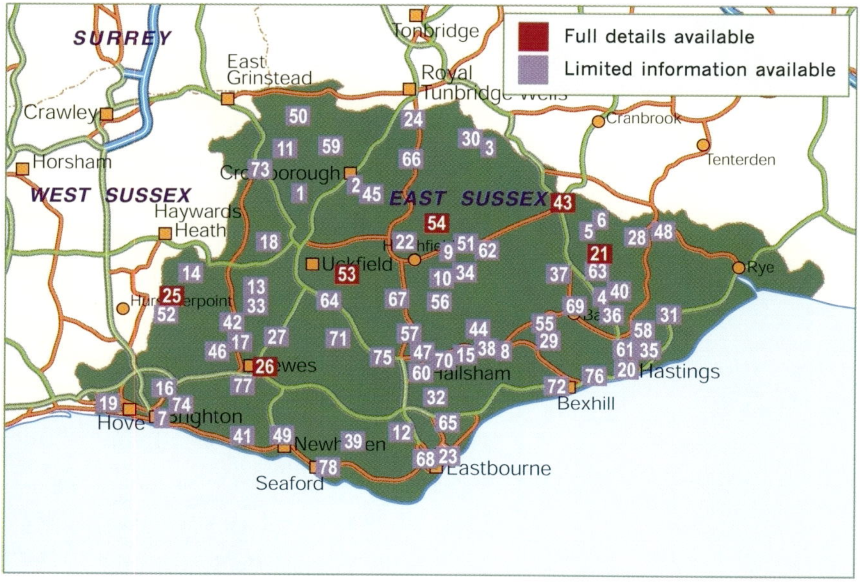

1 Ashdown Forest Garden Nursery

Duddleswell, Duddleswell, Uckfield,
East Sussex TN22 3JP
Tel: 01825 712300

2 B L Jordan & Son

Rockington Nurseries, Blackness Road,
Crowborough, East Sussex TN6 2NB
Tel: 01892 654083

3 Beals Barn Gardens

Bewlbridge Lane, Wadhurst, East Sussex TN5 6HJ
Tel: 01892 782657

4 Blackbrooks Garden Centre

Main A21, New Road, Sedlescombe, Battle,
East Sussex TN33 0RJ
Tel: 01424 870710

5 Bodiam Bonsai

Ewhurst Green, Ewhurst Green, Robertsbridge,
East Sussex TN32 5RJ
Tel: 01580 830644
Specialities: Bonsai, Bonsai

6 Bodiam Nursery

Cowfield Cottage, Bodiam, Robertsbridge,
East Sussex TN32 5RA
Tel: 01580 830811

7 Bonsai Ko

45 Sydney St, Brighton, East Sussex BN1 4EP
Tel: 01273 621743
Specialities: Trees, Trees, Bonsai

8 Boreham Mill Nurseries

Boreham Street, Hailsham, East Sussex BN27 4SD
Tel: 01323 832103
Specialities:

9 Broad Oak Garden Centre

Burwash Road, Broad Oak, Heathfield,
East Sussex TN21 8UP
Tel: 01435 860890
Specialities: Restaurant/Coffee Shop

10 Cade Street Nursery

Cade St, Cade Street, Heathfield,
East Sussex TN21 9BU
Tel: 01435 862988
Specialities: Annuals/Biennials/Bedding

11 Casablanca Nursery

Shepherds Hill, Colemans Hatch, Hartfield,
East Sussex TN7 4HN
Tel: 01342 824064

12 Chalk Farm Nursery

Coopers Hill, Willingdon, Eastbourne,
East Sussex BN20 9JD
Tel: 01323 503322
Specialities: Herbaceous Perennials

13 Chubbs Nursery

Brickyard Farm, Town Littleworth, Cooksbridge,
Lewes, East Sussex BN8 4TD
Tel: 01273 400218

14 Coldharbour Nursery

Hundred Acre Lane, Wivelsfield Green, Haywards
Heath, East Sussex RH17 7RS
Tel: 01444 471034
Specialities: Hanging Baskets, Shrubs

15 Coopers Croft Nurseries

New Rd, Herstmonceux, Hailsham,
East Sussex BN27 1PX
Tel: 01323 832151

16 Covers

Home Farm Road, Brighton, East Sussex BN1 9HU
Tel: 01273 607044
Specialities: Landscape Gardening, Paving

17 Covers

Cooksbridge, Cooksbridge, Lewes,
East Sussex BN8 4TJ
Tel: 01273 476133
Specialities: Fencing

18 Denniker Fruit Farm and Nursery

Fletching, Uckfield, East Sussex TN22 3SH
Tel: 01825 723596

19 Discount Hydroponics

320 Portland Road, Hove, East Sussex BN3 5LP
Tel: 01273 422210
Specialities: Hydroponics

20 Discount Hydroponics

1 Bus Buildings, Waldegrave Street, Hastings,
East Sussex TN34 1SJ
Tel: 01424 428186
Specialities: Hydroponics

21 Ed's Nursery & Sunflower Garden

Cripps Corner Road, Staplecross,
East Sussex FN32 5QA
Tel/Fax: 01580 830701

Specialities:

Roses; Climbers; Herbaceous

Products and Services:

- Seasonal Displays
- Disabled Facilities
- Display Gardens
- Fruit Trees
- Garden Construction
- Garden Design
- Garden Products
- Large Car Park
- Plants

Directions:

From the A21 south of Robertsbridge, go east on the B2089 to Vinehall Street. Turn left to Cripps Corner, left again on the B2165 and continue for 2 miles. The Nursery is on the left hand side.

Description:

Owner Ed Rogers comes from a farming family and has worked in horticulture for more than 30 years. He took over the present site in 1992 and over the years has extended and improved it considerably. Within the 2 ½ acre site are displayed many varieties of shrubs, conifers and herebaceous plants; climbers – especially clematis, honeysuckle, akebia and jasmine; and an extensive range of roses, including old-fashioned ones. Ornamental and fruits trees, compost and seeds, and terracotta and glazed pots are also here in abundance.

22 English Woodlands

Herring Lane, Cross In Hand, Heathfield,
East Sussex TN21 0UG
Tel: 01435 862992
Specialities: Shrubs, Trees

23 Esk Plants

Courtlands Rd, Courtlands Road, Eastbourne, East Sussex BN22 8TR
Tel: 01323 722922

24 Flower Power

Frant Road, Tunbridge Wells, East Sussex TN3 9HB
Tel: 01892 510190

25 Garden Pride Garden Centre

Common Lane, Ditchling, Sussex BN6 8TN
Tel: 01273 846844 Fax: 01273 845540
e-mail: info@garden-pride.com
website: www.garden-pride.com

SEE ADVERTISEMENT ABOVE

Specialities:

Plants; Gifts; Furniture

Products and Services:

- Barbeques
- Books and Stationery
- Clothing
- Credit Cards Accepted
- Disabled Facilities
- Dried & Artificial Flowers
- Fencing
- Garden Furniture
- Garden Products
- Giftware
- Greenhouses and Sheds
- Houseplants
- Information Desk
- Large Car Park
- Paving
- Plants
- Restaurant/Coffee Shop
- Seeds and Bulbs
- Trees

Directions:

From the A275 about 3 miles north of Lewes take the B2116 west to Ditchling (approx. 5 miles). Turn right on the B2112 through Ditchling village and the Garden Centre is 2 miles north of the village, on the right

26 Giganteum

The Yard, North Court, Lewes, East Sussex BN7 2AR
Tel: 01273 480777
e-mail: giganteum@hotmail.co.uk

Specialities:

Garden Landscaping & Design; Hardy Exotics; Topiary

Products and Services:

- Seasonal Displays
- Credit Cards Accepted
- Fruit Trees
- Garden Construction
- Garden Design
- Giftware Tokens
- Houseplants
- Information Desk
- Plant Guarantee
- Plants
- Seeds and Bulbs
- Trees

Directions:

Head east on the A27 to Lewes via the Cuilfail Tunnel. After the tunnel take first exit on the

roundabout; at second roundabout, take 1st exit then turn right into Harvey's Way, then turn left into North Court.

Description:
Giganteum really captures the imagination with a catalogue of inspirational "statement plants" that are interesting even in the depths of winter. Bamboos, palms, topiary, succulents, ferns, trees and shrubs are all available to buy, or to hire for those special occasions.

Giganteum also offers a consultation service tailored to the clients' needs, whether it be expert advice on existing gardens and the plants within, or a full design service aimed at working with the unique opportunities each space presents. Giganteum vouchers are also available for that special gift. The shop is open from 9.30am to 5pm on Tuesdays, Thursdays, Fridays and Saturdays. Private appointments can be made on Mondays and Wednesdays.

27 Goldcliff Nurseries

Uckfield Road, Ringmer, Lewes,
East Sussex BN8 5RY
Tel: 01273 813968
Specialities: Shrubs

28 Great Dixter Nursery

High Park Close, Northiam, Rye,
East Sussex TN31 6PH
Tel: 01797 253107

29 Great Park Farm Nursery

Great Park Farm, Catsfield, Battle,
East Sussex TN33 9DT
Tel: 01424 772531

30 Greencap Nursery

Sleepers Stile Road, Cousley Wood, Wadhurst,
East Sussex TN5 6QX
Tel: 01892 782685

31 Harborough Nurseries

The Thorne, Guestling, Hastings,
East Sussex TN35 4LU
Tel: 01424 814220

32 Hillier Garden Centre Eastbourne

Hailsham Road, Stone Cross, Pevensey,
East Sussex BN24 5BS
Tel: 01323 763240

33 House of Plants

Town Littleworth, Cooksbridge, Lewes,
East Sussex BN8 4TD
Tel: 01273 400001
Specialities: Houseplants

34 Hugletts Farm

Hugletts Lane, Heathfield, East Sussex TN21 9BY
Tel: 01435 862318
Specialities: Black Plants Speciality, Minature Bulbs

35 Ivy House Nursery

Bramble Wood, Ivyhouse Lane, Hastings,
East Sussex TN35 4NN
Tel: 01424 754763

36 Kent Street Nurseries

Kent Street, Sedlescombe, Battle,
East Sussex TN33 0SF
Tel: 01424 751134
Specialities: Geranium

37 Kipling Plants Ltd

Crossing Farm, Mountfield, Robertsbridge,
East Sussex TN32 5LR
Tel: 01580 882161

38 Lime Cross Nursery

Herstmonceux, Herstmonceux, Hailsham,
East Sussex BN27 4RS
Tel: 01323 833229
Specialities: Conifers, Shrubs, Trees

39 Litlington Nursery

Litlington, Polegate, East Sussex BN26 5RB
Tel: 01323 871211
Specialities: Pots, Alkaline Plants

40 The Lodge Nursery

Cottage Lane, Westfield, Hastings,
East Sussex TN35 4RP
Tel: 01424 870953

41 Louvain Organic Nursery

66 Telscombe Road, Peacehaven,
East Sussex BN10 7UB
Tel: 01273 584156

Merriments Gardens is a unique combination of a large plant nursery, garden shop, gift shop, licensed tearoom, wild bird care centre and one of the best gardens in the south of England. It has been featured on BBC-TV's Gardener's World, and in most major gardening magazines. The well-loved nursery grows an extensive range of perennials, shrubs and trees. The owner takes pride in the wonderful selection of unusual plants, many of which can be seen growing in the garden. Plants of exceptional quality, combined with friendly and knowledgeable staff, make this cherished nursery a paradise for plant lovers and the ultimate experience for discerning garden lovers. In the gift shop, alongside a good range of garden accessories, there's a wide choice of unusual items. Take a few minutes to stroll to the bird hide in the garden and watch the birds feeding. Then visit the extensive wild bird care centre for all you will need to care for the birds in your own garden. Finally, why not relax in the comfort of the tearoom which serves a selection of delicious snacks, lunches and cream teas.

- Books and Stationery
- Seasonal Displays
- Credit Cards Accepted
- Display Gardens
- Fruit Trees
- Giftware
- Large Car Park
- Plants
- Restaurant/Coffee Shop
- Seeds and Bulbs
- Trees

AS FEATURED ON TV'S GARDENER'S WORLD ......

MERRIMENTS GARDENS

Hawkhurst Road, Hurst Green, East Sussex TN19 7RA
Tel: 01580 860666 Fax: 01580 860324
e-mail: info@merriments.co.uk
website: www.merriments.co.uk

42 Mcbean's Orchids

Mcbeans Nursery, Resting Oak Hill, Cooksbridge, Lewes, East Sussex BN8 4PR
Tel: 01273 400228
Specialities: Orchids

43 Merriments Gardens

Hawkhurst Road, Hurst Green, East Sussex TN19 7RA
Tel: 01580 860666 Fax: 01580 860324
e-mail: info@merriments.co.uk
website: www.merriments.co.uk

SEE ADVERTISEMENT PAGE 268

Specialities:

Perennials; shrubs; Trees

Products and Services:

- Books and Stationery
- Seasonal Displays
- Credit Cards Accepted
- Display Gardens
- Fruit Trees
- Giftware
- Large Car Park
- Plants
- Restaurant/Coffee Shop
- Seeds and Bulbs
- Trees

Directions:

Merriments Gardens are situated on the A229, just off the A21 at Hurst Green

44 Merryweather's Herbs

Merryweathers Farm, Chilsham Lane, Herstmonceux, Hailsham, East Sussex BN27 4QH
Tel: 01323 833316
Specialities: Herbs, Wildlife Plants

45 Millbrook Garden Company

Tubwell Lane, Crowborough, East Sussex TN6 3RJ
Tel: 01892 663822
Specialities: Annuals/Biennials/Bedding

46 Mount Harry Trees

Mount Harry House, Ditchling Road, Offham, Lewes, East Sussex BN7 3QW
Tel: 01273 474456
Specialities: Specimen Trees, Trees

47 New Road Nurseries

New Rd, Herstmonceux, Hailsham, East Sussex BN27 4HU
Tel: 01323 846577

48 The Old Rose Nursery

Beales Lane, Northiam, Rye, East Sussex TN31 6QY
Tel: 01797 252680
Specialities: Rose

49 Paradise Park Garden Centre

Avis Road, Newhaven, East Sussex BN9 0DH
Tel: 01273 512123

50 Perryhill Nurseries Ltd

Edenbridge Road, Hartfield, East Sussex TN7 4JP
Tel: 01892 770377

51 Petals for Plants Ltd

Burwash Road, Broad Oak, Heathfield, East Sussex TN21 8XG
Tel: 01435 884111

52 Peter Eastwood Plants Ltd

Beardsfield Nursery, Common Lane, Ditchling, Hassocks, East Sussex BN6 8TN
Tel: 01273 845163

53 Pure Plants

Blackboys Nursery, Blackboys, Uckfield, East Sussex TN22 5LS
Tel: 01825 890858 Fax: 01825 890878
e-mail: info@pureplants.com
website: www.pureplants.com

Specialities:

Trees; Perennials; Grasses

Products and Services:

- Fruit Trees
- Garden Design
- Large Car Park
- Plants
- Trees

Directions:

Pure Plants is located between the market towns of Heathfield and Uckfield, on the western edge of Blackboys village on the B2102

Description:

Since opening in the spring of 2003, Pure Plants has gone from strength to strength by providing an extremely wide range of plants grown to the highest standards almost entirely without using peat. The area under cultivation has more than doubled in the last 3 years. The sales area has a well-designed layout and extensive information is provided on all plants. Seasonal plant combinations are located

throughout the nursery as well as planted areas for ideas and inspiration. Uniquely, the nursery has 2 acres of field grown trees and shrubs which can be viewed and specimens reserved. The owner of Pure Plants, Brian Fidler, trained and practised as a landscape architect and garden designer before moving into horticulture more than 10 years ago. His wife Fiona has worked in horticulture for over 25 years and has a vast range of knowledge and experience. She is skilled at grafting the many unusual trees and shrubs the nursery offers.

54 Rapkyns Nursery

Street End Lane, Broad Oak, Near Heathfield, East Sussex TN21 8UB
Tel/Fax: 01825 830065
e-mail: rapkyns@homecall.co.uk

Specialities:

Cottage Garden Plants; Perennials; Climbers; Grasses & Shrubs

Directions:

The nursery is exactly 2 miles down Street End Lane towards Mayfield off the A265 in Broad Oak near Heathfield.

Description:

Open Tuesdays, Thursdays and Fridays, March to October inclusive. This family-run nursery stocks an exciting range of over 1500 varieties of many unusual plants, all home-grown. Friendly and knowledgeable staff are always at hand to provide information and advice.

55 Redcoat Orchard

Main Road, Catsfield, Nr Battle, East Sussex TN33 9DU
Tel: 01424 772685

56 River Mead Nursery

Vines Cross, Heathfield, East Sussex TN21 9HB
Tel: 01435 813353

57 Rosslow Roses

North Street Farm, North Street, Hellingly, Hailsham, East Sussex BN27 4DZ
Tel: 01323 440888
Specialities: Rose, Containerised Roses

58 Rotherview Nursery

Ivyhouse Lane, Hastings, East Sussex TN35 4NP
Tel: 01424 756228
Specialities: Camellia

59 Royal Mires Nursery

Lye Green, Crowborough, East Sussex TN6 1UU
Tel: 01892 668850

60 Scamardella

Larrita, Amberstone, Hailsham, East Sussex BN27 1PE
Tel: 01323 847675
Specialities: Herbaceous Perennials, Shrubs

61 Scollays

400 The Ridge, Hastings, East Sussex TN34 2RR
Tel: 01424 753072
Specialities: Landscape Gardening

62 Shanks Nursery & Plant Centre

Old Orchard Nursery, Heathfield Road, Burwash Common, Etchingham, East Sussex TN19 7NE
Tel: 01435 882060

63 Staplecross Shrub Centre

Cripps Corner Road, Staplecross, Robertsbridge, East Sussex TN32 5QR
Tel: 01580 830678

64 Staverton Nursery & Farm Shop

Eastbourne Road, Halland, Lewes, East Sussex BN8 6PU
Tel: 01825 840249

65 Stone Cross Garden Centre

Rattle Road, Stone Cross, Pevensey, East Sussex BN24 5EB
Tel: 01323 763250
Specialities: Rose

66 Sussex Country Gardens

Mark Cross, Crowborough, East Sussex TN6 3PJ
Tel: 01892 852828
Specialities: Garden Furniture

67 Thorpe Gardens

Little London Road, Horam, Heathfield, East Sussex TN21 0BG
Tel: 01435 812455
Specialities: Annuals/Biennials/Bedding

68 Tutts Barn Nursery

Tutts Barn Lane, Eastbourne, East Sussex BN22 8XR
Tel: 01323 734064
Specialities: Shrubs, Annuals/Biennials/Bedding

69 Uckham Lane Nurseries

Uckham Lane, Battle, East Sussex TN33 0LY
Tel: 01424 772919

70 Usual & Unusual Plants

Onslow House, Magham Down, Hailsham, East Sussex BN27 1PL
Tel: 01323 840967
Specialities: Exotics, Herbaceous Perennials

71 Wallace Plants

Lewes Road, Laughton, Lewes, East Sussex BN8 6BN
Tel: 01323 811729
Specialities: Grasses, Annuals/Biennials/Bedding, Vegetables

72 Warburtons & The Pette Shoppe

41-43 Station Road, Bexhill-on-Sea, East Sussex TN40 1RG
Tel: 01424 211560

73 Wych Cross Nurseries & Garden Centre

Wych Cross, Wych Cross, Forest Row, East Sussex RH18 5JW
Tel: 01342 822705
Specialities: Rose

74 Wyevale Garden Centre (Brighton)

Warren Road, Brighton, East Sussex BN2 9XX
Tel: 01273 607888

75 Wyevale Garden Centre (Hailsham)

A22 Lower Dicker, Lower Dicker, Hailsham, East Sussex BN27 4BJ
Tel: 01323 844834

76 Wyevale Garden Centre (Hastings)

Bexhill Road, St. Leonards-on-Sea, East Sussex TN38 8AR
Tel: 01424 443414

77 Wyevale Garden Centre (Lewes)

Newhaven Road, Kingston, Lewes, East Sussex BN7 3NE
Tel: 01273 473510

78 Wynnes Garden Centre

53 Sutton Drove, Seaford, East Sussex BN25 3NQ
Tel: 01323 891292

SUSSEX - WEST

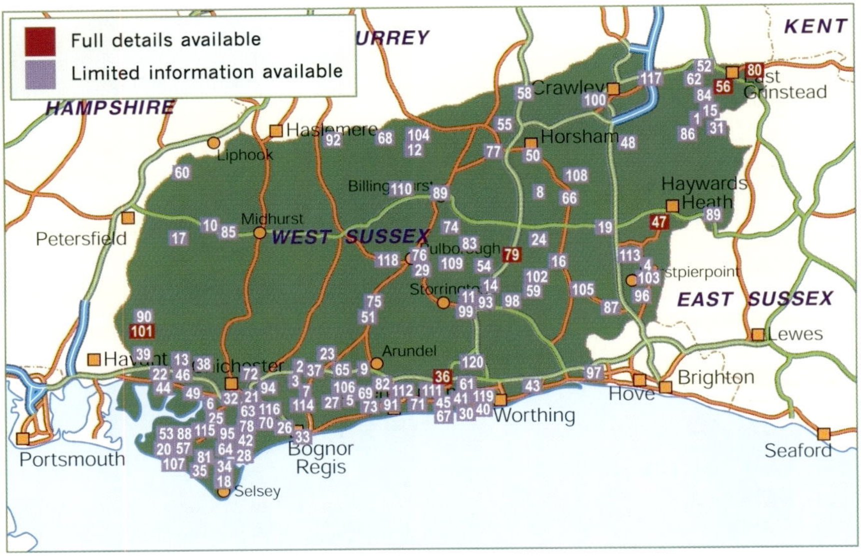

1 ADT

Whitestone Farm, Selsfield Road, West Hoathly,
East Grinstead, West Sussex RH19 4QN
Tel: 01342 810748

2 Aldingbourne Country Centre

Blackmill Lane, Norton, Chichester,
West Sussex PO18 0JP
Tel: 01243 542075

3 Aldingbourne Nurseries

Church Road, Aldingbourne, Chichester,
West Sussex PO20 3TU
Tel: 01243 544941

4 Allwoods

Summerfield Nursery, London Road, Hassocks,
West Sussex BN6 9NB
Tel: 01273 844229
Specialities: Dianthus (Carnation/Pinks)

5 Angels Nursery

Yapton Rd, Barnham, Bognor Regis,
West Sussex PO22 0AY
Tel: 01243 553250
Specialities: Hanging Baskets, Annuals/Biennials/Bedding, Fuchsia

6 Apuldram Roses

Appledram La South, Chichester,
West Sussex PO20 7EG
Tel: 01243 785769
Specialities: Rose

7 Architectural Plants

Lidsey Road, Woodgate, Nr Chichester,
West Sussex PO20 3SU
Tel: 01243 545008
Specialities: Bamboos, Seaside Speciality

8 Architectural Plants

Cooks Farm, Nuthurst Street, Nuthurst, Horsham,
West Sussex RH13 6LH
Tel: 01403 891772

9 Arundel Arboretum

Scotland Barn, Chichester Road, Arundel,
West Sussex BN18 0AD
Tel: 01903 883251

10 Aylings Garden Centre

Trotton, Petersfield, West Sussex GU31 5ES
Tel: 01730 813621
Specialities: Herbaceous Perennials, Trees, Greenhouses and Sheds

11 Barnard's

Rock Road, Washington, Pulborough,
West Sussex RH20 3BH
Tel: 01903 892320

12 Barnsfold Nurseries

Loxwood Road, Tismans Common, Rudgwick,
Horsham, West Sussex RH12 3BP
Tel: 01403 822493
Specialities: Annuals/Biennials/Bedding

13 Beachlands Nursery

Newells Lane, West Ashling, Chichester,
West Sussex PO18 8DD
Tel: 01243 573117

14 Big Plant Nursery

Hole Street, Ashington, Pulborough,
West Sussex RH20 3DE
Tel: 01903 891466

15 Birch Farm Nursery

W E Th Ingwersen Ltd., Gravetye, East Grinstead,
West Sussex RH19 4LE
Tel: 01342 810236
Specialities: Rockery Plants, Annuals/Biennials/Bedding

16 Birchfield Nursery

Kidders Lane, London Road, Henfield,
West Sussex BN5 9AB
Tel: 01273 494058

17 Birchfleet Nurseries

Greenfields Close, Nyewood, Petersfield,
West Sussex GU31 5JQ
Tel: 01730 821636
Specialities: Acer, Fagus (Beech), Quercus (Oak)

18 Bird Pond Nursery

Selsey Rd, Sidlesham, Chichester,
West Sussex PO20 7NF
Tel: 01243 641212

19 Bolney Nursery

Cowfold Road, Bolney, Haywards Heath,
West Sussex RH17 5QR
Tel: 01444 881784

20 Bramber Plant Centre

Chichester Road, West Wittering, Chichester,
West Sussex PO20 8QA
Tel: 01243 512004

21 Brick Kiln Nurseries

Bognor Road, Merston, Chichester,
West Sussex PO20 1EJ
Tel: 01243 531700

22 Broad Road Nurseries

Broad Road, Hambrook, Chichester,
West Sussex PO18 8RF
Tel: 01243 572889
Specialities: Annuals/Biennials/Bedding

23 C.R Upton

4 Top Road, Slindon, Arundel,
West Sussex BN18 0RP
Tel: 01243 814219
Specialities: Pumpkins

24 Camelia Botnar Homes & Gardens

Littleworth Lane, Partridge Green, Horsham,
West Sussex RH13 8NA
Tel: 01403 864556

25 Cedar Garden Centre

Birdham Road, Chichester, West Sussex PO20 7EQ
Tel: 01243 782666
Specialities: Specimen Plants, Trees

26 Chalcroft Nurseries

Chalcraft Lane, Bognor Regis,
West Sussex PO21 5TS
Tel: 01243 863346

27 Chapmans Nurseries

38 Hill Lane, Barnham, Bognor Regis,
West Sussex PO22 0BL
Tel: 01243 553706

28 Chicester Roses

Chaider Farm, Chalder Lane, Sidlesham, Chichester,
West Sussex PO20 7RN
Tel: 01243 641219
Specialities: Rose

29 The Citrus Centre

Mare Hill Nurseries, West Mare Lane, Pulborough,
West Sussex RH20 2EA
Tel: 01798 872786

30 Country Fayre

Littlehampton Road, Ferring, Worthing,
West Sussex BN12 6PN
Tel: 01903 245123

31 Courtlands Nurseries

Chilling Street, Sharpthorne, East Grinstead,
West Sussex RH19 4JF
Tel: 01342 810780

32 Covers

Quarry Lane, Chichester, West Sussex PO19 8PE
Tel: 01243 785141

33 Covers

Station Yard, Lyon Street West, Bognor Regis,
West Sussex PO21 1BT
Tel: 01243 863100

34 Cow Lane Gardens

6A Cow Lane, Sidlesham, Chichester,
West Sussex PO20 7LN
Tel: 01243 641106

35 Cuckoo's Nest Nursery

125 Third Avenue, Batchmere, Chichester,
West Sussex PO20 7LB
Tel: 01243 512541

36 Culberry Nursery

Dappers Lane, Angmering, West Sussex BN16 4EW
Tel: 01903 784107 Fax: 01903 779404
e-mail: mnjarvis@btinternet.com
website: www.culberry.com

Specialities:

Herbs; Vegetable Plants; Cottage Garden Plants

Products and Services:

- Seasonal Displays
- Disabled Facilities
- Display Gardens
- Houseplants
- Information Desk
- Car Park
- Plant Guarantee
- Plants
- Trees

Directions:

From the A27 west of Worthing take the A280 to Angmering village. Turn right onto Dappers Lane – the Nursery is ½ mile on the right.

37 Denmans Garden

Denmans Lane, Fontwell, Arundel,
West Sussex BN18 0SU
Tel: 01243 542808
Specialities: Houseplants, Restaurant/Coffee Shop

38 East Ashling Nurseries

Lye Lane, East Ashling, Chichester,
West Sussex PO18 9DD
Tel: 01243 575523

39 Ellesmere Nurseries

Ellesmere Orchard, Emsworth,
West Sussex PO10 8TP
Tel: 01243 371302

40 Ferring Country Centre

Rife Way, Ferring, Worthing, West Sussex BN12 5JZ
Tel: 01903 245078
Specialities: Vegetables, Hanging Baskets, Rose

41 Ferring Nurseries

Littlehampton Rd, Ferring, Worthing,
West Sussex BN12 6PN
Tel: 01903 241122
Specialities: Shrubs, Annuals/Biennials/Bedding

42 Fisher Common Nursery

Fisher Lane, North Mundham, Chichester,
West Sussex PO20 1YU
Tel: 01243 783889

43 Gardner & Scardifield

1-3 South Street, Lancing, West Sussex BN15 8AE
Tel: 01903 764964

44 Greenacre Nursery

Main Road, Chidham, Chichester,
West Sussex PO18 8TP
Tel: 01243 572441
Specialities: Annuals/Biennials/Bedding

45 Haskins Garden Centre

Roundstone Bypass, Angmering, Littlehampton,
West Sussex BN16 4BD
Tel: 01903 777666

46 Haywards Carnations

Ruhamah, Newells Lane, West Ashling, Chichester,
West Sussex PO18 8DF
Tel: 01243 576700
Specialities: Dianthus (Carnations/Pinks)

47 Heaselands Garden Nursery

The Old Lodge, Isaacs Lane, Haywards Heath,
West Sussex RH16 4SA
Tel/Fax: 01444 458084
e-mail: headgardener@heaselandsnursery.co.uk
website: www.heaselandsnursery.co.uk

SEE ADVERTISEMENT PAGE 275

Specialities:

Hardy Hybrid Rhododendrons; Deciduous Azaleas

Description:

Heaselands Garden Nursery offers more than 80 varieties of Hardy Hybrid Rhododendrons for sale to landscapers, nurseries and the general public. Heaselands 40-acre site provides a haven for countless unusual and often generally believed unavailable garden plants. Heaselands hosts two National Collections: Exbury/Knaphill and Mollis Azalea types. The nursery is open by appointment only.

Directions:

The nursery is on the A273 between Burgess Hill and Haywards Heath at Heasewood Farm Offices

48 High Beeches Nursery Ltd

High Beeches Lane, Handcross,
West Sussex RH17 6HQ
Tel: 01444 401398

49 Hillier Garden Centre Chichester

Main Road, Bosham, Chichester,
West Sussex PO18 8FL
Tel: 01243 573696

50 Hillier Garden Centre Horsham

Brighton Road, Horsham, West Sussex RH13 6QA
Tel: 01403 210113

51 Hillside Nurseries

Bury Common, Bury, Pulborough,
West Sussex RH20 1NR
Tel: 01798 831325
Specialities: Hanging Baskets, Annuals/Biennials/Bedding

52 Hillway

Crawley Down Road, Felbridge, East Grinstead,
West Sussex RH19 2PS
Tel: 01342 324950

53 Holdens Nursery

Shipton Green, Itchenor, Chichester,
West Sussex PO20 7BZ
Tel: 01243 512375
Specialities: Shrubs, Trees

54 Holly Gate Cactus Nursery

Billingshurst Road, Ashington, Pulborough,
West Sussex RH20 3BB
Tel: 01903 892930

55 Hot Plant Co

Warnham Court Gardens, Warnham Court,
Warnham, Horsham, West Sussex RH12 3RU
Tel: 01403 750558
Specialities: Exotics

56 Imberhorne Lane Nursery

Imberhorne Lane, East Grinstead,
West Sussex RH19 1TZ
Tel: 01342 321175 Fax: 01342 317477
e-mail: gardenexpert@btinternet.com
website: www.camelliasonline.co.uk

Specialities:

Camellias; Chinese Tree Peonies; Rhododendrons; Azaleas & Acers

Products and Services:

- Seasonal Displays
- Credit Cards Accepted
- Environmentally friendly composts and mulches
- Fruit Trees
- Information Desk
- Large Car Park

Directions:

From the north: From J6 of M25 take A22 towards Eastbourne. After the junction with the A264 at the next ttraffic lights, turn right into Imberhorne Lane – the nursery is 1.6m on the left

From the south: From A22 at Felbridge, turn left into Imberhorne Lane.

Description:

Owner Michael Dongray qualified in horticulture in 1973 and established his own wholesale nursery in 1986. In 1996 this became retail, and in 2006 included Mail Order. He is now a leading specialist of camellias with over 400 varieties. On site is a dedicated breeder of camellias whose knowledge, gained over 35 years, is often called upon by the RHS. The nursery has 67 varieties of Tree Peonies, 40 varieties of Japanese Acers, 200 varieties of rhododendrons and azaleas, and an extensive range of shrubs, climbers, bamboos, herbaceous and seasonal bedding plants. Always popular are the eco-friendly mulches and soil improvers available at competitive prices, as used by Michael on the nursery.

57 Kelly's Nursery

Bell Lane, Birdham, Chichester,
West Sussex PO20 7HY
Tel: 01243 513148
Specialities: Annuals/Biennials/Bedding, Hanging Baskets

58 Kingsfold Terra Décor

Dorking Road, Kingsfold, Horsham,
West Sussex RH12 3SD
Tel: 01306 628055
Specialities: Annuals/Biennials/Bedding, Pots

59 Lancasters Nursery

Bungalow Lancaster Nursery, West End Lane,
Henfield, West Sussex BN5 9RG
Tel: 01273 493913

60 Langley Boxwood Nursery Ltd

Rake, Liss, West Sussex GU33 7JN
Tel: 01730 894467
Specialities: Rare and Unusual Plants

61 Lansdowne Nursery

Littlehampton Road, Ferring, Worthing,
West Sussex BN12 6PB
Tel: 01903 241241

62 Laurence Hobbs Orchids Ltd

Bailiffs Cottage, Hophurst Lane, Crawley Down,
Crawley, West Sussex RH10 4LN
Tel: 01342 715142
Specialities: Orchids

63 Leythorne Nursery

Vinnetrow Road, Runcton, Chichester,
West Sussex PO20 1QB
Tel: 01243 782117

64 Little Oak Bonsai

Street End Lane, Sidlesham, Chichester,
West Sussex PO20 7RG
Tel: 01243 641302
Specialities: Bonsai, Rose, Hanging Baskets

65 Littlewoods Nursery

Yapton Lane, Walberton, Arundel,
West Sussex BN18 0LS
Tel: 01243 554952

66 Loder Plants

Camellia Grove Nursery, Market Garden, Brighton Road, Lower Beeding, Horsham, West Sussex RH13 6PP
Tel: 01403 891143
Specialities: Rhododendron (Inc Azalea)

67 Lower Trees Nursery

Roundstone-by-Pass, Angmering, Littlehampton, West Sussex BN16 4BD
Tel: 01903 770457

68 Loxwood Conifer Nurseries

Guildford Rd, Loxwood, Billingshurst, West Sussex RH14 0SA
Tel: 01403 753389
Specialities: Conifers

69 Manor Nurseries

Pagham Road, Runcton, Chichester, West Sussex PO20 1LJ
Tel: 01243 781734

70 Manor Nurseries

Yapton Road, Barnham, Bognor Regis, West Sussex PO22 0BQ
Tel: 01243 552028

71 Manor Nursery

High Street, Angmering, Littlehampton, West Sussex BN16 4AW
Tel: 01903 786977
Specialities: Herbaceous Perennials, Shrubs

72 Maudlin Nursery

Stane Street, Westhampnett, Chichester, West Sussex PO18 0PA
Tel: 01243 773024

73 Mayfield Nurseries

Yapton Road, Barnham, Bognor Regis, West Sussex PO22 0BJ
Tel: 01243 551235

74 Mayfield Nursery

West Chiltington Lane, Broadford Bridge, Billingshurst, West Sussex RH14 9EA
Tel: 01403 741224

75 Mill Farm Trees

Bury Mill Farm, Bury Gate, Bury, Pulborough, West Sussex RH20 1NN
Tel: 01798 831008
Specialities: Hedging, Trees

76 Murrells Nurseries

Broomers Hill Lane, Pulborough, West Sussex RH20 2DU
Tel: 01798 875508

77 Newbridge Nurseries

Billingshurst Road, Broadbridge Heath, Horsham, West Sussex RH12 3LN
Tel: 01403 272686

78 North Farm Nursery

Church Road, North Mundham, Chichester, West Sussex PO20 1JU
Tel: 01243 775947

79 Old Barn Nurseries

Dial Post (A24), nr Horsham, West Sussex RH13 8NR
Tel: 01403 710000 Fax: 01403 710010
e-mail: info@old-barn.co.uk
website: www.old-barn.co.uk

Specialities:

Planteria; Furniture/Barbecues; Restaurant

Products and Services:

- Aquatics
- Barbeques
- Books and Stationery
- Childrens Play Area
- Seasonal Displays
- Clothing
- Credit Cards Accepted
- Cut Flowers
- Disabled Facilities
- Dried & Artificial Flowers
- Farm Shop
- Fencing
- Fireworks
- Floristry
- Fruit Trees
- Garden Furniture
- Garden Machinery
- Garden Products
- Giftware
- Greenhouses and Sheds
- Houseplants
- Information Desk
- Large Car Park
- Paving
- Pet Products
- Plant Guarantee
- Plants
- Restaurant/Coffee Shop
- Seeds and Bulbs
- Trees

Directions:

Located on the A24, 8 miles south of Horsham and 12 miles north from the coast.

Description:

Old Barn Nurseries were established in 1991 on the site of a disused nursery and began with new glasshouses, a restaurant and planteria. More recently, the owners have added their own growing nursery, a children's play area and a farm shop department. Highlights of the nurseries include a superb range of garden furniture, specimen houseplants (including 25ft high banana plants), beautiful giftware and an amazing Christmas display. At the core of the nurseries is one of the largest planterias in the South East. Many of the plants are grown on site with more than 2000 varieties, 3000 specimen plants and 40 varieties of Acers. Between them, the staff at the Old Barn have more than 250 years of horticultural knowledge and

are always keen to pass this on. The nursery also offers a 3-year Plant Guarantee. The Old Barn restaurant occupies a lovingly restored 18th century barn which has a warm and friendly atmosphere. Many customers visit day in, day out, be it rain or shine. Some even come for breakfast, lunch and then tea!

80 Orchard Nursery

Holtye Road, East Grinstead, West Sussex RH19 3PP
Tel: 01342 311657
e-mail: orchardnursery@btconnect.com
website: www.orchardnursery.co.uk
Specialities: Hanging Baskets, Herbaceous Perennials

81 Orchid Answers Ltd

113 Second Avenue, Batchmere, Chichester, West Sussex PO20 7LF
Tel: 01243 511322
Specialities: Orchids

82 P Perella

Glenville Chalet, Old Mead Road, Wick, Littlehampton, West Sussex BN17 7PU
Tel: 01903 713305

83 The Plant Co

Coolham Road, West Chiltington, Pulborough, West Sussex RH20 2LH
Tel: 01403 740100

84 Pots & Pithoi

The Barns, East Street, Turners Hill, Crawley, West Sussex RH10 4QQ
Tel: 01342 714793
Specialities: Pots, Aquatics

85 Rotherhill Nurseries and Garden Centre

School Lane, Stedham, Midhurst, West Sussex GU29 0PA
Tel: 01730 813687

86 Royal Botanic Gardens Kew

Wakehurst Place, Ardingly, Haywards Heath, West Sussex RH17 6TN
Tel: 01444 894050

87 Rushfields Plant Centre

Henfield Road, Poynings, Brighton, West Sussex BN45 7AY
Tel: 01273 857445
Specialities: Rose

88 Russells Garden Centre

Main Road, Birdham, Chichester, West Sussex PO20 7BY
Tel: 01243 512525
Specialities: Garden Furniture

89 Scaynes Hill Nursery

Lewes Road, Scaynes Hill, Haywards Heath, West Sussex RH17 7PG
Tel: 01444 831673

90 The Secret Garden Nursery

The Walled Garden, Stansted Park, Rowland's Castle, West Sussex PO9 6DX
Tel: 023 9241 3993

91 Seymores Garden Centre

St Martins Road, Littlehampton, West Sussex BN17 6EU
Tel: 01903 713046

92 Shillinglee Gardens

Shillinglee, Godalming, West Sussex GU8 4SZ
Tel: 01428 708156

93 Shoots Garden Centre

London Road, Washington, Pulborough, West Sussex RH20 3BP
Tel: 01903 892006
Specialities: Garden Products, Annuals/Biennials/Bedding, Pet Products

94 Shopwhyke Nurseries

Tangmere Road, Shopwhyke, Chichester, West Sussex PO20 2BL
Tel: 01243 783123
Specialities: Annuals/Biennials/Bedding

95 Sidlesham Basket & Bedding Plant Nursery

Street End Road, Sidlesham Common, Chichester, West Sussex PO20 7QD
Tel: 01243 641044
Specialities: Annuals/Biennials/Bedding, Hanging Baskets

96 South Downs Garden Centre

Brighton Road, Hassocks, West Sussex BN6 9LY
Tel: 01273 845232

97 Southwick Nursery

15 Roman Way, Southwick, Brighton,
West Sussex BN42 4TN
Tel: 01273 592865

98 Spithandle Nurseries

Spithandle Lane, Wiston, Steyning,
West Sussex BN44 3DY
Tel: 01903 816299
Specialities: Annuals/Biennials/Bedding, Hanging Baskets, Xmas Trees

99 Spring Garden Nursery

Spring Gardens, Washington, Pulborough,
West Sussex RH20 3BS
Tel: 01903 892225
Specialities: Homegrown

100 Squires Garden Centre (Crawley)

Horsham Road, Crawley, West Sussex RH11 8PL
Tel: 01293 522101

101 Stansted Park Garden Centre

Stansted Park, Rowlands Castle,
Hampshire PO9 6DX
Tel: 023 9241 3090 Fax: 023 0241 3308
website: www.stanstedparkgardencentre.net

SEE ADVERTISEMENT PAGE 280

Specialities:

Unusual Plants; Children's Workshops; Feature Weekends

Products and Services:

- Barbeques
- Books & Stationery
- Camping Equipment
- Childrens Play Area
- Seasonal Displays
- Clothing
- Conservatories
- Credit Cards Accepted
- Disabled Facilities
- Display Gardens
- Fencing
- Fruit Trees
- Garden Furniture
- Garden Machinery
- Greenhouses and Sheds
- Giftware
- Houseplants
- Information Desk
- Large Car Park
- Paving
- Plant Guarantee
- Restaurant/Coffee Shop
- Seeds and Bulbs
- Trees

Directions:

The garden centre is just 10 miles west of Chichester, and about 5 miles from the junction of the M27 / A27 / A3(M). Follow the Stansted House brown heritage signs from the A3 (Emsworth and Rowlands Castle B2149) or A27 (Havant).

102 Stonepit Nurseries Ltd

Stone Pit Lane, Henfield, West Sussex BN5 9QU
Tel: 01273 493016
Specialities: Annuals/Biennials/Bedding

103 Stonepound Nursery

Brighton Road, Hassocks, West Sussex BN6 9NB
Tel: 01273 842551

104 Sunnyside Nurseries

Loxwood Road, Tismans Common, Rudgwick,
Horsham, West Sussex RH12 3BT
Tel: 01403 822896

105 Swains Farm Shop

Brighton Road, Woodmancote, Henfield,
West Sussex BN5 9RP
Tel: 01273 494582

106 T H Beck & Sons

Sunnyside, Yapton Road, Barnham, Bognor Regis,
West Sussex PO22 0AX
Tel: 01243 553062

107 Tawny Nurseries

Bell Lane, Birdham, Chichester,
West Sussex PO20 7HY
Tel: 01243 512333

108 Terra Decor

Handcross Road, Plummers Plain, Horsham,
West Sussex RH13 6NX
Tel: 01403 891969
Specialities: Pots

109 The Village Nurseries

Sinnocks, West Chiltington, Pulborough,
West Sussex RH20 2JX
Tel: 01798 813040
Specialities: Annuals/Biennials/Bedding

110 Wallabies Nursery

Kirdford Road, Wisborough Green, Billingshurst,
West Sussex RH14 0DD
Tel: 01403 700147

111 Warwick Nurseries

The Acre, Dappers Lane, Angmering, Worthing,
West Sussex BN16 4EN
Tel: 07970 202039
Specialities: Exotics, Trees, Shrubs

Stansted Park Garden Centre

Stansted Park, Rowlands Castle,
Hampshire PO9 6DX
Tel: 023 9241 3090 Fax: 023 0241 3308
website: www.stanstedparkgardencentre.net

Specialities:

Unusual Plants; Children's Workshops; Feature Weekends

Description:

Set in 1500 acres of parkland at Stansted Park, one of the South's finest stately homes, Stansted Park Garden Centre has evolved from a green field site in 1999 to what is now a large garden centre with everything for the garden and home. Gardeners will find an extensive range of outdoor plants, a garden shop offering everything a gardener will need from seeds to tools, and a houseplant and gift area bursting with gift ideas. The centre also stocks landscaping materials - everything from pots and planters to paving and sheds. Other attractions include a home interiors department, a tea room, aboretum, light railway, sculpture garden and fabulous Stansted House itself.

Directions:

The garden centre is just 10 miles west of Chichester, and about 5 miles from the junction of the M27 / A27 / A3(M). Follow the Stansted House brown heritage signs from the A3 (Emsworth and Rowlands Castle B2149) or A27 (Havant).

Products and Services:

Barbecues
Books and Stationery
Camping Equipment
Children's Play Area
Seasonal Displays
Clothing
Conservatories
Credit Cards Accepted
Disabled Facilities
Display Gardens
Dried and Artificial Flowers
Fencing
Fruit Trees
Garden Construction
Garden Design
Garden Furniture
Garden Products
Giftware
Greenhouses and Sheds
Houseplants
Information Desk
Large Car Park
Paving
Plant Guarantee
Plants
Restaurant/Coffee Shop
Seeds and Bulbs
Trees

112 Westholme Nurseries

Toddington Lane, Wick, Littlehampton,
West Sussex BN17 7PP
Tel: 01903 714845

113 Windmill Nursery

High Hatch Lane, Hurstpierpoint, Hassocks,
West Sussex BN6 9LH
Tel: 01273 834696
Specialities: Annuals/Biennials/Bedding, Xmas Trees

114 Wings Nursery

Lidsey Road, Woodgate, Nr Chichester,
West Sussex PO20 3SU
Tel: 01243 544804

115 Wophams Lane Nursery

Wophams Lane, Birdham, Chichester,
West Sussex PO20 7BS
Tel: 01243 512862
Specialities: Annuals/Biennials/Bedding

116 Wyevale Garden Centre (Chichester)

Bognor Road, Merston, Chichester,
West Sussex PO20 1EG
Tel: 01243 789276

117 Wyevale Garden Centre (Crawley)

Copthorne Road, Crawley, West Sussex RH10 3PD
Tel: 01293 883311

118 Wyevale Garden Centre (Pulborough)

Stopham Road, Pulborough,
West Sussex RH20 1DS
Tel: 01798 872981

119 Wyevale Garden Centre (Worthing Ferring)

Littlehampton Road, Ferring, Worthing,
West Sussex BN12 6PG
Tel: 01903 242003

120 Wyevale Garden Centre (Worthing Findon)

Rogers Lane, Findon, Worthing,
West Sussex BN14 0RE
Tel: 01903 874111

TYNE AND WEAR

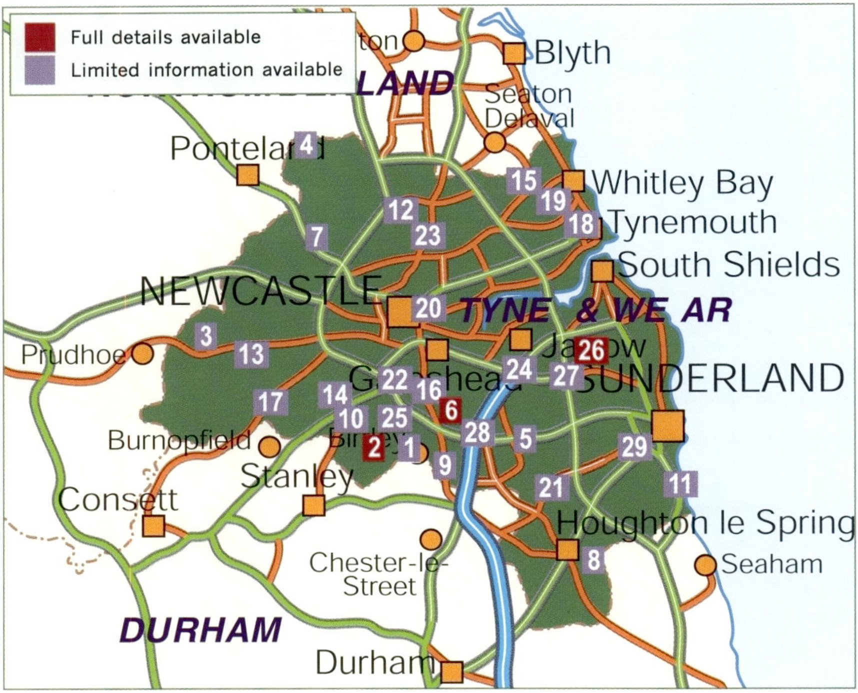

1 Angel Koi Aquatics

53 Greenford, Gateshead,
Tyne And Wear NE11 0TW
Tel: 01914 109985
Specialities: Aquatics

2 Birkheads Secret Gardens & Nursery

Nr Hedley Wood, Sunniside, Newcastle Upon Tyne,
Tyne And Wear NE16 5EL
Tel: 01207 232262 Mobile: 07778 447920
e-mail: birkheads,nursery@virgin.net
website: www.birkheadsnursery.co.uk
Specialities: Unusual and Rare Hardy Plants

3 Bradley Gardens and Nursery

Sled Lane, Wylam, Tyne and Wear NE41 8JH
Tel: 01661 852176
Specialities: Herbaceous Perennials, Farm Shop

4 Capital Gardens Ltd

Horton Grange, Berwick Hill Road, Seaton Burn,
Newcastle Upon Tyne, Tyne And Wear NE13 6BU
Tel: 01661 824631

5 Clay's Garden Centre

The Peel Centre, Peel Retail Park, Washington,
Tyne And Wear NE37 2PA
Tel: 01914 177777

6 Cottage Garden Nursery

Hunters Lodge, Galloping Green Road, Eighton
Banks, Gateshead, Tyne And Wear NE9 7XD
Tel: 01914 820059
e-mail: petershotton@aol.com
website: www.cottagegardennursery.co.uk
Specialities: Cottage Garden Plants

7 Cowells Garden Centre

Main Road, Woolsington, Newcastle Upon Tyne,
Tyne And Wear NE13 8BW
Tel: 01912 863403
Specialities: Outdoor Plantarium, Annuals/Biennials/Bedding, Garden Furniture

8 Deancroft Plant Nursery

Deancroft, Houghton Le Spring,
Tyne And Wear DH5 8LX
Tel: 01915 842141
Specialities: Annuals/Biennials/Bedding, Herbaceous Perennials

9 Dobbies Garden World

Durham Road, Birtley, Chester Le Street,
Tyne and Wear DH3 2PF
Tel: 01914 102556

10 Douglas Bros

Lingey Fine Gardens, Pennyfine Road, Sunniside,
Newcastle Upon Tyne, Tyne And Wear NE16 5EP
Tel: 01914 887487
Specialities: Annuals/Biennials/Bedding, Herbaceous Perennials

11 Emilies Nursery

Ryhope Grange Farm, Ryhope Road, Sunderland,
Tyne And Wear SR2 9SU
Tel: 01915 143777
Specialities: Annuals/Biennials/Bedding, Herbaceous Perennials

12 Gosforth Park Koi, Aquatic and Garden Centre

High Gosforth Park, Newcastle Upon Tyne,
Tyne And Wear NE3 5EJ
Tel: 01912 365775
Specialities: Aquatics, Garden Furniture

13 Greenside Nursery

Stephens Hall Farm, Lead Road, Greenside, Ryton,
Tyne And Wear NE40 4JE
Tel: 01914 136030
Specialities: Annuals/Biennials/Bedding, Shrubs

14 Haddrieck Hall Estates

Cheviot View, Streetgate, Sunniside,
Newcastle Upon Tyne, Tyne And Wear NE16 5LQ
Tel: 01914 881201

15 Hawthorn Nurseries

Earsdon, Whitley Bay, Tyne And Wear NE25 9TQ
Tel: 01912 528777
Specialities: Annuals/Biennials/Bedding, Shrubs

16 Home & Garden

529-531 Durham Road, Gateshead,
Tyne And Wear NE9 5EY
Tel: 01914 200185

17 Lintzford Bridge Garden Nursery

Lintzford Road, Rowlands Gill,
Tyne And Wear NE39 1DG
Tel: 01207 549202
Specialities: Annuals/Biennials/Bedding

18 Manor Road Nurseries

Manor Road, North Shields,
Tyne And Wear NE30 4RH
Tel: 01912 577264
Specialities: Herbaceous Perennials, Shrubs

19 Monkseaton Pet & Garden Supplies

40 Cauldwell Lane, Whitley Bay,
Tyne And Wear NE25 8LN
Tel: 01912 527869
Specialities: Pet Products, Garden Furniture

20 Ouseburn Valley Garden Centre

Stepney Road, Newcastle Upon Tyne,
Tyne And Wear NE2 1TZ
Tel: 01912 611616

21 Penshaw Nurseries

East Barnwell Farm, Penshaw, Houghton Le Spring,
Tyne And Wear DH4 7NJ
Tel: 01915 345241
Specialities: Annuals/Biennials/Bedding, Shrubs

22 Peter Barratts Garden Centre

No 1 Princess Park, Team Valley Trading Estate,
Gateshead, Tyne And Wear NE11 0NF
Tel: 01914 958800

23 Peter Barratts Garden Centre

Gosforth Park, Newcastle Upon Tyne,
Tyne And Wear NE3 5EP
Tel: 01912 367111

24 Red Fox Nurseries

Wardley, Gateshead, Tyne And Wear NE10 8YS
Tel: 01914 380821
Specialities: Annuals/Biennials/Bedding, Herbaceous Perennials

5 Badger Nurseries Ltd

Birmingham Road, Mappleborough Green, Studley, Warwickshire B80 7DL
Tel: 01527 852631
Specialities: Herbaceous Perennials, Garden Products

6 Blenheim Nursery

Blenheim, Stratford Road, Bidford-on-Avon, Alcester, Warwickshire B50 4LU
Tel: 01789 490448
Specialities: Hanging Baskets, Annuals/Biennials/Bedding, Shrubs

7 Blooms of Bressingham

Straight Mile, Bourton, Rugby, Warwickshire CV23 9QQ
Tel: 01788 812830
Specialities: Annuals/Biennials/Bedding, Herbaceous Perennials, Shrubs

8 Bridge Nursery

Tomlow Road, Napton, Nr Southam, Warwickshire CV47 8HX
Tel: 01926 812737
Specialities: Grasses, Bamboos, Herbaceous Perennials

9 Brookside Aquatics

Brookside, Gaydon Road, Bishops Itchington, Southam, Warwickshire CV47 2QZ
Tel: 01926 612320
Specialities: Aquatics, Water Products and Features

10 The Bungalow Nurseries

55 London Road, Stretton on Dunsmore, Rugby, Warwickshire CV23 9JB
Tel: 01788 810505
Specialities: Hanging Baskets, Ponsettias, Geranium

11 Castle Nurseries

Birmingham Road, Mappleborough Green, Studley, Warwickshire B80 7DF
Tel: 01527 853689

12 Charlecote Garden Store

Charlecote, Warwick, Warwickshire CV35 9ER
Tel: 01789 841842

13 Charlecote Plant & Flowers

Dog Kennel Close, Charlecote, Warwick, Warwickshire CV35 9EW
Tel: 01789 470947
Specialities: Herbaceous Perennials

14 Cheapside Nurseries

5 Cheapside Cottages, Honeybourne Road, Bickmarsh Bidford-on-Avon, Alcester, Warwickshire B50 4PB
Tel: 01789 778048
Specialities: Annuals/Biennials/Bedding, Alpines/Rock Plants, Conifers

15 Compton Lane Nurseries

Little Compton, Moreton-in-Marsh, Warwickshire GL56 0SJ
Tel: 01608 674578
Specialities: Alpines/Rock Plants, Woodland Plants

16 Crabtree Garden Centre

3 Stratford Road, Bidford-on-Avon, Alcester, Warwickshire B50 4LL
Tel: 01789 490390
Specialities: Annuals/Biennials/Bedding, Shrubs, Herbaceous Perennials

17 Crackley Lane Nursery

Crackley Lane, Kenilworth, Warwickshire CV8 2JW
Tel: 024 7646 7390

18 Dobbies Garden World

Nuneaton Road, Mancetter, Atherstone, Warwickshire CV9 1RF
Tel: 01827 715511

19 Duncan's Pet & Garden Centre

10 Warwick Road, Wellesbourne, Warwick, Warwickshire CV35 9ND
Tel: 01789 841847
Specialities: Pet Products

20 Durno Nurseries

Old Holly Lane, Atherstone,
Warwickshire CV9 2HD
Tel: 01827 713233
Specialities: Alpines/Rock Plants

21 Earlswood Nurseries and Garden Centre

Forshaw Heath Road, Earlswood, Solihull,
Warwickshire B94 5JU
Tel: 01564 702314
Specialities: Restaurant/Coffee Shop, Aquatics, Fencing

22 Fancy Plants

9 Meadow Close, Stretton on Dunsmore, Rugby,
Warwickshire CV23 9NL
Tel: 024 7654 3010

23 Farnborough Garden Centre

Southam Road, Farnborough, Banbury,
Warwickshire OX17 1EL
Tel: 01295 690479

24 Golls Nurseries

Pitchhill, Pitchill, Evesham,
Warwickshire WR11 8SN
Tel: 01386 871740

25 Greenleaf Farmshop & Nursery

Banbury Road, Southam,
Warwickshire CV47 2BL
Tel/Fax: 01926 815287
e-mail: greenleaf.nursery@virgin.net

Specialities:

Bedding Plants; Hanging Baskets; Herbaceous Perennials

Description:

Peter Healey, the owner of Greenleaf, has been a Nurseryman for some 30 years and is very knowledgeable and helpful. In addition to a wide selection of plants, seeds and bulbs, the farmshop offers an extensive choice of locally-grown vegetables.

Directions:

Greenleaf is on the A423, 2 miles south of Southam and 10 miles north of Banbury.

26 Halwyn Nurseries

7 Stratford Street, Nuneaton,
Warwickshire CV11 5BS
Tel: 024 7638 7232

27 The Hiller Garden & Plant Centre

Dunnington Heath Farm, Dunnington, Alcester,
Warwickshire B49 5PD
Tel: 01789 491342
Specialities: Rose, Climbers, Herbaceous Perennials

28 Hillier Garden Centre Redditch

Henley Road, Mappleborough Green, Studley,
Warwickshire B80 7DR
Tel: 01527 852266

29 Hilltop Garden Centre

Shilton Lane, Shilton, Coventry,
Warwickshire CV7 9LH
Tel: 024 7661 4752

30 Hintons Nurseries

Coventry Road, Guys Cliffe, Warwick,
Warwickshire CV34 5YL
Tel: 01926 492273
Specialities: Herbaceous Perennials, Shrubs, Fruit Bushes

31 Hoar Park Garden Centre

Ansley, Nuneaton, Warwickshire CV10 0QU
Tel: 024 7639 6011
Specialities: Herbaceous Perennials, Shrubs, Annuals/Biennials/Bedding

32 Hurrans Garden Centre

Myton Road, Leamington Spa,
Warwickshire CV31 3PB
Tel: 01926 881122
Specialities: Shrubs, Herbaceous Perennials, Conifers

33 Littlehurst Nurseries Ltd

Bennetts Road North, Corley, Coventry,
Warwickshire CV7 8BG
Tel: 01676 540178
Specialities: Herbaceous Perennials, Alpines/Rock Plants

34 Lodge Farm Plants & Wildflowers

Case Lane, Five Ways, Hatton, Warwick, Warwickshire CV35 7JD
Tel: 01926 484649
Specialities: Fruit Bushes, Wild Flowers

35 Medda Place Market Garden

Coventry Road, Thurlaston, Rugby, Warwickshire CV23 9JR
Tel: 01788 810382
Specialities: Hanging Baskets, Annuals/Biennials/Bedding, Trees

36 The National Herb Centre

Banbury Road, Warmington, nr Banbury, Oxfordshire OX17 1DF
Tel: 01295 690999 Fax: 01295 690034
e-mail: info@herbcentre.co.uk
website: www.herbcentre.co.uk

Specialities:

Herbs; Bistro & Delicatessen; Nature Trail & Gardens

Description:

The Centre has more than 500 varieties of culinary, medicinal and scented herbs available. The Garden & Gift Shop sells an extensive range of accessories for the garden, home and outdoor living, and there's also a Bistro & Delicatessen.

Directions:

The Centre is on the B4100, about 3 miles north of Banbury, just 15 minutes from Junction 12 of the M40.

37 New Inn Lane Nurseries

New Inn Lane, Pitchill, Evesham, Warwickshire WR11 8SN
Tel: 01386 870073
Specialities: Hanging Baskets, Fuchsia

38 Newland Nursery

Coppice Lane, Middleton, Tamworth, Warwickshire B78 2BU
Tel: 01213 087197
Specialities: Shrubs, Trees, Annuals/Biennials/Bedding

39 Newton Regis Garden Store

Shuttington Lane, Shuttington, Tamworth, Warwickshire B79 0HA
Tel: 01827 830888

40 The Oaks Nurseries

326 Weddington Road, Nuneaton, Warwickshire CV10 0EU
Tel: 024 7638 4550

41 Orchard Nurseries

Duck Lane, Welford On Avon, Stratford-upon-Avon, Warwickshire CV37 8QH
Tel: 01789 750317

42 Planters Garden Centre

Woodlands Farm, Freasley, Tamworth, Warwickshire B78 2EY
Tel: 01827 251511
Specialities: Annuals/Biennials/Bedding, Shrubs, Herbaceous Perennials

43 Russells Garden Centre

Mill Hill, Baginton, Coventry, Warwickshire CV8 3AG
Tel: 024 7630 3627

44 Ryton Nurseries

Leamington Road, Ryton On Dunsmore, Coventry, Warwickshire CV8 3FN
Tel: 024 7630 5511
Specialities: Greenhouses and Sheds, Houseplants

45 Smiths Nurseries

Stoneleigh Road, Baginton, Coventry, Warwickshire CV8 3BA
Tel: 024 7630 3382
Specialities: Bamboos

46 Spernal Ash Nurseries

Alcester Road, Studley, Warwickshire B80 7PD
Tel: 01527 854538

47 Springfield Nurseries

Hertford Road, Alcester, Warwickshire B49 6AT
Tel: 01789 762648
Specialities: Annuals/Biennials/Bedding, Hanging Baskets

48 Stratford Garden Centre

Campden Road, Clifford Chambers,
Stratford upon Avon, Warwickshire CV37 8LW
Tel: 01789 205745 Fax: 01789 261843
e-mail: mail@stratfordgardencentre.co.uk
website: www.stratfordgardencentre.co.uk

SEE ADVERTISEMENT ABOVE

Specialities:

Garden Furniture; Specimen Plants; Shrubs & Roses

Products and Services:

- Aquatics
- Barbeques
- Books and Stationery
- Children's Play Area
- Seasonal Displays
- Clothing
- Conservatories
- Credit Cards Accepted
- Disabled Facilities
- Dried & Artificial Flowers
- Farm Shop/Delicatessen
- Fencing
- Fruit Trees
- Garden Furniture
- Garden Products
- Gift Vouchers
- Giftware
- Greenhouses and Sheds
- Houseplants
- Information Desk
- Large Car Park
- Paving
- Plant Guarantee
- Plants
- Restaurant/Coffee Shop
- Seeds and Bulbs
- Trees

Directions:

From Stratford town centre take the A3400 towards Shipston on Stour, then the B4632. 2 miles south of the town.

49 Tebbs Garden Centre

59-69 Hilmorton Road, Rugby,
Warwickshire CV22 5AE
Tel: 01788 542997

Specialities: Houseplants, Annuals/Biennials/Bedding, Herbaceous Perennials

50 Whitacre Garden Centre

Tamworth Road, Nether Whitacre, Coleshill,
Warwickshire B46 2DP
Tel: 01675 481306
e-mail: mail@whitacregardencentre.co.uk
website: www.whitacregardencentre.co.uk

Specialities:

Garden Furniture; Perennials & Shrubs; Garden Products; Cafe Camellia

Products and Services:

- Aquatics
- Barbeques
- Books and Stationery
- Childrens Play Area

- Seasonal Displays
- Clothing
- Credit Cards Accepted
- Disabled Facilities
- Fencing
- Fruit Trees
- Garden Construction
- Garden Furniture
- Garden Products
- Gift Vouchers
- Giftware
- Houseplants
- Information Desk
- Large Car Park
- Plant Guarantee
- Plants
- Restaurant/Coffee Shop
- Seeds and Bulbs
- Trees

Directions:

From Junction 9 of the M42, take the A4097 towards Kingsbury and Tamworth. At the second roundabout, take the B4098 towards Coventry and Nether Whitacre. The garden centre is about 2.5 miles on the right hand side of the road.

Description:

The centre, established 25 years ago, is now under the same ownership as the Stratford upon Avon Garden Centre. A member of the Horticultural Trades & Garden Centre Association, the 3-acre site is very well laid out with plants clearly identified and with helpful staff happy to advise and answer questions. There's a full range of plants throughout the year, a wide choice of garden furniture, barbecues and accessories, tools and all garden needs, composts, seeds, glassware, pottery, candles and fragrances. The Cafe Camellia offers excellent snacks and meals, freshly prepared each day.

51 Withybrook Nurseries

Overstone Road, Withybrook, Coventry, Warwickshire CV7 9LU

Tel: 01455 220297

Specialities: Conifers

52 Wyevale Garden Centre (Church Lawford)

Kings Newnham Road, Church Lawford, Rugby, Warwickshire CV23 9EP

Tel: 024 7654 2319

53 Wyevale Garden Centre (Stratford-Upon-Avon)

Warwick Road, Black Hill, Stratford-upon-Avon, Warwickshire CV37 0QW

Tel: 01789 734200

WEST MIDLANDS

1 A Caldicott & Son

50 Lovelace Avenue, Solihull,
West Midlands B91 3JR
Tel: 01217 051747
Specialities: Pick Your Own, Annuals/Biennials/Bedding

2 Akamba Garden Centre

Tythe Barn Lane, Bickens Heath, Birmingham,
West Midlands B90 1PH
Tel: 01217 060306
Specialities: Hardy Garden Palms, Exotics, Tropical

3 Ambleside Garden Centre

432 Norton Lane, Earlswood, Solihull,
West Midlands B94 5LR
Tel: 01564 703553
Specialities: Nymphaea (Water Lilies), Restaurant/Coffee Shop

4 Ashfurlong Nursery

Tamworth Road, Four Oaks, Sutton Coldfield,
West Midlands B75 6JL
Tel: 01213 080530

5 Beacon Nurseries

Rowan Road, Sedgley, Dudley,
West Midlands DY3 1LT
Tel: 01902 882933

6 BFS Garden Centre

92 Stechford Lane, Birmingham,
West Midlands B8 2AN
Tel: 01217 833075
Specialities: Ornamentals

7 Blooms of Bressingham

Kenilworth Road, Hampton-in-Arden, Solihull,
West Midlands B92 0LP
Tel: 01675 442866

8 Braceys Nurseries and Garden Centre

Catherine-De-Barnes Lane, Bickenhill, Solihull,
West Midlands B92 0DB
Tel: 01675 442587
Specialities: Restaurant/Coffee Shop, Annuals/Biennials/Bedding

33 Park Lane Garden Centre

Park Lane, Aston, Birmingham,
West Midlands B6 5DL
Tel: 01216 857127

34 Pet & Garden Centre

126 Church Rd, Yardley, Birmingham,
West Midlands B25 8UT
Tel: 01217 832735

35 Pets & Gardens

372-376 Birmingham Road, Sutton Coldfield,
West Midlands B72 1YH
Tel: 01213 731491

36 Rathbone Garden Centre

Umberslade Road, Selly Oak,
Birmingham B29 7SQ
Tel: 0121 472 2233 Fax: 0121 471 3465
e-mail: enquiries@rathbone.co.uk
website: www.rathbone.co.uk

Specialities:

Choice and unusual shrubs; Ferns and grasses; Vegetable Plants

Products and Services:

- Garden Products
- Houseplants
- Information Desk
- Plant Guarantee
- Plants
- Seeds and Bulbs

Directions:

Rathbone Garden Centre is located off the A38 Bristol Road at Selly Oak, 3 miles south of the city centre. Take the signs to the hospital, turn left at traffic lights, go past the hospital car park, over the railway bridge then take the 2nd right into Umberslade Road, The centre is situated on the right after about 600 metres.

Description:

A truly urban garden centre, Rathbone Garden Centre developed in 1994 as part of a horticultural training centre for young persons with learning difficulties. Since that time it has continued to develop and now provides not only a training facility but also a retail outlet for the plants produced by the students.

In 2000 the centre underwent a redesign and rebuild to accommodate an extended range of plants. Currently it carries a comprehensive range of plants with an emphasis on the unusual.

This is a charity plant centre offering good service, an extensive range and competitive prices. All profits are used to help train students with learning difficulties.

37 Seasons Change

Three Crowns, Sutton Road, Walsall,
West Midlands WS9 0QH
Tel: 01922 454557

38 Shirley Plant Centre

102-106 Haslucks Green Road, Shirley, Solihull,
West Midlands B90 2EJ
Tel: 01217 332140

39 Shrubbers World Garden Centre

47 Walsall Street, Wednesbury,
West Midlands WS10 9HA
Tel: 01215 053558
Specialities: Garden Design

40 Two Gates Garden Market

Oldnall Road, Halesowen, West Midlands B63 2JP
Tel: 01384 567582

41 Valley Nurseries

297 Erdington Road, Walsall,
West Midlands WS9 0SB
Tel: 01922 456060

42 Water In The Garden

Meriden Mill Farm, Kenilworth Road, Meriden,
Coventry, West Midlands CV7 7LJ
Tel: 01675 442564
Specialities: Nymphaea (Water Lilies)

43 Webbs of Bearwood (O J Webb & Son)

64-66 Three Shires Oak Road, Smethwick,
West Midlands B67 5BU
Tel: 01214 292033

44 Whites Flower & Plant Centre

126 Church Rd, Bradmore, Wolverhampton,
West Midlands WV3 7EN
Tel: 01902 333422

45 William Wheat & Son

The Flower & Garden Centre, Chester Road,
Little Aston, Walsall, West Midlands WS9 0PU
Tel: 01213 537123

46 Woolmans Garden Centre

72-74 Solihull Road, Shirley, Solihull,
West Midlands B90 3HN
Tel: 01217 443094

47 Wyevale Garden Centre (Bournville)

Maple Road, Birmingham, West Midlands B30 2AE
Tel: 01214 720303

48 Wyevale Garden Centre (Coventry)

Brownshill Green Road, Coventry,
West Midlands CV5 9PB
Tel: 024 7633 3998

49 Wyndley Garden Centre

Warwick Road, Knowle, Solihull,
West Midlands B93 0DX
Tel: 01564 777106

50 York Supplies

57 Waterloo Road, Kings Heath, Birmingham,
West Midlands B14 7SG
Tel: 01214 443342

WILTSHIRE

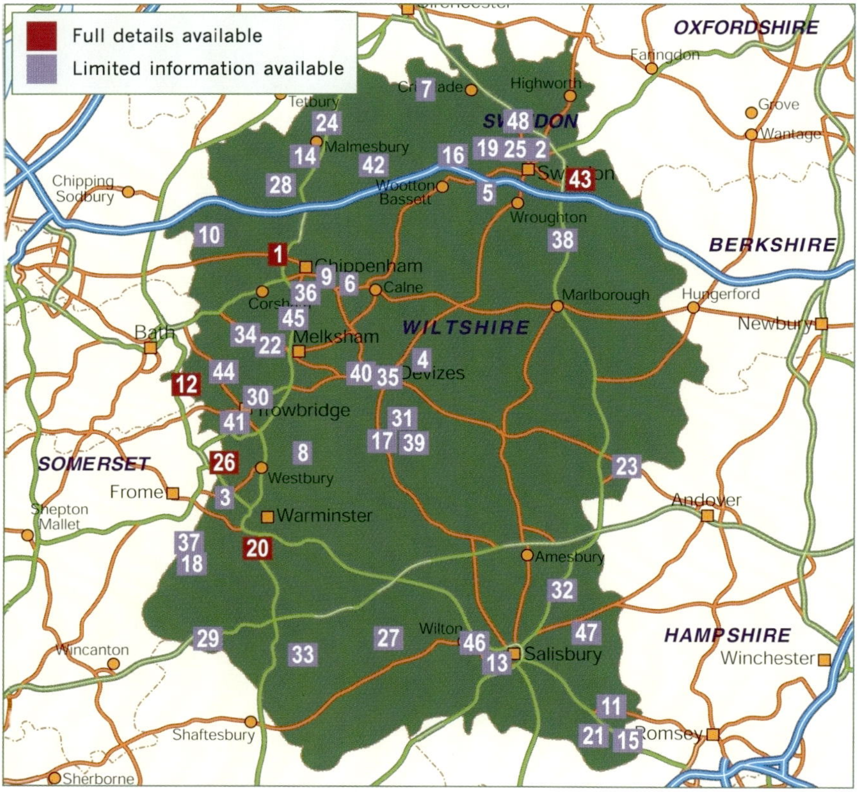

1 Allington Plants & Garden Shop

Allington Bar Farm, Allington, Chippenham, Wiltshire SN14 6LJ
Tel: 01249 659700

Specialities:

Herbaceous Perennials; Shrubs; David Austin Roses

Products and Services:

- Seasonal Displays
- Credit Cards Accepted
- Disabled Facilities
- Farm Shop
- Fireworks
- Fruit Trees
- Garden Products
- Giftware
- Information Desk
- Large Car Park
- Pet Products
- Plant Guarantee
- Plants
- Pots,Sundials etc
- Seeds and Bulbs
- Trees

Directions:

M4-J17 South on A350 to A420. West on A420 then third of a mile on left to Allington Plants and Garden Shop.

Description:

Visitors to the Allington Plants & Garden Shop will receive a warm welcome from owner Brenda Meade, whose passion for horticulture is evident. The site opened in early 2006 but Brenda has been involved

in horticulture for some 14 years and is extremely knowledgeable about all aspects of plant nurture.

Within the well laid out ½ acre site you will find not just an extensive range of plants, alpines, perennials and shrubs but a wide choice of pots, metalware, baskets, conservatory items and topiary. Brenda sells David Austin roses. She also stocks King's Seeds and Francini seeds from Italy.

Other popular lines include hanging baskets and planted pots prepared to order. There's also a large selection of new and secondhand tools.

2 Argyle Garden Supplies

Argyle Commercial Centre, Argyle Street, Swindon, Wiltshire SN2 8AR
Tel: 01793 520827
Specialities: Cut Flowers, Pet Products, Shrubs

3 Barters Plant Centre

High Street, Chapmanslade, Westbury, Wiltshire BA13 4AL
Tel: 01373 832694
Specialities: Garden Products, Herbaceous Perennials, Hanging Baskets

4 Belmont House Nursery

Little Horton, Devizes, Wiltshire SN10 3LJ
Tel: 01380 860510
Specialities: Garden Products

5 Blooms of Bressingham

Hay Lane, Studley, Swindon, Wiltshire SN4 9QT
Tel: 01793 852736

6 Blounts Court Nurseries

Studley, Calne, Wiltshire SN11 9NH
Tel: 01249 812103

7 Bowley Plants

Church Farm, North End, Ashton Keynes, Swindon, Wiltshire SN6 6QR
Tel: 01285 640352
Specialities: Rhododendron (Inc Azalea)

8 Bratton Plant Centre

Fitzroy Farm, Bratton, Westbury, Wiltshire BA13 4RP
Tel: 01380 830219
Specialities: Hanging Baskets, Herbaceous Perennials, Shrubs

9 Classic Baskets Ltd

3 Ludlow Close, Chippenham, Wiltshire SN15 3UG
Tel: 07799 330892
Specialities: Hanging Baskets

10 The Classic Gardener

West Kington, Chippenham, Wiltshire SN14 7JQ
Tel: 01249 783880
Specialities: Annuals/Biennials/Bedding

11 Courtens Garden Centre

Romsey Road, Whiteparish, Salisbury, Wiltshire SP5 2SD
Tel: 01794 884489

12 Downside Nurseries

Upper Westwood, Bradford-on-Avon, Wiltshire BA15 2DE
Tel: 01225 862392 Fax: 01225 866902

Specialities:

Herbaceous Perennials; Unusual Plants

Products and Services:

- Open 7 Days All Year
- Large Car Park
- 90% Plants Home Grown
- Associated Café 1 Mile
- Credit Cards Accepted

Directions:

From the B3109 follow the brown signs.
From the A36 take the Freshford turning, nursery 2 miles.

Description:

Downside Nurseries is a family-run business in a pleasant rural setting. There are more than 2 acres of well kept nursery to walk around. The majority of the high quality plants are home grown. They include herbaceous perennials, roses, shrubs, trees and climbers, Summer and Autumn bedding,

vegetable plants and herbs, along with hanging baskets and containers. Owners Lorraine and Richard Young aim to keep as wide a range as possible and are always on the look-out for new and unusual plants. Their associated café, the Mad Hatter, is 1 mile away in picturesque Avoncliff.

13 Flowerland Garden Centre

Netherhampton, Netherhampton, Salisbury, Wiltshire SP2 8PR
Tel: 01722 743206
Specialities: Bulbs, Rose, Annuals/Biennials/Bedding

14 Foxley Road Nurseries

Foxley Road, Malmesbury, Wiltshire SN16 0JQ
Tel: 01666 822171
Specialities: Annuals/Biennials/Bedding, Herbaceous Perennials

15 Golden Acres Nursery

Park Farm, Southampton Road, Landford, Salisbury, Wiltshire SP5 2BE
Tel: 01794 390319

16 Great Field Garden Centre

Greatfield, Swindon, Wiltshire SN4 8EQ
Tel: 01793 770381

17 Heather Bank Nursery

1 High Street, Littleton Panell, Devizes, Wiltshire SN10 4EL
Tel: 01380 812739
Specialities: Herbaceous Perennials

18 The Herbary

161 Chapel Street, Horningsham, Warminster, Wiltshire BA12 7LU
Tel: 01985 844442
Specialities: Vegetables, Herbs, Seeds

19 John Toomer & Sons Ltd

Stone Lane, Lydiard Millicent, Swindon, Wiltshire SN5 3LD
Tel: 01793 771766

20 Lakeside Garden Centre

Crockerton, Warminster, Wiltshire BA12 8AP
Tel: 01985 217413
website: www.fonthill-lakeside.com

SEE ADVERTISEMENT PAGE 299

Specialities:

Trees & Shrubs, Aquatics, Stoneware

Products and Services:

- Aquatics
- Barbeques
- Books
- Seasonal Displays
- Credit Cards Accepted
- Disabled Facilities
- Display Gardens
- Dried & Artificial Flowers
- Fencing
- Floristry
- Fruit Trees
- Garden Furniture
- Garden Products
- Giftware
- Greenhouses and Sheds
- Houseplants
- Information Desk
- Large Car Park
- Paving
- Plant Guarantee
- Plants
- Restaurant/Coffee Shop
- Seeds and Bulbs
- Trees

Directions:

Lakeside Garden Centre can be found in the village of Crockerton 2 miles south of Warminster on the A350.

21 Landford Trees

Landford Lodge, Landford, Salisbury, Wiltshire SP5 2EH
Tel: 01794 390808
Specialities: Trees

22 Lowden Nurseries

Folly Lane, Shaw, Melksham, Wiltshire SN12 8EZ
Tel: 01225 702345
Specialities: Annuals/Biennials/Bedding

23 Ludgershall Garden Centre

25 Astor Crescent, Ludgershall, Andover, Wiltshire SP11 9RG
Tel: 01264 791421
Specialities: Annuals/Biennials/Bedding, Aquatics, Water Products and Features

24 Malmesbury Garden Centre

Crudwell Road, Malmesbury, Wiltshire SN16 9JL
Tel: 01666 822809
Specialities: Annuals/Biennials/Bedding, Garden Furniture

25 Manor Garden Centre

Cheney Manor Industrial Estate, Swindon, Wiltshire SN2 2QJ
Tel: 01793 538400
Specialities: Annuals/Biennials/Bedding, Shrubs

26 The Mead Nursery

Brokerswood, nr Westbury, Wiltshire BA13 4EG
Tel: 01373 859990
website: www.themeadnursery.co.uk

Specialities:

Alpines; Herbaceous; Pot Grown Bulbs

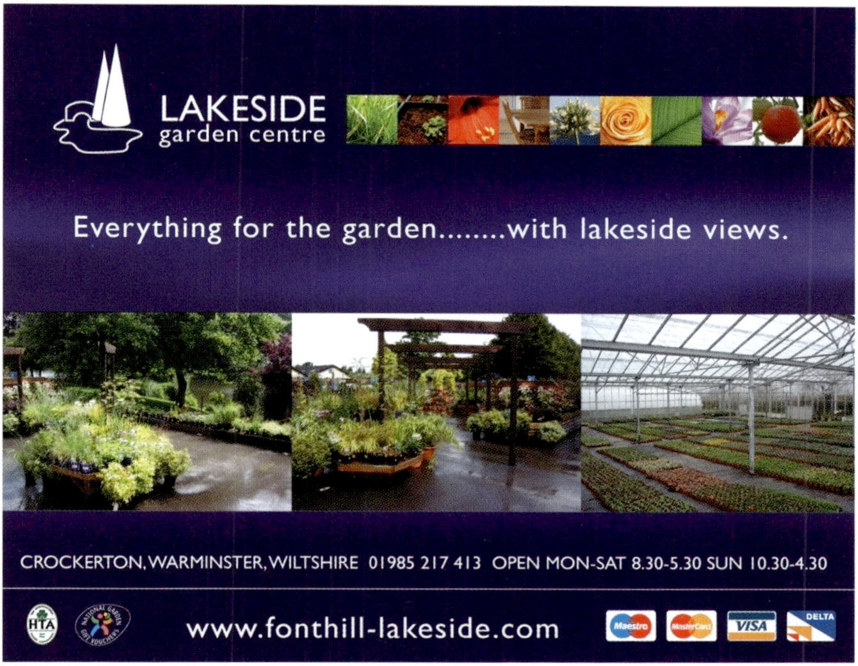

Products and Services:

- Childrens Play Area
- Credit Cards Accepted
- Disabled Facilities
- Display Gardens
- Information Desk
- Large Car Park
- Plants

Directions:

From Westbury take the A350 towards Warminster, then the A36 towards Bath. In Standerwick, turn right at The Bell pub to Rudge. In Rudge, turn right at the Full Moon pub.

Description:

If you love plants, you will love The Mead. Owned and run by Stephen and Emma Lewis-Dale, The Mead specialises in growing perennials, alpines, grasses and bulbs. As soon as you enter, you will see why The Mead has such a reputation for warmth of welcome and the sensitivity with which the nursery is managed environmentally. All plants are grown in a special peat-free compost mix and the peaceful rural setting of the nursery provides a haven for wildlife. The wide variety of plants available are grown on site and you will get the best professional advice to ensure your growing success at home. Within the site is a mature garden laid out to give ideas on colours and design, with raised alpine beds, a sink garden, herbaceous borders and a wildlife pond.

Opening times:

February to mid October, Wednesday - Saturday & Bank Holiday Mondays 9am - 5pm, Sundays 12 - 5pm. (Closed Easter Sunday).

27 Nadder Valley Nurseries

Park Farm, Teffont Magna, Salisbury, Wiltshire SP3 5RR
Tel: 01722 716985
Specialities: Garden Products

28 Natural Selection

1 Station Cottages, Hullavington, Chippenham, Wiltshire SN14 6ET
Tel: 01666 837369

29 North Street Nurseries

North Street, Mere, Warminster,
Wiltshire BA12 6HH
Tel: 01747 860259
Specialities: Houseplants, Cut Flowers, Annuals/Biennials/Bedding

30 Palmer Gardens Plant Centre

Islington, Trowbridge, Wiltshire BA14 8QJ
Tel: 01225 768977
Specialities: Rose, Shrubs, Bulbs

31 Parham Bungalow Plants

Parham Lane, Market Lavington, Devizes,
Wiltshire SN10 4QA
Tel: 01380 812605
Specialities: Alpines/Rock Plants

32 Porton Garden Centre

Porton, Porton, Salisbury, Wiltshire SP4 0LA
Tel: 08707 701011
Specialities: Annuals/Biennials/Bedding, Garden Furniture

33 Pythouse Organics

Pythouse Walled Garden, West Hatch, Tisbury,
Salisbury, Wiltshire SP3 6PA
Tel: 01747 870444
Specialities: Farm Shop

34 Rookery Nursery

Cottles Lane, Atworth, Melksham,
Wiltshire SN12 8HU
Tel: 07850 328756
Specialities: Garden Products

35 Roses od Devizes

4 The Green, Estcourt Street, Devizes,
Wiltshire SN10 1LH
Tel: 01380 721575
Specialities: Garden Furniture, Garden Machinery

36 Showell Nurseries

Showell, Chippenham, Wiltshire SN15 2NU
Tel: 01249 652487
Specialities: Annuals/Biennials/Bedding, Herbaceous Perennials

37 Simpsons Seeds

The Walled Garden Nursery, Horningsham,
Warminster, Wiltshire BA12 7NQ
Tel: 01985 845004
Specialities: Seeds, Vegetables

38 Sonia Wright Plants

Buckerfields Nursery, Ogbourne St. George,
Marlborough, Wiltshire SN8 1TF
Tel: 01672 841065

39 Superior Plants

Fiddington Hill Nurseries, Market Lavington,
Devizes, Wiltshire SN10 4BU
Tel: 01380 813210
Specialities: Hanging Baskets, Annuals/Biennials/Bedding, Herbaceous Perennials

40 Townsends Garden Centre

Bath Road, Devizes, Wiltshire SN10 2BB
Tel: 01380 723722
Specialities: Annuals/Biennials/Bedding, Garden Furniture

41 Trowbridge Garden Centre

288 Frome Road, Trowbridge, Wiltshire BA14 0DT
Tel: 01225 763927
Specialities: Garden Furniture, Annuals/Biennials/Bedding, Alpines/Rock Plants

42 The Walled Garden Nursery

Brinkworth, Chippenham, Wiltshire SN15 5DF
Tel: 01666 826637
Specialities: Herbaceous Perennials, Climbers, Clematis

43 Wanborough Herb Nursery

Callas Hill, Upper Wanborough, Swindon, Wiltshire
Tel: 01793 790327
e-mail: biggs@wanboroughherbnursery.fsnet.co.uk

Specialities:

Herbs; Hardy Perennials; Plant Searches

Products and Services:

- Display Gardens
- Garden Design
- Plants

Directions:

From Junction 15 of the M4 take the 3rd left at the roundabout on the A419 (signposted Wanborough). Continue along this road to the crossroads, turn right (signposted Fox Hill) and the nursery is 100 metres on the left.

Description:

Wanborough Herb Nursery is a small family-run nursery set in an area of outstanding natural beauty. Owner Peter Biggs has created gardens to show how many of the plants may be used. Once established, the plants in the gardens are not watered and the use of chemicals and pesticides is shunned. If you have children with you, they can view the rare breed rabbits and a small flock of chickens and ducks

whilst you browse amongst the ever-changing specialist range of herbs and unusual plants. Most of the plants are grown in Warrior compost which is produced from Wiltshire's green waste. Due to lack of space, the nursery propagates in small quantities so if you are coming from a distance for something in particular, you are advised to phone in advance to check availability. The nursery offers a garden design service and welcomes group visits by appointment.

44 Westdale Nurseries

Holt Road, Bradford-on-Avon, Wiltshire BA15 1TS
Tel: 01225 863258
Specialities: Garden Furniture, Garden Products, Pet Products

45 Whitehall Garden Centre

Corsham Road, Lacock, Chippenham, Wiltshire SN15 2LZ
Tel: 01249 730204

46 Wilton House Garden Centre

Salisbury Road, Wilton Estate, Wilton, Salisbury, Wiltshire SP2 0BJ
Tel: 01722 746746

47 Woodspring Fuchsia Nursery

Weston Lane, Winterslow, Salisbury, Wiltshire SP5 1RL
Tel: 01980 862075
Specialities: Fuchsia

48 Wyevale Garden Centre (Swindon)

Hyde Road, Swindon, Wiltshire SN2 7SE
Tel: 01793 822224

WORCESTERSHIRE

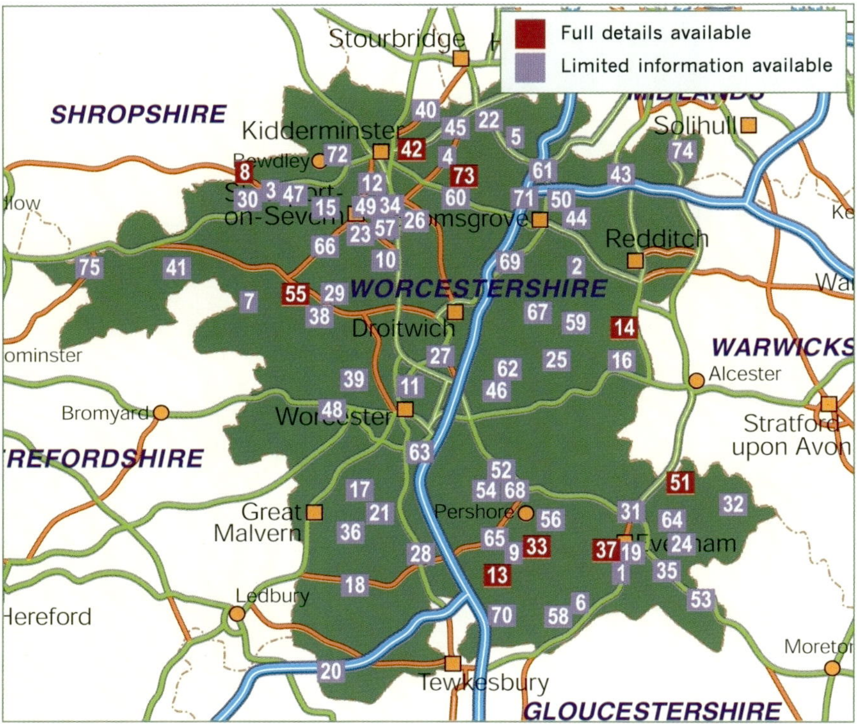

1 Abbey Road Nurseries	

1 Abbey Road Nurseries

Abbey Road, Evesham, Worcestershire WR11 4SB
Tel: 01386 442882
Specialities: Herbaceous Perennials

2 Banks Green Nurseries

Banks Green, Upper Bentley, Redditch, Worcestershire B97 5SU
Tel: 01527 404144
Specialities: Annuals/Biennials/Bedding, Herbaceous Perennials

3 Barncroft Nursery

Callow Hill, Rock, Kidderminster, Worcestershire DY14 9XB
Tel: 01299 266021

4 Barnett Hill Garden & Leisure Ltd

Worcester Road, Clent, Stourbridge, Worcestershire DY9 0EE
Tel: 01562 700308

5 Beechcroft Nurseries & Garden Centre

Madeley Road, Belbroughton, Stourbridge, Worcestershire DY9 9XA
Tel: 01562 710358

6 Belfield Nurseries

Cheltenham Road (A46), Ashton-under-Hill, Evesham, Worcestershire WR11 7QT
Tel: 01386 881870
Specialities: Grasses

7 Big Leaf

Old Post Office, Shelsley Walsh, Worcester, Worcestershire WR6 6RP
Tel: 01886 812838

8 Bill White Nurseries & Plant Centre

Cleobury Road, Far Forest, nr Bewdley, Worcestershire DY14 9EB
Tel: 01299 266326 Fax: 01299 266797
e-mail: flowerpot.bill@btinternet.com
website: www.plantsline.co.uk

Specialities:

Hanging Baskets; Vegetable Plants; Herbs

Description:

Bill White Nurseries opened in 1988 growing container herbs for the Garden Centre market. It moved to its present site in 1994 and has developed its range of plants over the years to now supply quality herbs, ferns, wildflowers, alpines and heathers. In September 1998 the nursery launched its special collection of large plants and chamomile lawn to provide the "Instant Garden" effect – no messing, no fussing, just lay like turf!

Directions:

The nurseries are on the A4117, 4 miles west of Bewdley.

9 Birlingham Nurseries & Garden Centre

Upper Ends, Birlingham, Pershore, Worcestershire WR10 3AB
Tel: 01386 750668
Specialities: Trees, Shrubs, Vegetable

10 Bishops Wood Nursery Plant Centre

Bishops Wood Lane, Crossway Green, Stourport-on-Severn, Worcestershire DY13 9SE
Tel: 01299 251208

11 Blooms of Bressingham

Droitwich Road, Claines, Worcester, Worcestershire WR3 7SW
Tel: 01905 451231

12 Blue Cedars

Wilden Top Road, Stourport-on-Severn, Worcestershire DY13 9JL
Tel: 01299 826813
Specialities: Annuals/Biennials/Bedding

13 Blue Gecko Plants

Tewkesbury Road, Eckington, nr Pershore, Worcestershire WR10 3DE
Tel/Fax: 01386 751802
e-mail: cath@bluegeckoplants.com
website: www.bluegeckoplants.com

Specialities:

Hardy Shrubs; Herbaceous Perennials; Ornamental Trees

Products and Services:

- Credit Cards Accepted
- Giftware
- Information & Advice
- Plant Guarantee
- Display Gardens
- Large Car Park

Directions:

Blue Gecko Plants

From Pershore take the A4104 south for about 3 miles then turn left on the B4080 towards Eckington. Continue through Eckington and the nursery is just south of the village.

Description:

Blue Gecko Plants is located in the village of Eckington, close to the elegant market town of Pershore. Since it opened in June 2006, this friendly, family-run business has rapidly gained a reputation for offering high quality plants presented in a stylish environment. The majority of plants are grown on site by Catherine Thorpe, the nursery owner. An interesting range of hardy shrubs, climbers and perennials is available, along with ornamental trees, bamboos, herbs and grasses. The nursery is particularly noted for the number of Camellia varieties on offer as well as an extensive seasonal selection of vegetable plants. Help and advice is always available, and there is also a small gift shop at the nursery.

14 Botany Bay Nurseries

Edgioake Lane, Astwood Bank, Redditch,
Worcestershire B96 6BG
Tel: 01527 893885 Fax: 01527 892835
e-mail: jamie@botany-bay.co.uk
website: www.botanybaynurseries.co.uk

Specialities:

Geraniums & Fuchsias; Shrubs & Plants; Wildlife Products

Products and Services:

- Seasonal Displays
- Credit Cards Accepted
- Disabled Facilities
- Fruit Trees
- Garden Products
- Gift Vouchers
- Houseplants
- Large Car Park
- Plants
- Seeds and Bulbs
- Trees

Directions:

The nurseries are on the B4092, 3 miles south of Redditch. Turn right for Inkberrow off the A441.

15 Bournewood Nursery

Lickhill Rd North, Stourport-on-Severn,
Worcestershire DY13 8RU
Tel: 01299 828988

16 Bouts Cottage Nurseries

Bouts Lane, Inkberrow, Worcester,
Worcestershire WR7 4HP
Tel: 01386 792923
Specialities: Old Fashioned Violas

17 Brockamin Plants

Brockamin, Old Hills, Callow End, Worcester,
Worcestershire WR2 4TQ
Tel: 01905 830370
Specialities: Geranium, Annuals/Biennials/Bedding

18 Brookend House

Drake Street, Welland, Malvern,
Worcestershire WR13 6LN
Tel: 01684 594005
Specialities: Nerines

19 Carlton Nurseries

Rudge Rd, Evesham, Worcestershire WR11 4JR
Tel: 01386 423888

20 Chris Pattison Nursery

Brookend, Pendock, Gloucester,
Worcestershire GL19 3PL
Tel: 01531 650480
Specialities: Alpines/Rock Plants, Shrubs

21 Chris's Cottage Garden Plants

Clevelode, Malvern, Worcestershire WR13 6PD
Tel: 01684 311243

22 Clent Nurseries

Bromsgrove Road, Clent, Stourbridge,
Worcestershire DY9 9QA
Tel: 01562 885071
Specialities: Ornamentals, Water Products and Features, Garden Furniture

23 Cooks Garden Centre

26 Worcester Road, Stourport-on-Severn,
Worcestershire DY13 9PB
Tel: 01299 826169

24 Cotswold Garden Flowers

Sands Lane, Badsey, Evesham,
Worcestershire WR11 7EZ
Tel: 01386 833849
Specialities: Herbaceous Perennials

25 Cranesbill Nursery

White Cottage, Earls Common Road, Stock Green,
Redditch, Worcestershire B96 6SZ
Tel: 01386 792414
Specialities: Geranium, Herbaceous Perennials

26 Creative Landscapes Gardens & Nurseries

Whitlenge Gardens, Whitlenge Lane, Hartlebury,
Kidderminster, Worcestershire DY10 4HD
Tel: 01299 250720
Specialities: Display Gardens, Shrubs

27 Davids Nurseries

Ash Lane, Martin Hussingtree, Worcester,
Worcestershire WR3 8TB
Tel: 01905 458558

28 Earls Croome Garden Centre

Worcester Road, Earls Croome, Upton-upon-Severn,
Worcestershire WR8 9DA
Tel: 01684 592143

29 Eastgrove Cottage Garden Nursery

Sankyns Green, Nr Shrawley, Little Witley,
Worcester, Worcestershire WR6 6LQ
Tel: 01299 896389

30 Evergreen Conifer Centre

Tenbury Road, Rock, Kidderminster, Worcestershire DY14 9RB
Tel: 01299 266581
Specialities: Conifers, Japanese

31 Evesham Garden Centre

Twyford, Evesham, Worcestershire WR11 4TP
Tel: 01386 761888

32 Fibrex Nurseries Ltd

Honeybourne Road, Pebworth, Stratford-upon-Avon, Worcestershire CV37 8XP
Tel: 01789 720788
Specialities: Ferns, Pelargonium, Hedera (Ivy)

33 Four Acres Nurseries

Great Comberton Road, Pensham, nr Pershore, Worcestershire WR10 3DY
Tel: 01386 550357 e-mail:four.acres@ic24.net

Specialities:

Herbaceous Perennials; Hanging Baskets; Summer Patio Plants

Products and Services:

- Seasonal Displays
- Compost
- Credit Cards Accepted
- Fruit Trees
- Garden Products
- Gift Vouchers
- Large Car Park
- Plant Guarantee
- Plants
- Pots
- Trees

Directions:

South of Pershore town centre, turn right to "The Combertons" after crossing the River Avon. The nurseries are half a mile along this road.

34 Fuchsiavale Nurseries

Worcester Road, Kidderminster, Worcestershire DY11 7SB
Tel: 01299 251162
Specialities: Fuchsia

35 Garden Styles

99 Pitchers Hill, Wickhamford, Evesham, Worcestershire WR11 7RT
Tel: 01386 831199

36 Grange Farm Nurseries

Rectory Lane, Guarlford, Malvern, Worcestershire WR13 6NT
Tel: 01684 562544

37 Harrell's Hardy Plants

Rudge Road, Evesham, Worcestershire WR11 6JR
Tel: 01386 443077
Mobiles: 07799577120 or 07733446606
e-mail: mail@harrellshardyplants.co.uk
website: www.harrellshardyplants.co.uk

Specialities:

Hemerocallis, Unusual Perennials, Grasses

Description:

The nursery was featured on Channel Four and *Gardener's World* in 2007. Sisters, Liz and Kate, have a nursery and acre of garden in an idyllic setting. The planting offers colour and interest most of the year and provides inspiration for many visitors. There is a comprehensive range of plants for sale.
Open Sundays 10am - 12noon, by appointment at most other times. Groups are welcome and can book refreshments.

Directions:

From the A4184 turn into Queen's Road opposite the Catholic Church. Turn right at the end, then left onto Rudge Road. 150 yards on the right a lane leads to the nursery.

38 Haven Nurseries

Bank Road, Little Witley, Worcester, Worcestershire WR6 6LS
Tel: 01886 888674

39 Heath Nurseries

Martley Road, Lower Broadheath, Worcester, Worcestershire WR2 6QG
Tel: 01905 640392

40 Highdown Nursery

1 Highdown Cottages, Sugar Loaf Lane, Iverley, Kidderminster, Worcestershire DY10 3PA
Tel: 01562 700821

41 Highwood Nurseries

Highwood, Eastham, Tenbury Wells, Worcestershire WR15 8PA
Tel: 01584 781333
Specialities: Trees, Shrubs, Garden Design

42 Hodge Hill Garden Centre

Birmingham Road, Kidderminster, Worcestershire DY10 3NR
Tel: 01562 822058

Specialities:

Bedding Plants & Hanging Baskets; Cut Flowers; Farm Shop

Description:

Established for more than 50 years, Hodge Hill is best for its cut flowers and bedding plants which are grown on site. There's also a florist's on site and a farm shop selling fresh locally grown produce.

Directions:

Hodge Hill is on the A456, 2 miles east of Kidderminster.

43 Hopwood Plant Centre Limited

Annandale, Ash Lane, Alvechurch, Birmingham, Worcestershire B48 7TT
Tel: 01214 454022

44 Hurrans

Alcester Road, Burcot, Bromsgrove, Worcestershire B60 1PW
Tel: 01527 873470
Specialities: Restaurant/Coffee Shop

45 Hurrans Garden Centre

Kidderminster Road South, Hagley, Stourbridge, Worcestershire DY9 0JB
Tel: 01562 700511

46 Hyde Cottage Nursery

Hyde Cottage, Church Road, Crowle, Worcester, Worcestershire WR7 4AT
Tel: 01905 381632

47 Ice Alpines

Lye Head, Bewdley, Worcestershire DY12 2UW
Tel: 01299 269219
Specialities: Alpines/Rock Plants, Herbaceous Perennials

48 Laylocks Nurseries & Garden Centre Limited

Bromyard Road, Cotheridge, Worcester, Worcestershire WR6 5LP
Tel: 01905 429212

49 Leapgate Nursery

Leapgate, Wilden, Stourport-on-Severn, Worcestershire DY13 9JH
Tel: 07742 115047

50 Little Heath Garden Centre

Littleheath Lane, Lickey End, Bromsgrove, Worcestershire B60 1HY
Tel: 01527 878174
Specialities: Aquatics, Garden Design, Restaurant/ Coffee Shop

51 Mill Lane Nurseries

Mill Lane, Cleeve Prior, nr Evesham, Worcestershire WR11 8JZ
Tel: 01789 773231

Specialities:

Summer & Winter Hanging Baskets; Patio Plants; Unusual Plants

Products and Services:

- Seasonal Displays
- Plants
- Garden Products
- Houseplants
- Large Car Park
- Plant Guarantee
- Credit Cards Accepted

Directions:

Cleeve Prior is 10 miles SW of Stratford-upon-Avon and 5 miles NE od Evesham. Mill Lane Nurseries are signposted off the B4085 in the village, and are about 300 yards along Mill Lane.

52 Neil Jackson's Nurseries

Worcester Road, Peopleton, Pershore, Worcestershire WR10 2DY
Tel: 01905 841042
Specialities: Conifers, Ornamentals

53 Orchard View Nursery

Orchard View, Evesham Road, Broadway, Worcestershire WR12 7PA
Tel: 01386 852346

54 P G Boulton

Stonebow Road, Drakes Broughton, Pershore, Worcestershire WR10 2AP
Tel: 01905 840880

55 Perhill Plants

Worcester Road, Great Witley, Worcestershire WR6 6JT
Tel: 01299 896329 Fax: 01299 896990
e-mail: perhillp@aol.com
website: www.perhillplants.co.uk

Specialities:

Herbaceous Perennials; Alpines & Rockery; Herbs

Products and Services:

- Credit Cards Accepted
- Mail Order
- Plants

Directions:

Perhill Plants is on the A443, 11 miles northwest of Worcester.

Description:

Perhill Plants is a small family-run business which was established in January 1988 mainly as a wholesale concern supplying rarer, more unusual border perennials, alpines and herbs to garden centres. However, Duncan and Sarah Straw found that their passion for collecting and growing as wide a range of plants as possible – currently they have some 2500 species and cultivars – did not fit in with the modern approach of garden centres. So they decided to concentrate solely on their retail and mail order customers. All the plants they sell are propagated and grown on site by Duncan and Sarah, and their friendly, part-time staff. Coaches are welcome at the site and illustrated talks are given on a variety of subjects.

56 The Plant Centre at Pershore College

Avon Bank, Pershore, Worcestershire WR10 3JP
Tel: 01386 561385

57 Poollands Farm Nursery

Lower Poollands Farm, Titton, Stourport-on-Severn, Worcestershire DY13 9QT
Tel: 01299 829133

58 Rails End Nursery

Back Lane, Ashton-under-Hill, Evesham, Worcestershire WR11 7RG
Tel: 01386 881884

59 Red House Farm Garden and Nursery

Flying Horse Lane, Bradley Green, Redditch, Worcestershire B96 6QT
Tel: 01527 821269
Specialities: Fuchsia, Herbaceous Perennials, Conifers

60 Rowberry's Nursery & Farmshop

On A448, Lower Chaddesley, Kidderminster, Worcestershire DY10 4QN
Tel: 01562 777017

61 Singleton's @ Lydiate Ash Plant Centre

600 Birmingham Road, Lydiate Ash, Bromsgrove, Worcestershire B61 0HT
Tel: 01214 579977
Specialities: Homegrown

62 Springridge Nurseries

Springridge, Old Turnpike Road, Crowle, Worcester, Worcestershire WR7 4AG
Tel: 01905 381451
Specialities: Annuals/Biennials/Bedding

63 St Peters Garden Centre

Broomhall, Norton, Worcester, Worcestershire WR5 2NY
Tel: 01905 357595

64 Station Garden Nurseries

Station Road, Blackminster, Evesham, Worcestershire WR11 7TF
Tel: 01386 832698

65 Three Springs Nursery

Defford Road, Pershore, Worcestershire WR10 3AA
Tel: 01386 555476

66 Tudor Rose Nursery

Tudor Rose Cottage, Dunley, Stourport-on-Severn, Worcestershire DY13 0TT
Tel: 01299 879552

67 Vincent's

Woolmere Pastures, Droitwich Road, Hanbury, Bromsgrove, Worcestershire B60 4DD
Tel: 01527 821196
Specialities: Trees, Shrubs

68 Walcot Organic Nursery

Lower Walcot Farm, Walcot Lane, Drakes Broughton, Pershore, Worcestershire WR10 2AL
Tel: 01905 841587
Specialities: Organic Fruit Trees

69 Webbs Garden Centre

Worcester Road, Wychbold, Droitwich, Worcestershire WR9 0DG
Tel: 01527 860000

70 Westmancote Nurseries

Westmancote Nursery, Main Road, Bredon, Nr Tewkesbury, Worcestershire GL20 7QN
Tel: 01684 772362

71 Willow Brook Garden Centre

Stourbridge Road, Catshill, Bromsgrove, Worcestershire B61 0BL
Tel: 01527 835084
Specialities: Herbaceous Perennials

72 Woodland Cottage Nursery

The Cottage, Habberley Road, Bewdley, Worcestershire DY12 1LA
Tel: 01299 404403

73 Woodrow Nurseries

Woodrow, Chaddesley Corbett, Kidderminster, Worcestershire DY10 4QF
Tel: 01562 777235
website: www.woodrownurseries.com
Specialities: Trees, Herbaceous Perennials, Climbers

74 Wythall Nurseries Ltd

Silver Street, Wythall, Birmingham, Worcestershire B47 6LZ
Tel: 01564 826014
Specialities: Annuals/Biennials/Bedding

75 Zenith Nurseries

Oldwood Road, Tenbury Wells, Worcestershire WR15 8TA
Tel: 01584 810311
Specialities: Annuals/Biennials/Bedding, Herbaceous Perennials

YORKSHIRE - EAST

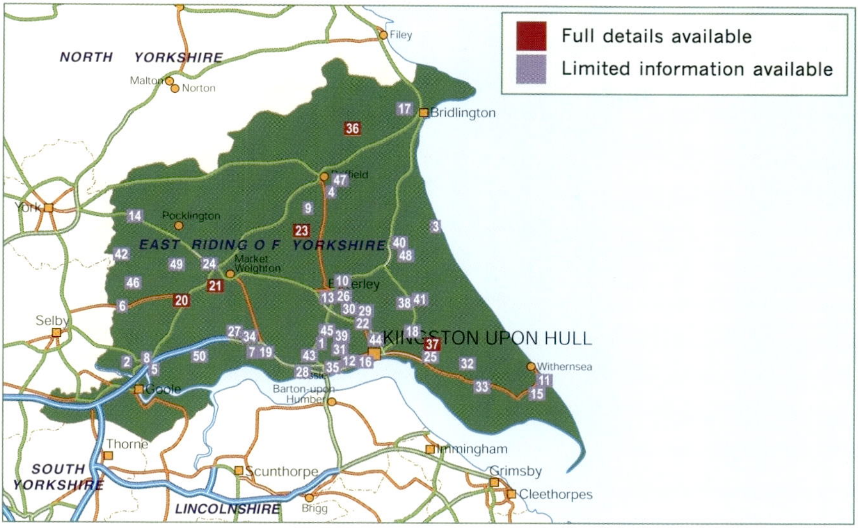

1 Allandale Nurseries

Great Gutter Lane West, Willerby, Hull,
East Yorkshire HU10 6DP
Tel: 01482 657305
Specialities: Garden Products

2 Asselby Nurseries

Barmby Marsh Road, Asselby, Goole,
East Yorkshire DN14 7HE
Tel: 01757 638463
Specialities: Annuals/Biennials/Bedding, Herbaceous Perennials, Shrubs

3 Atwick Mill Plant Centre

Atwick Road, Hornsea, East Yorkshire HU18 1EL
Tel: 01964 537971

4 Bell Mills Garden Centre

Skerne Road, Driffield, East Yorkshire YO25 6XL
Tel: 01377 254043
Specialities: Restaurant/Coffee Shop

5 Bellcross Nurseries

Bellcross Lane, Howden, Goole,
East Yorkshire DN14 7TQ
Tel: 01430 430284
Specialities: Garden Products

6 Box Hedging

49 Main Street, Bubwith, Selby,
East Yorkshire YO8 6LX
Tel: 01757 289972

7 Bramble Wood Nurseries

Stonepit Lane, South Cave, Brough,
East Yorkshire HU15 2BZ
Tel: 01430 425852
Specialities: Annuals/Biennials/Bedding, Herbaceous Perennials

8 California Gardens

Boothferry Road, Howden, Goole,
East Yorkshire DN14 7TF
Tel: 01430 430824
Specialities: Restaurant/Coffee Shop, Trees, Shrubs

9 Cranswick Garden Centre

Burn Butts Lane, Cranswick, Driffield,
East Yorkshire YO25 9JJ
Tel: 01377 270272

10 Croft Nurseries Garden Centre

245 Hull Bridge Road, Beverley,
East Yorkshire HU17 9RS
Tel: 01482 869522

11 Crossroads Nursery

North Leys Road, Hollym, Withernsea,
East Yorkshire HU19 2QN
Tel: 01964 614872
Specialities: Herbaceous Perennials, Shrubs, Annuals/Biennials/Bedding

12 Dairycoates Garden Supplies

Haddon St, Hawthorne Avenue, Hull,
East Yorkshire HU3 5LN
Tel: 01482 575407
Specialities: Garden Products

13 Dales Garden Centre Ltd

Long Lane, Beverley, East Yorkshire HU17 0RN
Tel: 01482 862513
Specialities: Garden Products

14 David Smith Garden Centre

Butts Lane, Hull Road, Wilberfoss, York,
East Yorkshire YO41 5NT
Tel: 01759 388863

15 Devine Nurseries

Patrington Road, Hollym, Withernsea,
East Yorkshire HU19 2QJ
Tel: 01964 613840

16 Dutch Imports Garden Centre

10-14 Alfred Street, Hull, East Yorkshire HU3 2DF
Tel: 01482 606658
Specialities: Garden Products, Water Products and Features, Garden Furniture

17 Eastfield Garden Centre Ltd

Easton Road, Easton, Bridlington,
East Yorkshire YO16 4XF
Tel: 01262 676285

18 Evergreens Garden Centre

119 Main Road, Bilton, Hull,
East Yorkshire HU11 4AB
Tel: 01482 813812
Specialities: Conifers, Garden Ornaments

19 Greens Garden Centre

Ring Beck, Ellerker, Brough,
East Yorkshire HU15 2DD
Tel: 01430 422256
Specialities: Xmas Trees, Trees, Shrubs

20 Greenside Up Plant Nursery

Back Lane, Holme-upon-Spalding-Moor, nr York,
East Yorkshire YO43 4AU
Tel: 01405 860119 Fax: 01405 860096
website: www.greensideup.co.uk

Specialities:

Bedding Plants; Soft Fruits; Hanging Baskets

Description:

Janice and Wayne Patrick took over here in 2005 and both offer a wealth of horticultural knowledge bedding plants, annuals and perennials – Wayne has been a specialist propagation expert for some 30 years. They also grow their own fruit and vegetables on site. If you are bringing children with you, they will be delighted with the small animal farm here which is home to several rare breed species.

Directions:

From Exit 37 of the M62 take the A614 north. After about 9 miles, in the village of Holme-upon-Spalding-Moor, turn left at the crossroads after the village sign.

21 Henleys Nurseries

Riverhead Farm, Market Weighton, York,
East Yorkshire YO43 4LX
Tel: 01430 872464 Fax: 01430 871401
e-mail: henleysnurseries@yahoo.co.uk
website: www.henleysnurseries.co.uk

Specialities:

Ornamental & Fruit Trees; Herbaceous Plants

Description:

Henleys Nurseries moved to its site in 1944 when Charles Henley purchased Riverhead Farm. It is currently run by his son and daughter-in-law, Richard and Judith Henley. The nursery occupies approx 23 acres of which 20 acres produces field grown trees and the remainder has polytunnels and standing areas for container grown stock.

Directions:

From the A1079 Market Weighton by-pass, follow the signs to Holme-on-Spalding Moor. The nursery is 1½ mile along on the left.

22 Ings Lane Nurseries & Water Garden Centre Ltd

Ings Lane, Dunswell, Hull, East Yorkshire HU6 0AL
Tel: 01482 850181
Specialities: Books and Stationery, Garden Products

23 Kilnwick Gardens

Church Lane, Kilnwick, nr Driffield,
East Yorkshire YO25 9JG

Specialities:

Bedding Plants & Baskets; Perennials; Cyclamen

Description:

Established more than 50 years ago, Kilnwick Gardens covers two acres and is open all year round for plant sales. Owner Richard Benwell is an award-winning grower and is noted particularly for his hardy perennials and bedding plants. Set within a walled garden, Kilnwick also specialises at Christmas time with its holly wreaths and cyclamen which make lovely presents. The gardens are closed on Wednesdays.

Directions:

From the A164, about 7 miles north of Beverley, turn left on minor road to Kilnwick (1 mile) The gardens are in the centre of the village.

24 Langlands Nurseries

York Road, Shiptonthorpe, York,
East Yorkshire YO43 3PN
Tel: 01430 873426
Specialities: Restaurant/Coffee Shop, Farm Shop, Garden Products

25 Middle Lane Nurseries

Middle Lane, Preston, Hull,
East Yorkshire HU12 8TQ
Tel: 01482 897138

26 Minster Plant Centre

Long Lane, Beverley, East Yorkshire HU17 0RN
Tel: 01482 874684

27 Mires Beck Nursery

Low Mill Lane, North Cave, Brough,
East Yorkshire HU15 2NR
Tel: 01430 421543
Specialities: Herbaceous Perennials, Wild Flowers

28 O.K Roses

131 Ferriby High Road, North Ferriby,
East Yorkshire HU14 3LA
Tel: 01482 634237
Specialities: Rose

29 Outlet Garden Centre

324 Hull Road, Woodmansey, Beverley,
East Yorkshire HU17 0RU
Tel: 01482 677518
Specialities: Annuals/Biennials/Bedding, Herbaceous Perennials, Hardy Garden Palms

30 Peter Barretts Garden Centre

Hull Rd, Woodmansey, Beverley,
East Yorkshire HU17 0RS
Tel: 01482 880002

31 Ponderosa Garden Centre

60 Woodlands Road, Hull, East Yorkshire HU5 5EF
Tel: 01482 352619
Specialities: Garden Products

32 Premier Koi

Springfield Water Garden Centre, Hedon Road,
Burstwick, Hull, East Yorkshire HU12 9HA
Tel: 01482 896770
Specialities: Conifers

33 Premier Plant Producers

Hull Road, Keyingham, Hull,
East Yorkshire HU12 9SU
Tel: 01964 622564

34 R A Marjoram

Woodside House, Low Road, Everthorpe, Brough,
East Yorkshire HU15 2AD
Tel: 01430 424493
Specialities: Pots, Hanging Baskets, Annuals/Biennials/Bedding

35 RHS Garden

83 Seaton Road, Hessle, East Yorkshire HU13 9EJ
Tel: 01482 646908
Specialities: Restaurant/Coffee Shop, Annuals/Biennials/Bedding

36 Roanne Nurseries

Driffield Road, Kilham, nr Driffield,
East Yorkshire YO25 4SP
Tel: 01262 420617

Specialities:

Hardy Fuchsias; Iris; Scented Pelargoniums

Description:

At this family-run nursery in the heart of the Yorkshire Wolds, owners Ron and Anne Mawer pride themselves on their selection of perennial plants, including some of the less common ones. All of them are grown here from seeds or cuttings. For the summer season they grow many varieties of fuchsias and pelargoniums for which they have won Large Gold, Gold or Silver medals at major shows.

Directions:

From the A166, about 4 miles northeast of Driffield, take the minor road left to Ruston Parva and Kilham. The Nurseries are located in the village, near the school.

Sandhill Nurseries & Garden Centre

Wyton Road, Preston, Hull HU12 8TY
Tel: 01482 898370 Fax: 01482 896099
e-mail: info@sandhill-gardencentre.co.uk
website: www.sandhill-gardencentre.co.uk

Opening Times 9am - 6pm

Thursday, open until 8pm
Sunday, 10.30am-4.30pm

Sandhill Nurseries and Garden Centre, based near Preston, east of Hull, is a family owned business and has been trading for over 60 years.

They now operate from a 7 acre site offering the largest selection of quality gardening products in East Yorkshire, supported by their 4 acre nursery which produces many of their plants and sits next to their 30 acre Country Park where they are currently planning further developments.

They pride themselves on the innovative displays and on the quality of service offered by their team of trained staff, who are always on hand to offer expert advice.

Sandhill offers something for everyone, aswell as fulfilling all your gardening needs we have an ever expanding gift room, a coffee shop, and a farm shop offer high quality, organic and specialist foods. To keep the kids happy we now have an animal farm and play area.

So for a truly inspirational experience, visit Sandhill Garden Centre.

Aquatics
Barbecues
Books and Stationery
Childrens Play Area
Seasonal Displays
Clothing
Conservatories
Credit Cards Accepted
Cut Flowers
Disabled Facilities
Display Gardens
Dried and Artificial Flowers
Farm Shop
Floristry
Fruit Trees
Garden Construction
Garden Machinery
Garden Products
Giftware
Greenhouses and Sheds
Houseplants
Information Desk
Large Car Park
Paving
Pet Products
Plant Guarantee
Plants
Restaurant/Coffee Shop
Seeds and Bulbs
Trees

37 Sandhill Nurseries & Garden Centre

Wyton Road, Preston, Hull, East Yorkshire HU12 8TY
Tel: 01482 898370 Fax: 01482 896099
e-mail: info@sandhill-gardencentre.co.uk
website: www.sandhill-gardencentre.co.uk

SEE ADVERTISEMENT PAGE 312

Specialities:

Bedding Plants & Perennials; Giftware; Farm Shop

Products and Services:

- Aquatics
- Barbecues
- Books and Stationery
- Childrens Play Area
- Seasonal Displays
- Clothing
- Conservatories
- Disabled Facilities
- Display Gardens
- Dried & Artificial Flowers
- Farm Shop
- Floristry
- Garden Construction
- Garden Machinery
- Garden Products
- Giftware
- Greenhouses and Sheds
- Houseplants
- Large Car Park
- Paving
- Pet Products
- Plants
- Restaurant/Coffee Shop
- Seeds and Bulbs
- Trees

Directions:

Preston is on the B1239, east of Hull and about 2 miles north of Hedon.

38 Skirlaugh Garden & Aquatics Centre

Hull Rd, Skirlaugh, Hull, East Yorkshire HU11 5AH
Tel: 01964 562413
Specialities: Aquatics, Garden Products

39 Southwood Farmshop & Nurseries

Ganton Avenue, Cottingham,
East Yorkshire HU16 5HG
Tel: 01482 845825
Specialities: Annuals/Biennials/Bedding, Herbaceous Perennials, Vegetables

40 Spring Garden Centre Ltd

Main Road, Sigglesthorne, Hull,
East Yorkshire HU11 5QL
Tel: 01964 543040

41 Stag View Nurseries

Crab Tree Lane, Old Ellerby, Hull,
East Yorkshire HU11 5AJ
Tel: 01482 815707
Specialities: Annuals/Biennials/Bedding, Shrubs

42 Storwood Nurseries

Hagg Lane, Storwood, York,
East Yorkshire YO42 4TF
Tel: 01759 318084

43 Swanland Nurseries

Beech Hill Road, Swanland, North Ferriby,
East Yorkshire HU14 3QY
Tel: 01482 633670
Specialities: Annuals/Biennials/Bedding, Shrubs

44 Thompson's Compost

181 Cleveland Street, Hull, East Yorkshire HU8 8AZ
Tel: 01482 221144
Specialities: Composts

45 Twinacre Nurseries Ltd

Castle Road, Cottingham, East Yorkshire HU16 5JF
Tel: 01482 841091
Specialities: Annuals/Biennials/Bedding, Herbaceous Perennials

46 Ukgardensupplies.Co.Uk

Bowland House, Ellerton, York,
East Yorkshire YO42 4PN
Tel: 08450 941043

47 Whin Hill Nurseries

Whin Hill Lock, Wansford Road, Wansford, Driffield,
East Yorkshire YO25 5NW
Tel: 01377 240399
Specialities: Garden Products, Hanging Baskets, Shrubs

48 White Cottage Alpines

Sunnyside Nurseries, Hornsea Road, Sigglesthorne,
Hull, East Yorkshire HU11 5QL
Tel: 01964 542692
Specialities: Annuals/Biennials/Bedding, Herbaceous Perennials

49 Woodland Nurseries

Thorpe Lane, Everingham, York,
East Yorkshire YO42 4JA
Tel: 01430 861997

50 Yorkshire Garden Centre

Scalby Lane, Gilberdyke, Brough,
East Yorkshire HU15 2UJ
Tel: 01430 440169

YORKSHIRE - NORTH

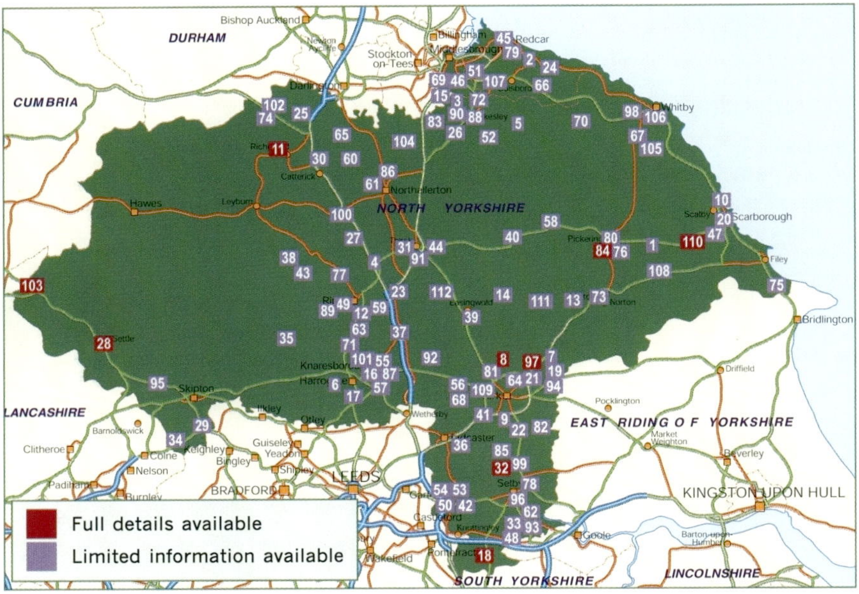

1 A & M Metcalfe

Byland Gardens, Peniston Lane, Ebberston,
Scarborough, North Yorkshire YO13 9NY
Tel: 01723 859101

2 Apple Orchard Nurseries

Marske Lane, Skelton-in-Cleveland,
Saltburn-by-The-Sea, North Yorkshire TS12 2HE
Tel: 01287 623349

3 Arcadia Nurseries

Brass Castle Lane, Nunthorpe, Middlesbrough,
North Yorkshire TS8 9EB
Tel: 01642 310782
Specialities: Garden Furniture, Annuals/Biennials/Bedding, Herbaceous Perennials

4 Baldersby Garden Centre

Ripon Rd, Baldersby, Thirsk,
North Yorkshire YO7 4PS
Tel: 01765 640775
Specialities: Geranium

5 Battersby Roses

Pear Tree Farm Cottages Battersby, Battersby,
Middlesbrough, North Yorkshire TS9 6LU
Tel: 01642 723402
Specialities: Rose

6 Bluecoat Wood Nursery

Otley Road, Beckwithshaw, Harrogate,
North Yorkshire HG3 1QL
Tel: 01423 522876
Specialities: Annuals/Biennials/Bedding, Shrubs

7 Breezy Knees Nurseries

Common Lane, Warthill, York,
North Yorkshire YO19 5XS
Tel: 01904 488800
Specialities: Annuals/Biennials/Bedding, Hanging Baskets

8 Brown's Nurseries

Corban Lane, Wigginton, nr York, North Yorkshire YO32 2RD
Tel/Fax: 01904 766266
website: www.brownsnurseries.co.uk

Specialities:

Hanging Baskets; Perennials; Ornamental Trees

Description:

Over the past 20 years 4 generations of the Brown family have been involved in this successful nursery specialising in hanging baskets and bedding plants. There's also an excellent range of perennials, shrubs and trees.

Directions:

From the A1237 York by-pass, take the B1363 north to Wigginton. In the village take the minor road on the left towards Shipton. The nurseries are on this road between Wigginton and Shipton.

9 Brunswick Organic Nursery

Appleton Road, Bishopthorpe, York, North Yorkshire YO23 2RF
Tel: 01904 701869
Specialities: Organic

10 Burniston Nursery

Coastal Road, Burniston, Scarborough, North Yorkshire YO13 0DB
Tel: 01723 871078
Specialities: Annuals/Biennials/Bedding, Herbaceous Perennials

11 Carl Johnson Landscapes & Aquatics Garden Centre

Gallowfields Industrial Estate, Borough Road, Richmond, North Yorkshire DL10 4SX
Tel: 01748 824058
e-mail: carl-johnson1@btconnect.com
website: www.richmondgardencentre.co.uk

Specialities:

Water Features, Fish, Plants.

Products and Services:

- Aquatics
- Credit Cards Accepted
- Display Gardens
- Fencing
- Garden Construction
- Garden Design
- Garden Furniture
- Garden Products
- Greenhouses and Sheds
- Information Desk
- Paving
- Plants
- Trees
- Water Features

Directions:

From Richmond follow signs to the Industrial Estate. Turn left into Green Howard Road. On the Industrial Estate Take first right onto Borough Road.The garden centre is on the right.

12 Cascade Garden & Aquatic Centre

Bishop Monkton, Harrogate, North Yorkshire HG3 3QD
Tel: 01765 677456

13 Castle Howard Plant Centre

Castle Howard, York, North Yorkshire YO60 7DA
Tel: 01653 648646
Specialities: Shrubs, Trees, Annuals/Biennials/Bedding

14 Cherry Hill Nurseries

Crayke Rd, Crayke Road, Brandsby, York, North Yorkshire YO61 4RR
Tel: 01347 888216

15 Cherry Hill Nurseries & Garden Centre

Stokesley Road, Hemlington, Middlesbrough, North Yorkshire TS8 9DY
Tel: 01642 590650
Specialities: Garden Furniture, Annuals/Biennials/Bedding, Herbaceous Perennials

16 The Christmas Tree Centre

8 Stephenson Close, Knaresborough, North Yorkshire HG5 8EG
Tel: 01423 862683
Specialities: Xmas Trees

17 Crimple Valley Garden Centre

Leeds Rd, Pannal, Harrogate, North Yorkshire HG3 1EW
Tel: 01423 872463
Specialities: Garden Furniture, Trees

18 Crown Garden Centre

Doncaster Road, Whitley, nr Doncaster,
North Yorkshire DN14 0JW
Tel: 01977 661 643 Fax: 01977 663 603
e-mail: smegson.crowngc@btinternet.com

Specialities:

Shrubs, Trees& Conifers; Poinsettias; Japanese Maples

Description:

The Megson family have operated this 4-acre site for more than 20 years, always carrying a wide range of garden plants, including grown on site conifers, seasonal bedding and poinsettias. They also carry a comprehensive range of garden products and accessories.

Directions:

From junction 34 of the M62 take the A19 south towards Doncaster. Drive through Whitley (1 mile) and the nursery is about 1 mile further south.

19 Dalby Nurseries

Warthill, York, North Yorkshire YO19 5XP
Tel: 01904 489788
Specialities: Annuals/Biennials/Bedding, Herbaceous Perennials

20 Dean's Garden Centre

Mere Valley, Seamer Road, Scarborough,
North Yorkshire YO12 4DT
Tel: 01723 361814

21 Dean's Garden Centre

Stockton Lane, Stockton On The Forest, York,
North Yorkshire YO32 9UE
Tel: 01904 400141

22 Deighton Nursery

York Rd, Deighton, York, North Yorkshire YO19 6HQ
Tel: 01904 728891

23 Dishforth Nursery Gardens

Topcliffe Road, Dishforth, Thirsk,
North Yorkshire YO7 3JT
Tel: 01845 578303
Specialities: Annuals/Biennials/Bedding, Herbaceous Perennials

24 Elmora Nurseries

Railway Cottages, Skelton-in-Cleveland,
Saltburn-by-The-Sea, North Yorkshire TS12 2AT
Tel: 01287 652105

25 Everards Nurseries

East Road, Melsonby, Richmond,
North Yorkshire DL10 5NF
Tel: 01325 718003

26 Fir Trees Pelargonium Nursery

Stokesley, Middlesbrough, North Yorkshire TS9 5LD
Tel: 01642 713066
Specialities: Trees, Pelargonium

27 Gables Nursery

Church Wynd, Burneston, Bedale,
North Yorkshire DL8 2JE
Tel: 01677 422228
Specialities: Shrubs, Herbaceous Perennials

28 The Garden Shed

Dugdales, Kirkgate, Settle,
North Yorkshire BD24 9DX
Tel: 01729 822337 Fax: 01729 822785
e-mail: dugdales1906@aol.com
website: www.dugdales-electrical.co.uk

Specialities:

Pots; Garden Furniture; Outdoor Lights

Description:

In 2006 this family-owned business celebrated 100 years of serving the gardeners of the Settle area. Their stock includes a wide range of garden products, pet products, pots galore, tools, seeds, garden furniture, gardening gear and gift items such as walking sticks.

Directions:

The Garden Shed is in the town centre with car parking opposite.

29 Gardeners Choice

Skipton Road, Cross Hills, Keighley,
North Yorkshire BD20 7DS
Tel: 01535 631630

30 Gatherley Garden Centre

Scorton Road, Brompton On Swale, Richmond,
North Yorkshire DL10 7EQ
Tel: 01748 818889

31 Gilbey's Plants

The Walled Garden, Cemetery Road, Thirsk,
North Yorkshire YO7 4DA
Tel: 01765 689927
Specialities: Annuals/Biennials/Bedding

32 Grange Side Nurseries

Thorpe Lane, Cawood, nr Selby,
North Yorkshire YO8 3SG
Tel/Fax: 01757 242953
website: www.grangesidenurseries.co.uk

Specialities:

Unusual Perennials; Seasonal Bedding Plants; Hanging Baskets

Description:

Grange Side Nurseries is a long established business with a good reputation for quality perennials, alpines, bedding plants and hanging baskets. Under new management since 2004 Steve and Di are always on hand for help and advice. A small farm shop is also now open selling many locally produced items.

Directions:

The nurseries are on the B1222 on the eastern side of Cawood village.

33 Green Garden Herbs

13 West Bank, Carlton, Goole,
North Yorkshire DN14 9PZ
Tel: 01405 860708
Specialities: Herbs

34 Green Syke Nursery

Colne Road, Cowling, Keighley,
North Yorkshire BD22 0NA
Tel: 01535 634024
Specialities: Annuals/Biennials/Bedding

35 Greenfingers

4 West Riding Terrace, Glasshouses, Harrogate,
North Yorkshire HG3 5QS
Tel: 01423 712048

36 Grimston Park Nurseries

Grimston Park, Grimston, Tadcaster,
North Yorkshire LS24 9DB
Tel: 01937 832188

37 Hardwicks Garden Centre

Stump Cross, Boroughbridge, York,
North Yorkshire YO51 9HU
Tel: 01423 323941

38 Healey Mill Nursery

Healey, Healey, Ripon, North Yorkshire HG4 4LH
Tel: 01765 689289

39 Hebdon & Sons

The Nurseries, Thirsk Road, Easingwold, York,
North Yorkshire YO61 3HL
Tel: 01347 824540
Specialities: Annuals/Biennials/Bedding

40 Helmsley Walled Garden Ltd

Cleveland Way, Helmsley, York,
North Yorkshire YO62 5AH
Tel: 01439 771427
Specialities: Restaurant/Coffee Shop, Clematis

41 Herbaceous Borders Nurseries

Chapel Lane, Askham Bryan, York,
North Yorkshire YO23 3RD
Tel: 01904 700771
Specialities: Herbaceous Perennials

42 Hillam Gardens Ltd

Hillam Common Lane, Hillam, Leeds,
North Yorkshire LS25 5HR
Tel: 01977 681209

43 The Himalayan Garden Co.

The Hutts, Hutts Lane, Grewelthorpe, Ripon,
North Yorkshire HG4 3DA
Tel: 01765 658009
Specialities: Rhododendron (Inc Azalea), Magnolia

44 Hollin Barn Nurseries

Sutton Road, Thirsk, North Yorkshire YO7 2ER
Tel: 01845 522961
Specialities: Annuals/Biennials/Bedding, Shrubs

45 The Home & Garden Barn

39-41 Esplanade, Redcar,
North Yorkshire TS10 3AG
Tel: 01642 489993

46 Ingles Nursery

44 Sidmouth Close, Middlesbrough,
North Yorkshire TS8 9DN
Tel: 01642 320871

47 Irton Garden Centre

Moor Lane, Irton, Scarborough,
North Yorkshire YO12 4RW
Tel: 01723 862978

48 John Owen Nurseries

20 West Bank, Carlton, Goole,
North Yorkshire DN14 9PZ
Tel: 01405 861415
Specialities: Trees, Shrubs

49 Larkhill Nurseries

Studley Road, Ripon, North Yorkshire HG4 2QR
Tel: 01765 602822
Specialities: Annuals/Biennials/Bedding, Herbaceous Perennials, Pots

50 Lumby Garden Centre

Selby Road, Lumby, South Milford, Leeds,
North Yorkshire LS25 5LE
Tel: 01977 682815

51 Lyndhurst Nurseries

42 Ormesby Bank, Ormesby, Middlesbrough,
North Yorkshire TS7 9EB
Tel: 01642 314547
Specialities: Annuals/Biennials/Bedding

52 Maynards Nursery

Great Broughton, Stokesley, Middlesbrough,
North Yorkshire TS9 7HU
Tel: 01642 778307
Specialities: Conifers, Hanging Baskets, Hedging

53 Milford Plants

Mill Nurseries, Common Lane, South Milford, Leeds,
North Yorkshire LS25 5BX
Tel: 01977 683623

54 Mill Farm Nursery

Mill Farm, Gorse Lane, South Milford, Leeds,
North Yorkshire LS25 6JR
Tel: 01977 682810
Specialities: Shrubs, Herbaceous Perennials

55 Miniature Bulbs UK

The Warren Estate, Greengate Drive, Knaresborough,
North Yorkshire HG5 9EN
Tel: 01423 542819
Specialities: Minature Bulbs

56 Moor Monkton Nurseries

Moor Monkton, Moor Monkton, York,
North Yorkshire YO26 8JJ
Tel: 01904 738770
Specialities: Bamboos, Ferns, Herbaceous Perennials

57 Moorland Nurseries & Garden Centre

Forest Moor Road, Knaresborough,
North Yorkshire HG5 8JY
Tel: 01423 866054
Specialities: Annuals/Biennials/Bedding

58 Moorside Nurseries

Piercy End, Kirkbymoorside, York,
North Yorkshire YO62 6DQ
Tel: 01751 431771
Specialities: Annuals/Biennials/Bedding, Herbaceous Perennials

59 Newby Hall Gardens

Newby Hall, Ripon, North Yorkshire HG4 5AE
Tel: 01423 322583

60 Norheath Nurseries

Streetlam, Northallerton, North Yorkshire DL7 0AH
Tel: 01325 378376
Specialities: Annuals/Biennials/Bedding, Herbaceous Perennials, Shrubs

61 Northdale Horticulture

Yafforth Road, Northallerton,
North Yorkshire DL7 0LQ
Tel: 01609 770269
Specialities: Annuals/Biennials/Bedding, Herbaceous Perennials

62 Oaktree Nursery

Mill Lane, Barlow, Selby, North Yorkshire YO8 8EY
Tel: 01757 618409
Specialities: Annuals/Biennials/Bedding

63 Orchard House Nursery

Ripon Road, Wormald Green, Harrogate,
North Yorkshire HG3 3NQ
Tel: 01765 677541
Specialities: Herbaceous Perennials

64 Orchard Nurseries

48 Hopgrove Lane South, Stockton on the South,
York, North Yorkshire YO32 9TG
Tel: 01904 421153
Specialities: Annuals/Biennials/Bedding, Iris

65 Paddock Farm Water Gardens

Dalton Gates, West Lane, Dalton On Tees,
Darlington, County Durham DL2 2PT
Tel: 01325 378286
Specialities: Water Products and Features, Garden Products, Trees

66 The Paddock Nursery

Lockwood Beck Road, Lingdale,
Saltburn-by-The-Sea, North Yorkshire TS12 3LF
Tel: 01287 652176
Specialities: Annuals/Biennials/Bedding, Herbaceous Perennials

67 Perry's Plants

The River Garden, The Carrs, Briggswath, Whitby,
North Yorkshire YO21 1RR
Tel: 01947 810329
Specialities: Annuals/Biennials/Bedding

68 Plant Sarus

Mayfield Nursery, Hessay Lane, Rufforth, York,
North Yorkshire YO23 3QQ
Tel: 01904 738077

69 Plantarama Garden Centre

Sandy Flatts Lane, Middlesbrough,
North Yorkshire TS5 7YN
Tel: 01642 320514

70 Poets Cottage Shrub Nursery

Lealholm, Whitby, North Yorkshire YO21 2AQ
Tel: 01947 897424
Specialities: Shrubs

71 R M J Sharpe

Nursery Garden, Ripon Road, South Stainley,
Harrogate, North Yorkshire HG3 3NA
Tel: 01423 770961
Specialities: Floristry

72 Rainbow Nurseries

Pannierman Lane, Nunthorpe, Middlesbrough,
North Yorkshire TS7 0NX
Tel: 01642 722234
Specialities: Annuals/Biennials/Bedding, Shrubs, Trees

73 Ralph Yates & Sons

The Garden Centre, Railway Street, Malton,
North Yorkshire YO17 7NR
Tel: 01653 605400
Specialities: Garden Furniture, Trees

74 Ravensworth Nurseries

Ravensworth, Richmond, North Yorkshire DL11 7HA
Tel: 01325 718370
Specialities: Annuals/Biennials/Bedding

75 Reighton Nursery

Hunmanby Road, Reighton, Filey,
North Yorkshire YO14 9RT
Tel: 01723 890359
Specialities: Shrubs, Annuals/Biennials/Bedding

76 Roger's Nurseries

Malton Road, Pickering, North Yorkshire YO18 7JW
Tel: 01751 472226
Specialities: Rose, Annuals/Biennials/Bedding

77 Royalmain Garden Centre

North Stainley, Ripon, North Yorkshire HG4 3HT
Tel: 01765 604385
Specialities: Garden Furniture, Trees, Shrubs

78 Selby Garden Centre Ltd

Hull Road, Osgodby, Selby,
North Yorkshire YO8 5HG
Tel: 01757 708658
Specialities: Garden Furniture, Trees

79 Simmons Nurseries

Redcar Road, Marske-by-The-Sea, Redcar,
North Yorkshire TS11 8LE
Tel: 01642 482387

80 Simpsons

Wold View House, Outgang Road, Pickering,
North Yorkshire YO18 7EL
Tel: 01751 473382
Specialities: Aquatics

81 Skelton Garden Centre

Shipton Road, Skelton, York,
North Yorkshire YO30 1XW
Tel: 01904 470072
Specialities: Restaurant/Coffee Shop, Giftware, Japanese Maples

82 Springwell Garden Centre

Main Street, Wheldrake, York,
North Yorkshire YO19 6AH
Tel: 01904 448733

83 Springwell Nurseries

Stainton Road, Seamer, Middlesbrough,
North Yorkshire TS9 5NA
Tel: 01642 710494
Specialities: Annuals/Biennials/Bedding, Shrubs

84 Steam and Moorland Garden Centre

Malton Road, Pickering, North Yorkshire YO18 7JW
Tel: 01751 471471 e-mail:
sales@hopkinsonandsons.com
website: www.steamandmoorland.com

Specialities:
Garden Furniture; Café/Coffee Shop; Mowers

Products and Services:

- Aquatics
- Barbecues
- Books and Stationery
- Children's Play Area
- Seasonal Displays
- Clothing
- Credit Cards Accepted
- Disabled Facilities
- Display Gardens
- Dried & Artificial Flowers
- Fencing
- Fireworks
- Garden Furniture
- Garden Machinery
- Garden Products
- Giftware
- Greenhouses and Sheds
- Houseplants
- Large Car Park
- Paving
- Pet Products
- Plant Guarantee
- Plants
- Restaurant/Coffee Shop
- Seeds and Bulbs

Directions:
The Centre is located on the A169, half a mile south of Pickering.

Description:
This purpose-built Garden Centre has just about everything the enthusiastic gardener could possibly need – there's even a fine selection of pagodas. The well laid out site contains thousands of healthy vigorous plants, shrubs and trees, as well as an extensive choice of quality garden furniture including sheds, picnic and regular tables, and much more. There's also a well-equipped children's play area. Inside the smart modern building large areas are devoted to mowing machines – including Hayter, Flymo and Husqvarna models – fertilisers, pots and planters, seeds and bulbs, garden tools, gardening clothes and barbecues. If it has anything to do with gardening, you'll find it here. That includes a section devoted to books. There's also an excellent coffee shop selling snacks, teas and light meals which is open from 10am to 4pm. The centre itself is open from 9am to 5.30pm, Monday to Saturday, and from 10am-4pm on Sunday. Adjacent to the Steam and Moorland Garden Centre, Rogers Nursery specialises in fruit trees and roses.

85 Stillingfleet Lodge Nurseries

Stewart Lane, Stillingfleet, York,
North Yorkshire YO19 6HP
Tel: 01904 728506
Specialities: Geranium, Foliage Plants

86 Strikes Garden Centre

Darlington Road, Northallerton,
North Yorkshire DL6 2PW
Tel: 01609 760970

87 Strikes Garden Centre

York Road, Knaresborough,
North Yorkshire HG5 0SP
Tel: 01423 865351

88 Strikes Garden Centre

The Roundabout, Stokesley, Middlesbrough,
North Yorkshire TS9 5HJ
Tel: 01642 710419

89 Studley Royal Garden Centre

Studley Park, Ripon, North Yorkshire HG4 3BB
Tel: 01765 604385
Specialities: Trees, Annuals/Biennials/Bedding

90 Tanton Nursery

Stokesley, Middlesbrough, North Yorkshire TS9 5JT
Tel: 01642 711883

91 Thirsk Garden Centre

Blakey Lane, Thirsk, North Yorkshire YO7 3AB
Tel: 01845 523032
Specialities: Restaurant/Coffee Shop, Annuals/Biennials/Bedding, Water Products and Features

92 Thorpe Trees Nursery

Thorpe Underwood, York,
North Yorkshire YO26 9SR
Tel: 01423 330977
Specialities: Trees

93 Too Hoots Nurseries

Hirst Road, Carlton, Goole,
North Yorkshire DN14 9PX
Tel: 01405 862854
Specialities: Restaurant/Coffee Shop, Giftware

94 Trevor Smith Landscapes Ltd

Weir Pond Nurseries, Main Street, Holtby, York,
North Yorkshire YO19 5UD
Tel: 01904 489998
Specialities: Landscape Gardening, Annuals/Biennials/Bedding

95 Twin Locks Garden Centre Ltd

Hellifield Road, Gargrave, Skipton,
North Yorkshire BD23 3NB
Tel: 01756 749812

96 Vertech Plants

Unit 13 Selby Business Park, Oakney Wood Road,
Selby, North Yorkshire YO8 8LZ
Tel: 01757 700346

97 Vertigrow Plant Nursery

Lawnswood, Malton Road (A64), York,
North Yorkshire YO32 9TL
Tel: 01904 400092 Fax: 01904 400846
e-mail: info@vertigrow.co.uk
website: www.vertigrow.co.uk

SEE ADVERTISEMENT PAGE 321

Specialities:

Herbaceous Perenials, Large Ornamental Trees, Fruit Trees

Products and Services:

- Aquatics
- Seasonal Displays
- Credit Cards Accepted
- Disabled Facilities
- Fencing
- Fruit Trees
- Garden Construction
- Garden Design
- Garden Products
- Houseplants
- Information Desk
- Large Car Park
- Paving
- Plant Guarantee
- Plants
- Restaurant/Coffee Shop
- Seeds and Bulbs
- Trees

Directions:

On A64 five miles north west of York bypass.

98 Victoria Farm

Victoria Farm, Guisborough Road, Whitby,
North Yorkshire YO21 1TL
Tel: 01947 825418

99 Villa Nurseries

York Road, Riccall, York, North Yorkshire YO19 6QG
Tel: 01757 248257

100 W Braithwaite & Sons (Floral Nurseries)

Bedale Road, Leeming Bar, Northallerton,
North Yorkshire DL7 9BG
Tel: 01677 422861
Specialities: Floristry

101 Wayside Nurseries

New Road, Scotton, Knaresborough,
North Yorkshire HG5 9HR
Tel: 01423 868211

102 West Layton Nurseries

Greystones, West Layton, Richmond,
North Yorkshire DL11 7PR
Tel: 01325 718555
Specialities: Annuals/Biennials/Bedding, Trees

103 Westhouse Nursery

A65 Road, Westhouse, Ingleton, Lancashire LA6 3NJ
Tel: 015242 41878 / 42652
website: www.westhousenursery.co.uk

Specialities:

Fuchsias; Bedding Plants; Alpines

Description:

Located in a beautiful part of the Yorkshire Dales National Park, Westhouse Nursery is stocked with hundreds of lovingly home-grown plants. The heated indoor nursery boasts over 400 varieties of fuchsias, as well as perennials, shrubs, hanging baskets and terracotta pots.

Directions:

The nursery is on the A65, six miles southeast of Kirkby Lonsdale and 13 miles from junction 36 of the M6

104 Whitegates Nursery

West Rounton, Northallerton,
North Yorkshire DL6 2LL
Tel: 01609 882355
Specialities: Annuals/Biennials/Bedding, Shrubs

105 Wildplant Services

Dean Hall, Littlebeck Lane, Sneaton, Whitby,
North Yorkshire YO22 5HY
Tel: 01947 810406

106 Williamsons Plant Centre

Ruswarp Lane, Whitby, North Yorkshire YO21 1ND
Tel: 01947 602611

107 Windlebridge Garden Nursery

Middlesbrough Road, Guisborough,
North Yorkshire TS14 6QS
Tel: 01287 635642
Specialities: Annuals/Biennials/Bedding, Shrubs

108 Woodland Nurseries

Scarborough Road, West Heslerton, Malton,
North Yorkshire YO17 8RH
Tel: 01944 728420

109 Wyevale Garden Centre (York)

Northfield Lane, Upper Poppleton, York,
North Yorkshire YO26 6QE
Tel: 01904 795920

110 Wykeham Mature Plants

The Walled Garden, Wykeham, nr Scarborough,
North Yorkshire YO13 9QS
Tel: 01723 862406 Fax: 01723 865643
e-mail: m.howe@wykeham.co.uk
website: www.wykehammatureplants.co.uk

Specialities:

Trees; Hedging; Shrubs

Description:

Wykeham Mature Plants has a long-established reputation for producing extra heavy and semi-mature trees, as well as instant hedging. To complement this, an excellent range of mature shrubs has now been added. The backbone of the shrub range is grown in 45-litre and 115-litre pots using environmentally friendly, peat-free compost. These plants are genuinely 'instant' shrubs which, once planted, look like they have been in place for years.

Directions:

From Scarborough, take the A170 west to the village of Wykeham (6 miles) and look for the signs to the left for Wykeham Mature Plants.

111 Yorkshire Lavender

Yorkshire Lavender Farm, Terrington, York,
North Yorkshire YO60 6PP
Tel: 01653 648008
Specialities: Lavender

112 Yorkshire Plants

Carr Lane Sessay, Sessay, Thirsk,
North Yorkshire YO7 3NG
Tel: 01845 501494
Specialities: Annuals/Biennials/Bedding, Herbaceous Perennials

YORKSHIRE - SOUTH

1 Abbeydale Garden Company

Abbeydale Road South, Dore, Sheffield,
South Yorkshire S17 3LB
Tel: 01142 369091

2 Bawtry Garden Centre

Great North Road, Bawtry, Doncaster,
South Yorkshire DN10 6AB
Tel: 01302 711639
Specialities: Aquatics, Giftware, Annuals/Biennials/Bedding

3 Birley Moor Garden Centre

Moor Valley, Mosborough, Sheffield,
South Yorkshire S20 5BB
Tel: 01142 480666
Specialities: Garden Furniture, Annuals/Biennials/Bedding, Herbaceous Perennials

4 Black Plants Nursery

35 Longfield Road, Sheffield,
South Yorkshire S10 1QW
Tel: 01142 681700
Specialities: Black Plants Speciality

5 Braithwell Nurseries

2 Holywell Cottages, Braithwell, Rotherham,
South Yorkshire S66 7AB
Tel: 01709 812093

6 Brambling House Alpines

119 Sheffield Road, Warmsworth, Doncaster,
South Yorkshire DN4 9QX
Tel: 01302 850730
Specialities: Alpines/Rock Plants

7 Bridge Acre Nurseries & Pet Centre

Bridge Acre, Broomhouse Lane, Doncaster,
South Yorkshire DN4 9BW
Tel: 01302 857333
Specialities: Pet Products, Trees

8 Burncross Bonsai

101 Holy Cross, Chapeltown, Sheffield,
South Yorkshire S35 1ZA
Tel: 01142 451658
Specialities: Bonsai

9 Burncross Nurseries

235 Burncross Road, Chapeltown, Sheffield, South Yorkshire S35 1RZ
Tel: 01142 570959
Specialities: Annuals/Biennials/Bedding, Herbaceous Perennials

10 Cannon Hall Garden Centre

Bark House Lane, Cawthorne, Barnsley, South Yorkshire S75 4AT
Tel: 01226 790785

11 Charlton Brook Nurseries

Springwood Lane, High Green, Sheffield, South Yorkshire S35 4JQ
Tel: 01142 845118
Specialities: Shrubs, Trees, Conifers

12 Clarksons Landscapes

4 Alverley Lane, Doncaster, South Yorkshire DN4 8QW
Tel: 01302 855088
Specialities: Landscape Gardening

13 Croft Nurseries

South Street, Mosborough, Sheffield, South Yorkshire S20 5DF
Tel: 01142 486541
Specialities: Annuals/Biennials/Bedding, Herbaceous Perennials

14 D T & J A Hobson

130 Aughton Road, Swallownest, Sheffield, South Yorkshire S26 4TH
Tel: 01142 872532

15 Dearne Valley Garden Centre Ltd

Station Road, Wath-upon-Dearne, Rotherham, South Yorkshire S63 7DQ
Tel: 01709 877363
Specialities: Garden Furniture

16 Dore More Nursery & Garden Centre

Brickhouse Lane, Sheffield, South Yorkshire S17 3DQ
Tel: 01142 368144

17 Fosters Garden Centre

Doncaster Road, Thrybergh, Rotherham, South Yorkshire S65 4BE
Tel: 01709 850337
Specialities: Annuals/Biennials/Bedding

18 G & P Plevey

Chappell Drive, Doncaster, South Yorkshire DN1 2RF
Tel: 01302 327311
Specialities: Annuals/Biennials/Bedding

19 Garden Care

17 Imperial Buildings, Rotherham, South Yorkshire S60 1PA
Tel: 01709 373158

20 Godfreys Gardens

Hardwick Lane, Aston, Sheffield, South Yorkshire S26 2BE
Tel: 01142 872447
Specialities: Landscape Gardening, Annuals/Biennials/Bedding

21 Greenhouse

401 Ecclesall Road, Sheffield, South Yorkshire S11 8PG
Tel: 01142 660384

22 Hagan's Nurseries

Branton Farm Nurseries, Doncaster Road, Branton, Doncaster, South Yorkshire DN3 3QP
Tel: 01302 538708
Specialities: Shrubs, Giftware

23 High Moor Nursery & Garden Services

Mansfield Rd, Killamarsh, Sheffield, South Yorkshire S21 2BW
Tel: 01142 280600

24 High Riggs Nursery

Upper Gate Road, Stannington, Sheffield, South Yorkshire S6 6BY
Tel: 01142 854061
Specialities: Trees, Shrubs

25 Home & Garden

237 Ringinglow Road, Sheffield, South Yorkshire S11 7PU
Tel: 01142 351325

26 Hoyland Plant Centre

54 Greenside Lane, Hoyland, Barnsley, South Yorkshire S74 9PZ
Tel: 01226 744466
Specialities: Bulbs

27 Loxley Nursery

Long Lane, Loxley, Sheffield,
South Yorkshire S6 6RL
Tel: 01142 337134
Specialities: Annuals/Biennials/Bedding, Herbaceous Perennials, Shrubs

28 Manor Gardens Nursery

Doncaster Road, Hooton Roberts, Rotherham,
South Yorkshire S65 4PF
Tel: 01709 852433
Specialities: Annuals/Biennials/Bedding, Herbaceous Perennials

29 Manor Nurseries

Braithwell Road, Maltby, Rotherham,
South Yorkshire S66 8AE
Tel: 01709 812070

30 Markham Grange Nursery

Long Lands Lane, Brodsworth, Doncaster,
South Yorkshire DN5 7XB
Tel: 01302 330430

31 More Hall Water Gardens

Manchester Road, Bolsterstone, Sheffield,
South Yorkshire S36 3ZP
Tel: 01142 883388
Specialities: Aquatics, Water Products and Features

32 Munsborough Nurseries

Munsbrough Lane, Greasbrough, Rotherham,
South Yorkshire S61 4NS
Tel: 01709 551056

33 Normandale Nurseries

66 Rodney Hill, Loxley, Sheffield,
South Yorkshire S6 6SG
Tel: 01142 344703
Specialities: Shrubs, Annuals/Biennials/Bedding

34 Oak Dene Nurseries

10 Back Lane West, Royston, Barnsley,
South Yorkshire S71 4SB
Tel: 01226 722253
Specialities: Cacti And Succulents

35 Pastures Garden Centre

Pastures Industrial Estate, Pastures Road,
Mexborough, South Yorkshire S64 0JJ
Tel: 01709 584100
Specialities: Garden Furniture, Annuals/Biennials/Bedding, Trees

36 Pennine Lavender Farm

Greenbottom Farm, Barnsley Road, Hoylandswaine,
Sheffield, South Yorkshire S36 7JA
Tel: 01226 763246
Specialities: Lavender

37 PSI Nursery

4 High Street, Braithwell, Rotherham,
South Yorkshire S66 7AL
Tel: 01709 790642
Specialities: Herbaceous Perennials, Rare and Unusual Plants, Trees

38 Rainbow Plants

Springvale Nursery, Sheffield Road, Penistone,
Sheffield, South Yorkshire S36 6HJ
Tel: 07798 691853
Specialities: Herbaceous Perennials, Rare and Unusual Plants

39 Rhinegold Garden Centre

West Lane, Loxley, Sheffield,
South Yorkshire S6 6RW
Tel: 01142 851487
Specialities: Annuals/Biennials/Bedding

40 Rouletta Nurseries

Rouletta Villa, Melton Road, Sprotbrough,
Doncaster, South Yorkshire DN5 7TQ
Tel: 01302 853727
Specialities: Dahlia

41 Scawsby Hall Nurseries

Barnsley Road, Scawsby, Doncaster,
South Yorkshire DN5 7UB
Tel: 01302 782585
Specialities: Shrubs, Herbaceous Perennials

42 Strikes Stringer's Garden Centre

Crookhill Road, Conisbrough, Doncaster,
South Yorkshire DN12 2AE
Tel: 01709 863228

43 Sunnybank Nursery

Sunnybank, Bawtry Road, Hatfield Woodhouse,
Doncaster, South Yorkshire DN7 6PH
Tel: 01302 842954
Specialities: Annuals/Biennials/Bedding, Herbaceous Perennials, Alpines/Rock Plants

44 Taylors Clematis Nurseries

Sutton Road, Sutton, Doncaster,
South Yorkshire DN6 9JZ
Tel: 01302 700716
Specialities: Clematis

45 Thorpe Salvin Nursery

73 Common Road, Thorpe Salvin, Worksop,
South Yorkshire S80 3JJ
Tel: 01909 515393
Specialities: Shrubs, Herbaceous Perennials

46 Tickhill Garden Centres

Bawtry Road, Tickhill, Doncaster,
South Yorkshire DN11 9EX
Tel: 01302 742134
Specialities: Garden Furniture

47 Tom Horsfield

Pot House Hamlet, Barnsley Road, Silkstone,
Barnsley, South Yorkshire S75 4JU
Tel: 01226 790441
Specialities: Rose

48 Valleyside Garden Centre

Bell Hagg, Sheffield, South Yorkshire S10 5PX
Tel: 01142 301925
Specialities: Barbeques, Garden Machinery

49 Walkers Nurseries Ltd

Mosham Road, Blaxton, Doncaster,
South Yorkshire DN9 3BA
Tel: 01302 770325
Specialities: Trees, Shrubs

50 Wentworth Garden Centre

Hague Lane, Wentworth, Rotherham,
South Yorkshire S62 7TF
Tel: 01226 744842
Specialities: Garden Furniture

51 Windmill Nursery

Whitley House, Royd Lane, Millhouse Green,
Sheffield, South Yorkshire S36 9NY
Tel: 01226 765599

52 Yew Tree Garden Centre

Doncaster Rd, Barnsley, South Yorkshire S71 5EZ
Tel: 01226 750500
Specialities: Garden Products, Garden Furniture

YORKSHIRE - WEST

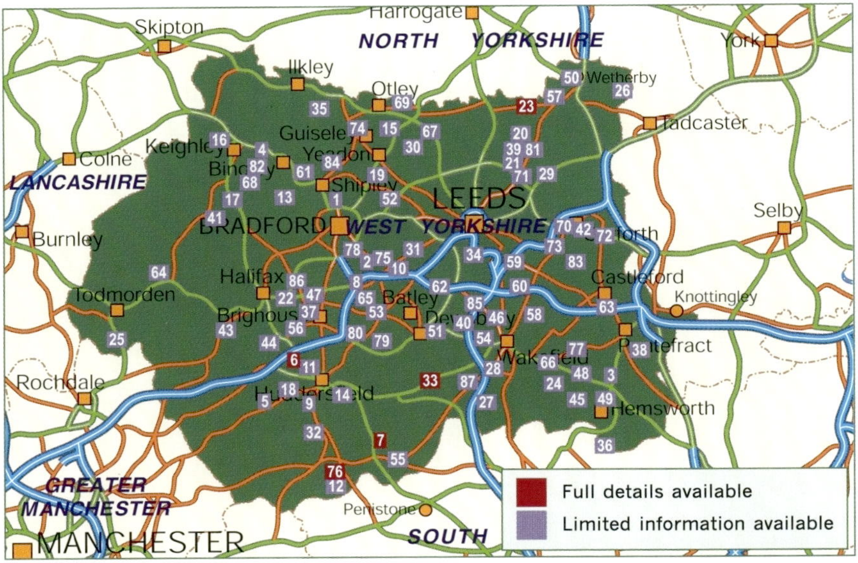

1 A C W Garden Centre

Canal Road, Bradford, West Yorkshire BD2 1AL
Tel: 01274 392344
Specialities: Restaurant/Coffee Shop, Garden Machinery

2 A J Keeling & Sons

Grange Nurseries, North View Road, Bradford, West Yorkshire BD4 6NS
Tel: 01274 682120

3 Ackworth Garden Centre

Barnsley Road, Ackworth, Pontefract, West Yorkshire WF7 7NB
Tel: 01977 612660
Specialities: Garden Furniture, Annuals/Biennials/Bedding, Restaurant/Coffee Shop

4 Acorn Garden Supplies

Units E-F, Castlefields Trading Estate, Crossflatts, Bingley, West Yorkshire BD16 2AF
Tel: 01274 563639

5 Aquatech

Unit 2E Spa Fields Industrial Estate, New Street, Slaithwaite, Huddersfield, West Yorkshire HD7 5BB
Tel: 01484 842632

6 Armitage's Birchencliffe Garden Centre

75 Birchencliffe Hill Road, Huddersfield, West Yorkshire HD3 3NJ
Tel: 01484 536010
e-mail: enquiries@armitages.com
website: www.armitages.com

SEE ADVERTISEMENT PAGE 329

Specialities:

Shrubs, Garden Furniture, Coffee Shop

Products and Services:

- Barbeques
- Books & Stationery
- Seasonal Displays
- Credit Cards Accepted
- Disabled Facilities
- Display Gardens
- Dried & Artificial Flowers
- Fencing
- Fruit Trees
- Garden Furniture
- Garden Products
- Giftware
- Greenhouses and Sheds
- Houseplants
- Information Desk
- Large Car Park
- Plant Guarantee
- Plants
- Restaurant/Coffee Shop
- Seeds and Bulbs
- Trees and Shrubs

Directions:

Off A629 south of M64 J24. Turn right after 1 mile into Birchencliffe Road.

7 Armitage's Pennine Garden Centre

Huddersfield Road, Shelley, Huddersfield,
West Yorkshire HD8 8LF
Tel: 01484 607248
e-mail: enquiries@armitages.com
website: www.armitages.com

SEE ADVERTISEMENT PAGE 329

Specialities:

Bedding, Shrubs, Aquatics, Coffee Shop

Products and Services:

- Aquatics
- Barbeques
- Books & Stationery
- Seasonal Displays
- Credit Cards Accepted
- Cut Flowers
- Disabled Facilities
- Display Gardens
- Dried & Artificial Flowers
- Fencing
- Fruit Trees
- Garden Furniture
- Garden Products
- Giftware
- Greenhouses and Sheds
- Houseplants
- Information Desk
- Large Car Park
- Plant Guarantee
- Plants
- Restaurant/Coffee Shop
- Seeds and Bulbs
- Trees and Shrubs

Directions:

On the B6116, 6 miles south of Huddersfield.

8 Beardsworths Ltd

Whitehall Road, Cleckheaton,
West Yorkshire BD19 6HG
Tel: 01274 871869

9 Beaumont Park Garden Centre

84 Meltham Road, Lockwood, Huddersfield,
West Yorkshire HD4 7BQ
Tel: 01484 667755
Specialities: Bamboos, Bonsai, Xmas Trees

10 Beeches Nurseries

31 Back Lane, Drighlington, Bradford,
West Yorkshire BD11 1LS
Tel: 01132 879077

11 Bridgewood Horticultural Nursery

The Stables, Bryan Road, Huddersfield,
West Yorkshire HD2 2AH
Tel: 01484 435272

12 Brooklands Nurseries

Scholes, New Mill, Holmfirth,
West Yorkshire HD9 1UJ
Tel: 01484 683891

13 Carlton Nurseries

Cottingley Moor Road, Bingley,
West Yorkshire BD16 1UU
Tel: 01274 496270

14 Castle Hill Nurseries

91 Northgate, Almondbury, Huddersfield,
West Yorkshire HD5 8RX
Tel: 01484 531868

15 Chevin Nurseries Ltd

Carlton Lane, East Carlton, Yeadon, Leeds,
West Yorkshire LS19 7BG
Tel: 01132 507626
Specialities: Annuals/Biennials/Bedding, Herbaceous Perennials

16 Cliffe Castle Nursery

Spring Gardens Lane, Keighley,
West Yorkshire BD20 6LH
Tel: 01535 606842

17 The Coley Nursery

Manywells Brow, Cullingworth, Bradford,
West Yorkshire BD13 5EA
Tel: 01274 834992
Specialities: Annuals/Biennials/Bedding, Herbaceous Perennials

18 Colne Valley Garden Centre

Scar Lane, Milnsbridge, Huddersfield,
West Yorkshire HD3 4QA
Tel: 01484 656164
Specialities: Restaurant/Coffee Shop, Cut Flowers, Pet Products

19 Craggwood Nurseries

Woodlands Drive, Rawdon, Leeds,
West Yorkshire LS19 6LQ
Tel: 01132 502016

20 D Pask

Brandon Lane, Leeds, West Yorkshire LS17 9JL
Tel: 01132 893044

21 David Stead Orchids

Greenscapes Horticultural Centre, Brandon Crescent, Leeds, West Yorkshire LS17 9JH
Tel: 01132 893933
Specialities: Orchids

22 Dove Cottage Nursery & Garden

23 Shibden Hall Road, Halifax,
West Yorkshire HX3 9XA
Tel: 01422 203553
Specialities: Herbaceous Perennials, Grasses

23 East Keswick Garden Centre

Harewood Road, East Keswick, nr Wetherby,
West Yorkshire LS17 9HG
Tel/Fax: 01937 573539
e-mail: drbowers@tiscali.co.uk
website: www.eastkeswickplantcentre.co.uk

Specialities:

Trees, Acers; Shrubs, Topiary; Conifers

Description:

East Keswick Garden Centre has been in business for some 20 years but has recently been taken over by David Bowers who has a wealth of horticultural experience and specialises in hardy outdoor nursery stock. The centre also offers many other plants along with an extensive range of garden pots.

Directions:

The centre is on the A659, 2 miles east of Harewood House and 2 miles from the A1 at Wetherby

24 First Impressions

4 Ash Street, New Crofton, Wakefield,
West Yorkshire WF4 1JL
Tel: 01924 865743

25 Gordon Rigg Nurseries

Calderbank, Rochdale Road, Todmorden,
West Yorkshire OL14 7TJ
Tel: 01706 813374

26 The Greenery Garden Centre

Thorp Arch Trading Estate, Thorp Arch, Wetherby,
West Yorkshire LS23 7BJ
Tel: 01937 844778
Specialities: Customer Service, Garden Products, Annuals/Biennials/Bedding

27 Hall Green Nurseries

Stoney Lane, Woolley, Wakefield,
West Yorkshire WF4 2LH
Tel: 01924 259589
Specialities: Annuals/Biennials/Bedding, Herbaceous Perennials

28 Hampsons Plant World

Denby Dale Road, Durkar, Wakefield,
West Yorkshire WF4 3AZ
Tel: 01924 298365
Specialities: Aquatics, Annuals/Biennials/Bedding, Restaurant/Coffee Shop

29 Hayes Garden World

York Road, Leeds, West Yorkshire LS15 4NF
Tel: 01132 731949

30 High Trees Garden Centre

Otley Old Road, Horsforth, Leeds,
West Yorkshire LS18 5HZ
Tel: 01132 587788

31 Hillcroft Nurseries

Back Lane, Leeds, West Yorkshire LS12 5HN
Tel: 01132 854928

32 Holmfield Nurseries

Long Lane, Honley, Holmfirth,
West Yorkshire HD9 6EA
Tel: 07976 523994

33 Horticentre

New Road, Overton, Nr Wakefield,
West Yorkshire. WF4 4RG
Tel: 01924 280306
website www.horticentre.co.uk

Specialities:

Fruit Trees, Ceramic Pots, Organic Composts.

Products and Services:

- Credit Cards Accepted
- Ceramic Pots
- Cut Flowers
- Fireworks
- Floristry
- Fruit Trees
- Garden Products
- Houseplants
- Information Desk
- Large Car Park
- Pet Products
- Plant Guarantee
- Plants
- Seeds and Bulbs
- Trees

Directions:

10 minutes from M1/M62. On A642 5 miles west of Wakefield, close to Coal Mining Museum.

34 Hydroponic & Growlight Centre Ltd

15 Lockwood Way, Leeds, West Yorkshire LS11 5TQ
Tel: 01132 706622
Specialities: Hydroponics

35 Ilkley Moor Garden Centre

Moor Road, Burley Woodhead, Burley - in-Wharfedale, Ilkley, West Yorkshire LS29 7BE
Tel: 01943 864303

36 J P Foss

Common Road, Brierley, Barnsley,
West Yorkshire S72 9EY
Tel: 01226 711338

37 Kershaw's Garden Shopping Centre

Halifax Road, Brighouse, West Yorkshire HD6 2QD
Tel: 01484 713435
Specialities: Restaurant/Coffee Shop, Aquatics

38 Kismet

Moor Lane, Carleton, Pontefract,
West Yorkshire WF8 3RX
Tel: 01977 703736

39 Leeds Lavender

Greenscapes Nursery, Brandon Crescent, Leeds,
West Yorkshire LS17 9JH
Tel: 01132 892922
Specialities: Lavender

40 Lindhill Nursery

Batley Road, Kirkhamgate, Wakefield,
West Yorkshire WF2 0SH
Tel: 01924 372433

41 Little Red Hen Nurseries

91 Denholme Road, Oxenhope, Keighley,
West Yorkshire BD22 9SJ
Tel: 01535 643786
Specialities: Annuals/Biennials/Bedding, Herbaceous Perennials, Shrubs

42 Millrace Nursery

84 Selby Road, Garforth, Leeds,
West Yorkshire LS25 1LP
Tel: 01132 869233
Specialities: Herbaceous Perennials

43 New Farm Nurseries

Longley Lane Noriand, Norland, Sowerby Bridge,
West Yorkshire HX6 3SA
Tel: 01422 834328
Specialities: Annuals/Biennials/Bedding, Herbaceous Perennials, Alpines/Rock Plants

44 Newbank Garden Centre

Stainland Road, Greetland, Halifax,
West Yorkshire HX4 8AN
Tel: 01422 371011
Specialities: Annuals/Biennials/Bedding, Garden Furniture

45 Newstead Garden & Bonsai Centre

Newstead Lane, Havercroft, Wakefield,
West Yorkshire WF4 2HW
Tel: 01977 610040
Specialities: Restaurant/Coffee Shop, Japanese

46 Newton Hill Alpines

335 Leeds Road, Wakefield,
West Yorkshire WF1 2JH
Tel: 01924 377056
Specialities: Alpines/Rock Plants

47 Nordgreen Nurseries

Leeds Road, Norwood Green, Halifax,
West Yorkshire HX3 8PT
Tel: 01274 690611

48 Nostell Priory Rose Gardens

Nostell Priory, Nostell, Wakefield,
West Yorkshire WF4 1QE
Tel: 01924 862248
Specialities: Rose

49 The Nursery

Pontefract Road, Ackworth, Pontefract,
West Yorkshire WF7 7DX
Tel: 01977 794387

50 The Nursery Shop

6 The Shambles, Wetherby,
West Yorkshire LS22 6NG
Tel: 01937 589422

51 Owl Lane Farm Nurseries

118 Owl Lane, Ossett, West Yorkshire WF5 9AU
Tel: 01924 276446
Specialities: Annuals/Biennials/Bedding, Herbaceous Perennials

52 Palmer Nurseries

Calverley Lane, Calverley, Pudsey,
West Yorkshire LS28 5QQ
Tel: 01132 563331

53 Park House Nurseries

Eddercliffe Crescent, Liversedge,
West Yorkshire WF15 6LR
Tel: 01274 861500

54 Planters Nursery

51 Flanshaw Lane, Wakefield,
West Yorkshire WF2 9JG
Tel: 01924 369500

55 Pots 'R' Us

6 Dale Close, Denby Dale, Huddersfield,
West Yorkshire HD8 8RJ
Tel: 01484 864354

56 R G R Garden Centre

Stand Lane, Elland Road, Brighouse,
West Yorkshire HD6 2RG
Tel: 01484 713256

57 Riverside Nurseries

Linton Common, Linton, Nr Wetherby,
West Yorkshire LS22 4JD
Tel: 01937 582598
Specialities: Annuals/Biennials/Bedding, Herbaceous Perennials

58 Roots Nursery & Garden Centre

Birkwood Rd, Normanton, West Yorkshire WF6 2JE
Tel: 01924 898100
Specialities: Annuals/Biennials/Bedding, Shrubs

59 Rothwell Garden Centre

251 Wood Lane, Rothwell, Leeds,
West Yorkshire LS26 0RJ
Tel: 01132 822227
Specialities: Fencing, Herbaceous Perennials, Trees

60 Royds Green Nursery

Sanderson Lane, Oulton, Leeds,
West Yorkshire LS26 8HB
Tel: 01133 934194

61 Saltair Garden Centre

Hirst Mill Crescent, Shipley,
West Yorkshire BD18 4DA
Tel: 01274 599445
Specialities: Garden Furniture, Annuals/Biennials/Bedding

62 Saxons

Topcliffe Lane, Morley, Leeds,
West Yorkshire LS27 0HL
Tel: 01132 527165

63 Seasons

Unit J2-J3 Carr Wood Industrial Estate, Carr Wood Road, Castleford, West Yorkshire WF10 4SB
Tel: 01977 519238

64 Slack Top Nurseries

Heptonstall, Hebden Bridge, West Yorkshire HX7 7HA
Tel: 01422 845348
Specialities: Alpines/Rock Plants, Annuals/Biennials/Bedding

65 Spen Nurseries

183 Spen Lane, Cleckheaton, West Yorkshire BD19 4AB
Tel: 01274 869282
Specialities: Garden Furniture

66 Spring Green Nurseries Ltd

Pontefract Road, Crofton, Wakefield, West Yorkshire WF4 1LW
Tel: 01924 863859
Specialities: Annuals/Biennials/Bedding

67 Springwood Pleiones

8 Tredgold Avenue, Bramhope, Leeds, West Yorkshire LS16 9BU
Tel: 01132 301158
Specialities: Orchids

68 Stephen H Smith Garden & Leisure

Aire Valley Branch, Harden, Bingley, West Yorkshire BD16 1BL
Tel: 01535 274653
Specialities: Shrubs

69 Stephen H Smith Garden & Leisure

Wharfe Valley Branch, Pool Road, Otley, West Yorkshire LS21 1DY
Tel: 01943 850050
Specialities: Garden Furniture, Restaurant/Coffee Shop, Trees

70 Strikes Garden Centre

Selby Road, Swillington Common, Leeds, West Yorkshire LS15 4LQ
Tel: 01132 862981

71 Strikes Garden Centre

Red Hall Lane, Leeds, West Yorkshire LS17 8NA
Tel: 01132 657839

72 Strikes Saville Brothers Garden Centre

Selby Road, Garforth, Leeds, West Yorkshire LS25 2AQ
Tel: 01132 862183

73 Swillington Nurseries

Goody Cross Lane, Swillington, Leeds, West Yorkshire LS26 8UY
Tel: 01132 862265
Specialities: Annuals/Biennials/Bedding

74 Swincar Nurseries

Chevin End Road, Guiseley, Leeds, West Yorkshire LS20 9DL
Tel: 01943 874614

75 Tong Garden Centre

Tong Village, Bradford, West Yorkshire BD4 0RY
Tel: 01132 853506

76 Totties Garden Centre

Downshutts Lane, Totties, Holmfirth, West Yorkshire HD9 1AU
Tel: 01484 683363

Specialities:

Herbaceous Perennials; Trees; Shrubs

Description:

Located in the heart of the *Last of the Summer Wine* country, Totties is a family-run establishment and stands 800ft up on the hillside facing north. If plants can thrive here, they will grow anywhere. Owners Steve Halifax and Sasche Brooks are currently planning to add a restaurant and coffee shop to the site's amenities. This should be up and running by 2008.

Directions:

Totties is located 1 mile from Holmfirth town centre

77 Victoria Garden Centre

Wakefield Road, Streethouse, Pontefract, West Yorkshire WF7 6BS
Tel: 01977 701560

78 Wagon House Farm Nursery and Garden Centre

209 Shetcliffe Lane, Bradford,
West Yorkshire BD4 6QJ
Tel: 01274 684400

79 Westfield Nurseries

Shill Bank Lane, Mirfield, West Yorkshire WF14 0DU
Tel: 01924 498259

80 Whiteleys Garden Centre

Whitegate, Leeds Road, Mirfield,
West Yorkshire WF14 0DQ
Tel: 01924 495944
Specialities: Garden Furniture, Trees, Shrubs

81 Wolinski Nursery

Bay Horse Lane, Leeds, West Yorkshire LS17 8SL
Tel: 01132 892873
Specialities: Hanging Baskets, Seasonal Plants

82 Woodbank Nurseries

Harden Road, Harden, Bingley,
West Yorkshire BD16 1BE
Tel: 01274 562971
Specialities: Restaurant/Coffee Shop, Aquatics, Shrubs

83 Woodend Nurseries

Preston Lane, Great Preston, Leeds,
West Yorkshire LS26 8AP
Tel: 01132 862888
Specialities: Annuals/Biennials/Bedding, Herbaceous Perennials, Shrubs

84 Woodward Nurseries

Barnsley Beck Grove, Baildon, Shipley,
West Yorkshire BD17 6NR
Tel: 01274 414789
Specialities: Annuals/Biennials/Bedding

85 Wyevale Garden Centre (Carr Gate)

Old Bradford Road, Carr Gate, Wakefield,
West Yorkshire WF2 0SY
Tel: 01924 823002

86 Wyevale Garden Centre (Halifax)

Denholmegate Road, Northowram, Halifax,
West Yorkshire HX3 7SB
Tel: 01422 206418

87 Zephyrwude Irises

48 Blacker Lane, Crigglestone, Wakefield,
West Yorkshire WF4 3EW
Tel: 01924 252101
Specialities: Iris

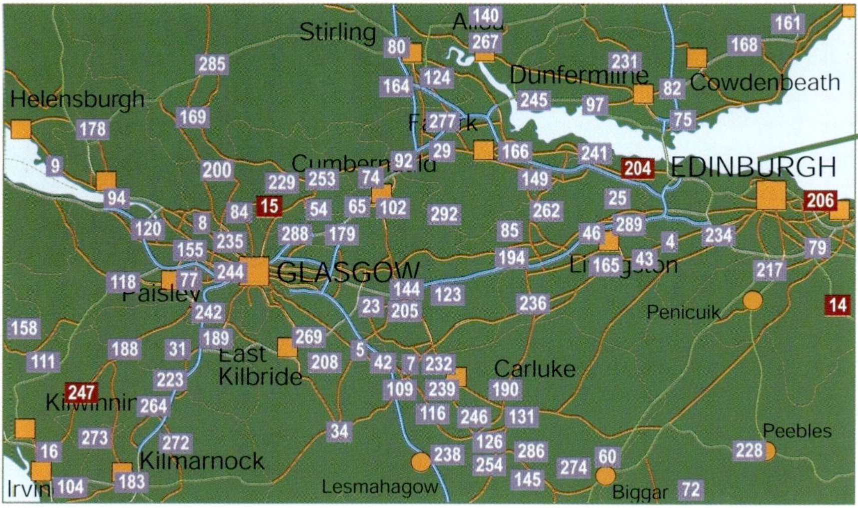

Full details available
Limited information available

SCOTLAND

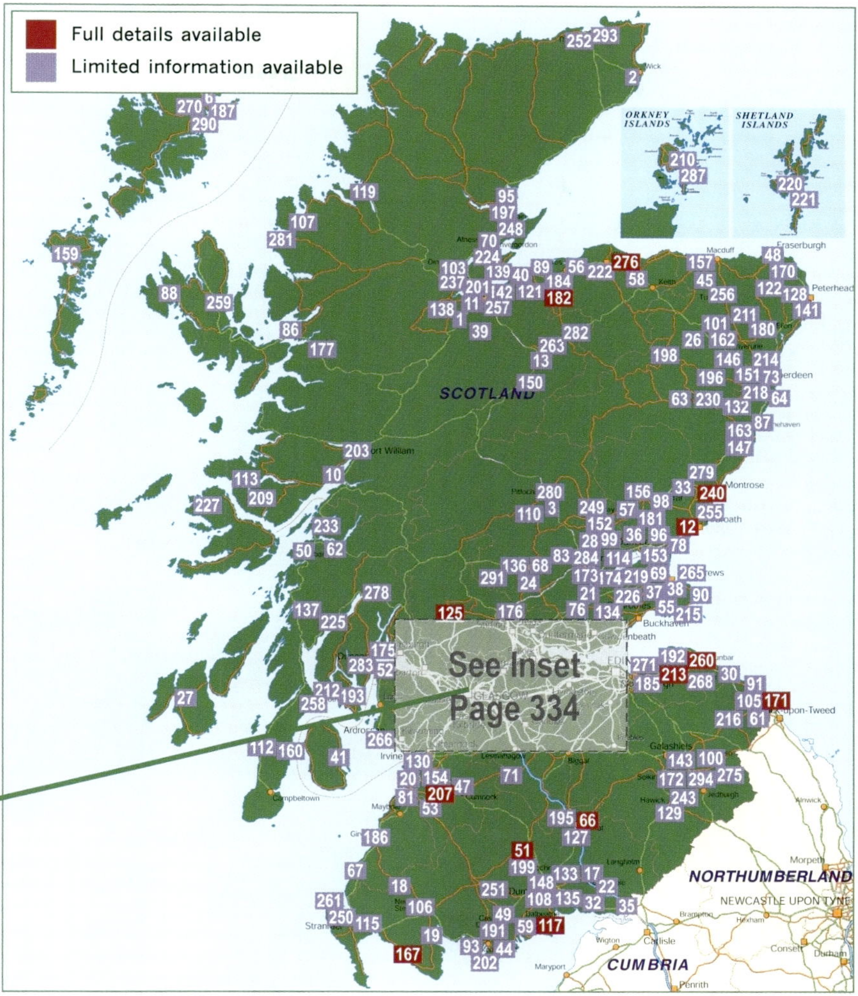

1 Abriachan Nurseries

Loch Ness Side, Inverness, Highland IV3 8LA
Tel: 01463 861232
Specialities: Herbaceous Perennials

2 Allan's of Gillock Ltd

Lochview, Wick, Caithness KW1 5SF
Tel: 01955 621609

3 Allium Garden Co

Ballinluig, Ballinluig, Pitlochry, Perthshire PH9 0LG
Tel: 01796 472358

4 Almondell Garden Centre

Raw Farm, East Calder, Livingston,
West Lothian EH53 0ET
Tel: 01506 880018
Specialities: Garden Furniture

5 Amazing Ponds & Gardens (Scotland) Ltd

Chatelherault House, Hamilton, Lanarkshire ML3 7UE
Tel: 01698 457700
Specialities: Garden Products, Water Products and Features

6 Anderson Road Nurseries

Anderson Rd, Stornoway, Isle of Lewis HS1 2PG
Tel: 01851 702379

7 Anderson's Garden Centre

Lanark Road, Larkhall, Lanarkshire ML9 2UB
Tel: 01698 883492

8 Anniesland Garden & Gift Centre

950A Crow Rd North, Glasgow, Lanarkshire G13 1JD
Tel: 01419 597050
Specialities: Giftware

9 Ardardan Walled Garden Nursery

Cardross, Cardross, Dumbarton, Dunbartonshire G82 5HD
Tel: 01389 849188
Specialities: Annuals/Biennials/Bedding

10 Ard-Daraich Shrub Nursery

Ardgour, Ardgour, Fort William, Inverness-Shire PH33 7AB
Tel: 01855 841248
Specialities: Shrubs

11 Ardfearn Nursery

Bunchrew, Inverness, Highland IV3 8RH
Tel: 01463 243250
Specialities: Alpines/Rock Plants, Shrubs, Trees

12 Ashbrook Nursery

Forfar Road, Arbroath DD11 3RB
Tel/Fax: 01241 873408
e-mail: ashbrooknursery@hotmail.co.uk website: www.ashbrooknursery.co.uk

Specialities:
Alpine and Herbaceous Perennials; Patio and Bedding Plants; Planted Baskets and Containers; Summer Posies.

Products and Services:
- Alpines
- Bulbs
- Compost & Fertilizers
- Cut Flowers
- Design & Planting Advice
- Gift Vouchers
- Herbaceous Perennials
- Information Desk
- Patio & Bedding
- Baskets & Planters
- Seasonal Displays
- Table Decorations
- Trees & Shrubs
- Christmas Trees
- Holly Wreaths
- Herb & Veg Plants

Directions:
The nursery is located on the northern outskirts of Arbroath, on the western side of the A933 Forfar to Brechin road, opposite the RM Condor marine base.

Description:
This is a family-run business where Francis, Joe and Anne Webster are supported by well qualified and enthusiastic staff who are always happy to make time to discuss customers' requirements. The majority of alpine, herbaceous, bedding and patio, herbs and vegetable plants sold here are grown on site. The Webster family offer a wide range of varieties – traditional and novel – not always available elsewhere. The nursery also stocks a good selection of trees and shrubs, sourced where possible from Scottish growers. The nursery features displays of complementary planting combinations and stocks an extensive choice of planted baskets and containers. Ashbrook won a silver medal for courtyard garden display at the Dundee Spring Flower Show. Throughout the year the nursery hosts various seasonal promotions – planters for Mother's Day, summer patio plants and cut flowers, and holly wreaths, table decorations and Christmas trees.

13 Auchgourish Gardens & Arboretum

Street of Kincardine, Boat of Garten, Inverness-Shire PH24 3BY
Tel: 01479 831464

14 Ballantyne Nursery

Shanks Garden, Gorebridge, Midlothian EH23 4LG
Tel: 01875 820859
e-mail: enquiries@ballantynenurseries.com
Specialities: Hanging Baskets, Window Boxes

15 Balmore Garden Centre

Old Balmore Road, Balmore, Glasgow G64 4AE
Tel:01360 620508 Fax: 01360 622655
e-mail: sales@balmoregardencentreltd.com
website: www.balmoregardenscentreltd.com

Specialities:

Hanging Baskets; Bulk Quantities Landscaping Materials; Bedding Plants

Products and Services:

- Aquatics
- Bulbs
- Conservatories
- Credit Cards Accepted
- Disabled Facilities
- Display Gardens
- Fruit Trees
- Garden Construction
- Garden Design
- Garden Furniture
- Garden Products
- Giftware
- Greenhouses and Sheds
- Information Desk
- Large Car Park
- Plants
- Restaurant/Coffee Shop
- Seasonal Displays, inc. Christmas
- Trees

Directions:

Situated in the village of Balmore on the A807.

Description:

The 2-acre site at Balmore Garden Centre stocks a wide variety of plants, bulbs, and aquatics as well as conservatories, greenhouses and sheds. The Centre also stocks garden furniture and offers garden construction and design services. A member of the HTA, owner Andrew Fraser also works closely with the Special Needs Association. A popular amenity at Balmore is Petals Tea Room which offers value-for-money full meals, open sandwiches, toasties, ploughman's and baked potatoes. Desserts include home-made lemon meringue

16 Bankhead Landscape Centre

26 Dovecot Lane, Kilwinning, Ayrshire KA13 6DX
Tel: 01294 558808
Specialities: Landscape Gardening

17 Bankhead Nursery

Bankhead, Lochmaben, Lockerbie, Dumfriesshire DG11 1RR
Tel: 01387 810459
Specialities: Herbaceous Perennials, Annuals/ Biennials/Bedding, Shrubs

18 Barwinnock Herbs

Barrhill, Girvan, Ayrshire KA26 0RB
Tel: 01465 821338
Specialities: Herbs

19 Bayview Nursery

19 Bank Street, Wigtown, Newton Stewart, Wigtownshire DG8 9HR
Tel: 01988 402040

20 Beauly Nursery

Mainholm, Ayrshire KA6 5HE
Tel: 01292 266053

21 Bellfield Organic Nursery

Ferryfield, Newburgh, Cupar, Fife KY14 6EW
Tel: 01738 850589

22 Bellhouse Plants

Johnsfield, Lockerbie, Dumfriesshire DG11 1SS
Tel: 01576 204323
Specialities: Herbaceous Perennials, Shrubs

23 Bellshill Garden Centre

Reema Road, Bellshill, Lanarkshire ML4 1RR
Tel: 01698 840000
Specialities: Garden Furniture

24 Bennybeg Nursery

Muthill Road, Crieff, Perthshire PH7 4HN
Tel: 01764 656345
Specialities: Annuals/Biennials/Bedding, Herbaceous Perennials, Trees

25 Binny Plants

Binny Estate, Ecclesmachan, Broxburn, West Lothian EH52 6NL
Tel: 07753 626116

26 Blossoms of Insch

18 Church Avenue, Insch, Aberdeenshire AB52 6JZ
Tel: 01464 820476

27 Blue Lupins

Islay House Square, Bridgend, Isle of Islay PA44 7NZ
Tel: 01496 810020

28 Bonhard Nursery

Murrayshall Road, Scone, Perth,
Perthshire PH2 7PQ
Tel: 01738 552791
Specialities: Trees, Shrubs

29 Bonnyview Nursery

Falkirk Road, Larbert, Stirlingshire FK5 3NQ
Tel: 01324 562207
Specialities: Hanging Baskets

30 Border Belles

Old Branxton, Dunbar, East Lothian EH42 1QT
Tel: 01368 840325

31 Botany Bay

5 Pilmuir Road, Newton Mearns, Glasgow,
Lanarkshire G77 6PS
Tel: 01416 162277
Specialities: Hanging Baskets, Annuals/Biennials/Bedding

32 Bowhill Nursery

Annan, Dumfriesshire DG12 5RP
Tel: 01461 700244

33 Brechin Castle Centre

Haughmuir, Brechin, Angus DD9 6RL
Tel: 01356 626813
Specialities: Restaurant/Coffee Shop, Giftware

34 Brian Young

Newtonside Nursery, Newton Road, Strathaven,
Lanarkshire ML10 6BQ
Tel: 01357 520895

35 Bridge End Nurseries

Gretna Green, Gretna, Dumfriesshire DG16 5HN
Tel: 01461 800612
Specialities: Annuals/Biennials/Bedding

36 Bridgefoot Nursery

Tigh An Dallach, Rosemill Road, Bridgefoot, Dundee,
Angus DD3 0RW
Tel: 01382 818922
Specialities: Annuals/Biennials/Bedding

37 Bridgend Garden Centre

Ladybank Road, South Bridge, Cupar,
Fife KY15 5HY
Tel: 01337 858293

38 Brighton Nurseries

The Smiddy, Kemback Bridge, Kemback, Cupar,
Fife KY15 5TP
Tel: 01334 654037

39 Brin Herb Nursery

Brin School Fields, Flichity, Inverness,
Highland IV2 6XD
Tel: 01808 521288

40 Broadley Garden Centre

Cawdor Road, Nairn, Highland IV12 5QU
Tel: 01667 452955
Specialities: Annuals/Biennials/Bedding, Herbaceous Perennials, Restaurant/Coffee Shop

41 Brodick Nursery Gardens

Pier Brae, Brodick, Isle of Arran, Ayrshire KA27 8AJ
Tel: 01770 302684
Specialities: Annuals/Biennials/Bedding, Herbaceous Perennials, Floristry

42 Brookside Homes & Gardens

Lanark Road End, Larkhall, Lanarkshire ML9 2UG
Tel: 01698 886464

43 Broompark Plant Centre Ltd

Broompark Farm, Mid Calder, Livingston,
West Lothian EH53 0EB
Tel: 01506 881488

44 Buckland Plants

Whinnieliggate, Kirkcudbright,
Dumfries & Galloway DG6 4XP
Tel: 01557 331323
Specialities: Herbaceous Perennials, Woodland Plants

45 Burn Grains

Burngrains Farm, Alvah, Banff,
Aberdeenshire AB45 3XX
Tel: 01466 780273

46 Bush Cottage Nursery

Wester Dechmont, Broxburn,
West Lothian EH52 6LS
Tel: 01506 811643

47 Butterworths' Organic Nursery

Garden Cottage, Auchinleck House Estate,
Auchinleck, Cumnock, Ayrshire KA18 2LR
Tel: 01290 551088
Specialities: Organic Trees

48 C & A West

64 Charlotte Street, Fraserburgh, Aberdeenshire AB43 9JE
Tel: 01346 511918

49 Cairnsmore Nursery

Chapmanton Road, Castle Douglas, Kirkcudbrightshire DG7 2NU
Tel: 01556 504819

50 Cakebread's Garden Centre

Old Tweedmill, Soroba Road, Oban, Argyll PA34 4HY
Tel: 01631 565141
Specialities: Annuals/Biennials/Bedding, Herbaceous Perennials

51 Capenoch Gardens

Penpont, Thornhill, Dumfriesshire DG3 4LZ
Tel/Fax: 01848 330423
e-mail: capenoch-gardens@virgin.net

Specialities:

Herbaceous Perennials, Shrubs & Trees, Seasonal Bedding

Products and Services:

- Fruit Trees
- Information Desk
- Car Park
- Plants
- Bulbs
- Trees

Directions:

From the A76 Kilmarnock to Dumfries road, 15 miles north of Dumfries, turn left at Thornhill onto the A702 to Penpont. Capenoch Gardens are on the left about a quarter mile outside Penpont.

52 Cardwell Nursery Garden Centre

Lunderston Bay, Cloch Road, Gourock, Renfrewshire PA19 1BB
Tel: 01475 521536

53 Cassillis Living Colour

Drongan, Ayr, Ayrshire KA6 6NQ
Tel: 01292 592252

54 Caulders Nursery

63 Kilsyth Road, Kirkintilloch, Glasgow, Lanarkshire G66 1QF
Tel: 01417 762001
Specialities: Annuals/Biennials/Bedding, Herbaceous Perennials, Customer Service

55 Chessington Nurseries Ltd

The Smiddy, Kemback Bridge, Kemback, Cupar, Fife KY15 5TP
Tel: 01334 653040

56 Christie Elite Nurseries

Forres, Bogton Road, Forres, Morayshire IV36 3TW
Tel: 01309 672633
Specialities: Hardy Woodland

57 Christie's Nursery

Downfield, Main Road, Westmuir, Kirriemuir, Angus DD8 5LP
Tel: 01575 572977
Specialities: Alpines/Rock Plants

58 Christies of Fochabers

The Nurseries, Fochabers, Morayshire IV32 7PF
Tel: 01343 820362

59 Clifton Place Garden Centre

318 High Street, Dalbeattie, Kirkcudbrightshire DG5 4DS
Tel: 01556 610777

60 Clydesdale Plants

13 Biggar Road, Libberton, Carnwath, Nr Biggar, Lanarkshire ML11 8LT
Tel: 01555 840783
Specialities: Herbaceous Perennials, Shrubs

61 Complete Gardens

Whitehouse, Chirnside, Duns, Berwickshire TD11 3XL
Tel: 01890 818339

62 Corachie Clematis

Corachie Farm, Taynuilt, Argyll PA35 1HY
Tel: 01866 822266
Specialities: Clematis

63 Coull Plants

10 Station Square, Aboyne, Aberdeenshire AB34 5HX
Tel: 01339 885700

64 Cove Bay Nurseries

Cove Bay, Aberdeen, Aberdeenshire AB12 3LA
Tel: 01224 897216

65 Craigend Nursery

Condorrat, Cumbernauld, Glasgow,
Lanarkshire G67 4HD
Tel: 01236 721822

66 Craigieburn Garden & Nursery

Craigieburn House, by Moffat,
Dumfriesshire DG10 9LF
Tel: 01683 221250 Fax: 01683 221721
e-mail: jameswheatcroft@hotmail.com
website: www.craigieburngarden.com

Specialities:
Himalayan Plants; Classic Herbaceous Plants; Woodland Plants

Description:
The nursery was established in 1996 by Janet Wheatcroft who has travelled extensively in Europe and the Himalayas. She has introduced a number of important plant varieties to the UK.

Charming and somewhat haphazard, Craigieburn Garden and Nursery is the antithesis of the normal garden centre. The nursery takes up most of a former kitchen garden on a fine, naturally sheltered part of the garden which is flanked by some fine 200-year-old beech trees. The Craigieburn runs close by, keeping the nursery plants fresh in the summer. Within the garden are the old bothies, tiny buildings which visitors are intrigued to discover once housed the gardener and his family. One of them now recreates an authentic Sherpa home, as lived in by Nepal's heroic mountaineers. The plants for sale cover half an acre, between 600 to 700 varieties of plants filling every corner of the nursery.

Directions:
Two miles from Moffat on the A708, just past the traffic signals. Look out for the large dark blue signs to CRAIGIEBURN NURSERY AND GARDEN. Go up the drive and park in the Stable Yard.

67 Craigiemains Home & Garden Centre

Main Street, Ballantrae, Girvan, Ayrshire KA26 0NB
Tel: 01465 831052
Specialities: Restaurant/Coffee Shop, Herbaceous Perennials

68 Crieff Plant Centre

Muthill Road, Crieff, Perthshire PH7 4HQ
Tel: 01764 652722

69 Cupar Garden Centre

Cupar Trading Estate, Cupar, Fife KY15 4SX
Tel: 01334 655766
Specialities: Garden Furniture, Greenhouses and Sheds, Barbeques

70 D J Ledingham

109 High Street, Invergordon, Ross-Shire IV18 0AB
Tel: 01349 854303

71 Daly Mclaughlan

Ayr Road, Glespin, Lanark, Lanarkshire ML11 0SF
Tel: 01555 850333

72 Dawyck Botanic Gardens

Stobo, Peebles, Borders EH45 9JU
Tel: 01721 760254
Specialities: Trees, Shrubs

73 Dobbies Garden Centre

Hazledene Road, Hazlehead, Aberdeen,
Aberdeenshire AB15 8QU
Tel: 01224 318658

74 Dobbies Garden Centre

Eastfield Road, Cumbernauld, Glasgow,
Lanarkshire G68 0EB
Tel: 01236 736100

75 Dobbies Garden Centre

Western Approach Road, Dalgety Bay, Dunfermline,
Fife KY11 9XP
Tel: 01383 823841

76 Dobbies Garden Centre

Turfhills, Kinross, Perthshire KY13 0NQ
Tel: 01577 863327

77 Dobbies Garden Centre

Hawkhead Road, Paisley, Renfrewshire PA2 7AD
Tel: 01418 875422

78 Dobbies Garden World

Ethiebeaton Park, Monifieth, Dundee,
Angus DD5 4HB
Tel: 01382 530333

79 Dobbies Garden World

Melville Nursery, Lasswade, Midlothian EH18 1AZ
Tel: 01316 631941

80 Dobbies Garden World

Drip Road, Drip Bridge, Stirling,
Stirlingshire FK9 4UF
Tel: 01786 458860

81 Dobbies Garden World

Old Toll, Ayr, Ayrshire KA6 5JJ
Tel: 01292 294750

82 Dobbies Garden World

Whimbrell Place, Dunfermline, Fife KY11 8EX
Tel: 01383 842757

83 Dobbies Garden World

Crieff Road, Perth, Perthshire PH1 3JJ
Tel: 01738 638555

84 Dobbies Milngavie

Boclair Road, Milngavie, Glasgow,
Lanarkshire G62 6EP
Tel: 01360 620721

85 Dow's Garden Supplies

Wester Hillhouse, Bathgate, West Lothian EH48 3AR
Tel: 01501 733931

86 Duncraig Nursery

3 Duncraig Square, Plockton, Ross-Shire IV52 8TZ
Tel: 01599 544760
Specialities: Landscape Gardening, Herbaceous Perennials, Annuals/Biennials/Bedding

87 Dunnottar Nurseries

1 Carronhall, Stonehaven,
Kincardineshire AB39 2QF
Tel: 01569 763422

88 Dunvegan Nursery

Strathord, Dunvegan, Isle of Skye,
Isle of Skye IV55 8GT
Tel: 01470 521425

89 Dyke Nurseries

Dyke, Forres, Morayshire IV36 2TF
Tel: 01309 641362

90 East Neuk Water Garden Centre

St. Andrews Road, Crail, Fife KY10 3UL
Tel: 01333 450530
Specialities: Aquatics, Water Products and Features

91 Edrom Nurseries

Coldingham, Eyemouth, Berwickshire TD14 5TZ
Tel: 01890 771386

92 Elainan Nursery

183 Glasgow Road, Dennyloanhead, Bonnybridge,
Stirlingshire FK4 1QY
Tel: 01324 840941

93 Elizabeth MacGregor

Ellenbank, Tongland Road, Kirkcudbright,
Dumfries & Galloway DG6 4UU
Tel: 01557 330620
Specialities: Cottage Garden Plants

94 Erskine Garden Centre

Erskine Hospital, Bishopton, Renfrewshire PA7 5PU
Tel: 01418 144626

95 Evelix Daffodils

Aird Asaig Evelix, Evelix, Dornoch,
Sutherland IV25 3NG
Tel: 01862 810715
Specialities: Narcissus (Daffodil)

96 Evergreen Nursery

Murroes Broughty Ferry, Dundee, Angus DD5 3PB
Tel: 01382 505446

97 Fairleys Garden Centre

53 Main Street, Cairneyhill, Dunfermline,
Fife KY12 8QT
Tel: 01383 880223
Specialities: Garden Furniture

98 Ferngreen Garden Centre

Kirriemuir Road, Forfar, Angus DD8 3TG
Tel: 01307 460012
Specialities: Annuals/Biennials/Bedding, Herbaceous Perennials, Shrubs

99 Ferngreen Nursery

Borelands, Coupar Angus, Blairgowrie,
Perthshire PH13 9LL
Tel: 01828 670651
Specialities: Annuals/Biennials/Bedding, Herbaceous Perennials, Shrubs

100 Floors Castle Garden Centre

Floors Castle, Kelso, Roxburghshire TD5 7RN
Tel: 01573 224530
Specialities: Landscape Gardening

101 Fordmouth Croft Ornamental Grass Nursery

Fordmouth Croft, Meikle Wartle, Inverurie, Aberdeenshire AB51 5BE
Tel: 01467 671519
Specialities: Grasses

102 Forsyth Garden Supplies

North Myvot, Cumbernauld, Glasgow, Lanarkshire G67 4HB
Tel: 01236 458357

103 Frank Nichols Garden Centre & Tearoom

Strathpeffer Road, Dingwall, Ross-Shire IV15 9QF
Tel: 01349 863854

104 Frasers Garden Centre

6 Ploughlands, Dundonald, Kilmarnock, Ayrshire KA2 9BT
Tel: 01563 850215

105 Fuscia Centre

Glendale Cottage Houndwood, Eyemouth, Berwickshire TD14 5TP
Tel: 01361 850207
Specialities: Fuchsia

106 Galloway Heathers Plant Centre

Holmpark Industrial Estate, Cumloden Road, Minnigaff, Newton Stewart, Wigtownshire DG8 6AP
Tel: 01671 402360
Specialities: Calluna/Erica (Heather)

107 Garden Cottage Nursery

Tournaig, Poolewe, Achnasheen, Ross-Shire IV22 2LH
Tel: 01445 781777
Specialities: Drought Resistant Plants

108 Garden Wise Plant & Garden Centre

Castle Douglas Road, Dumfries, Dumfries & Galloway DG2 8PP
Tel: 01387 262654

109 Garrion Bridges Garden Centre

Horseley Brae, Wishaw, Lanarkshire ML2 0RR
Tel: 01698 372288
Specialities: Annuals/Biennials/Bedding, Restaurant/Coffee Shop

110 Gatehouse Nursery

Gatehouse, Aberfeldy, Perthshire PH15 2EL
Tel: 01887 820472
Specialities: Annuals/Biennials/Bedding

111 Gateside Nursery

Gateside, Beith, Ayrshire KA15 2LR
Tel: 01505 502606
Specialities: Shrubs

112 Glenbarr Nursery

Glenbarr, Tarbert, Argyll PA29 6UT
Tel: 01583 421200
Specialities: Annuals/Biennials/Bedding, Herbaceous Perennials

113 Glenborrodale Nursery

Creag Darach, Glenborrodale, Acharacle, Argyll PH36 4JP
Tel: 01972 500205
Specialities: Annuals/Biennials/Bedding

114 Glendoick Garden Centre

Glendoick, Perth, Perthshire PH2 7NS
Tel: 01738 860260

115 Glenwhan Gardens Nursery

Glenwhan, Dunragit, Stranraer, Wigtownshire DG9 8PH
Tel: 01581 400222
Specialities: Annuals/Biennials/Bedding, Ornamentals

116 Goulding Grower Plant Centre

Dalpatrick Farm, Rosebank, Carluke, Lanarkshire ML8 5QA
Tel: 01555 860259

117 Gowanbrae Nursery

Mainsriddle, by Dumfries, Dumfries & Galloway DG2 8AG
Tel: 01387 780588
e-mail: gowanbrae.lyle@virgin.net
website: www.scotlands-garden.org.uk/gowanbrae.html
Specialities: Alpines, Perennials, Shrubs, Conifers, Rhododendrons

118 Gowanlea Nurseries

Barochan Road, Johnstone, Renfrewshire PA5 8YS
Tel: 01505 337222

119 Greenfingers

2 West Terrace, Ullapool, Ross-Shire IV26 2UU
Tel: 01854 612148

120 Greenhead Nursery

Old Greenock Road, Inchinnan, Renfrew, Argyll PA4 9PH
Tel: 01418 120121
Specialities: Seasonal Plants, Annuals/Biennials/Bedding, Rose

121 Greens Nurseries

New Fleenas Farm, Nairn, Highland IV12 5QN
Tel: 01667 452760
Specialities: Herbaceous Perennials

122 Greens Plants

Greens of Gaval, Fetterangus, Peterhead, Aberdeenshire AB42 4HB
Tel: 01771 623696

123 Greenside Nursery

High Street, Newhouse, Motherwell, Lanarkshire ML1 5ST
Tel: 01698 862600
Specialities: Annuals/Biennials/Bedding

124 Greenyards Garden Centre

Cowie Road, Bannockburn, Stirling, Stirlingshire FK7 8AD
Tel: 01786 817369
Specialities: Annuals/Biennials/Bedding, Composts, Garden Furniture

125 Guyana

Main Street, Aberfoyle, Stirling, Stirlingshire FK8 3UQ
Tel: 01877 382308 Fax: 01877 382128
e-mail: robert@guyana.ltf.uk
website: www.guyana.ltd.uk
Specialities: Alpines, Heathers, Shrubs, Perennials, Trees

126 Guyana

Crossford, Carluke, Lanarkshire ML8 5NJ
Tel: 01555 860033

127 Hammerlands Centre

Holme Road, Moffat, Dumfriesshire DG10 9QL
Tel: 01683 221220

128 Happy Plant Garden Centre

North Street, Mintlaw, Peterhead, Aberdeenshire AB42 5HH
Tel: 01771 623344
Specialities: Restaurant/Coffee Shop, Giftware, Herbaceous Perennials

129 Hawick Garden Centre

Earl Street, Hawick, Roxburghshire TD9 9PZ
Tel: 01450 371100

130 Hayes Garden Land

7 Tarbolton Road, Bogend, Symington, Kilmarnock, Ayrshire KA1 5PD
Tel: 01563 830204

131 Headspoint Garden Centre

Lanark Road, Braidwood, Carluke, Lanarkshire ML8 5PL
Tel: 01555 772303

132 Heather Centre

Glenisla, Drumoak, Banchory, Kincardineshire AB31 5AJ
Tel: 01330 811234
Specialities: Calluna/Erica (Heather)

133 Heathhall Garden Centre

Edinburgh Road, Dumfries, Dumfries & Galloway DG1 1SE
Tel: 01387 255636
Specialities: Garden Furniture

134 Hedgehogs Nursery

Crompton Road, Glenrothes, Fife KY6 2SF
Tel: 01592 772274

135 Hetland Garden Nursery

Carrutherstown, Carrutherstown, Dumfries, Dumfries & Galloway DG1 4JX
Tel: 01387 840632

136 Highland Heathers

Muirend, Comrie, Crieff, Perthshire PH6 2JA
Tel: 01764 670440
Specialities: Calluna/Erica (Heather)

137 Highland Heathers Ltd

Ford, Ford, Lochgilphead, Argyll PA31 8RH
Tel: 01546 810253
Specialities: Heathers

138 Highland Liliums

10 Loaneckheim, Loaneckheim Kiltarlity, Beauly, Inverness-Shire IV4 7JQ
Tel: 01463 741365

139 Highland Wildwoods

Blackstand, Fortrose, Ross-Shire IV10 8SW
Tel: 01381 621040
Specialities: Trees

140 Hillfoots Nurseries

Dollar Road, Tillicoultry, Clackmannanshire FK13 6PF
Tel: 01259 751819
Specialities: Annuals/Biennials/Bedding, Herbaceous Perennials

141 Hillhead Nurseries

Hillhead of Cocklaw, Blackhills, Peterhead, Aberdeenshire AB42 3LL
Tel: 01779 471688
Specialities: Herbaceous Perennials

142 Howdens Garden Centre

Stoneyfield, Inverness, Highland IV2 7PA
Tel: 01463 711134

143 Hundred Acre Wood Garden Centre

Annay Road, Melrose, Roxburghshire TD6 9LW
Tel: 01896 823717
Specialities: Fencing, Cut Flowers, Restaurant/Coffee Shop

144 Hygro Market

55 Main St, Holytown, Motherwell, Lanarkshire ML1 4TH
Tel: 01698 832720

145 Hyndford Nursery

Hyndford, Lanark, Lanarkshire ML11 8SJ
Tel: 01555 662608

146 Inverurie Garden & Machinery Centre

Oldmeldrum Road, Inverurie, Aberdeenshire AB51 0TP
Tel: 01467 621402
Specialities: Restaurant/Coffee Shop

147 Island Plants

Island of Criggie, Stonehaven, Kincardineshire AB39 2XY
Tel: 01569 740397

148 J Tweedie Fruit Trees

Maryfield Road Nursery, Terregles, Dumfries, Dumfries & Galloway DG2 9TH
Tel: 01387 720880
Specialities: Fruit Trees

149 J Woods & Sons

Parkhall Nursery, Maddiston, Falkirk, Stirlingshire FK2 0BN
Tel: 01324 712625
Specialities: Annuals/Biennials/Bedding

150 Jack Drake's Nursery

Aviemore, Inverness-Shire PH22 1QS
Tel: 01540 651287

151 James Cocker & Sons

Whitemyres, Lang Stracht, Aberdeen, Aberdeenshire AB15 6XH
Tel: 01224 313261
Specialities: Rose

152 James Mcintyre & Sons

Moness, Coupar Angus Road, Blairgowrie, Perthshire PH10 6UT
Tel: 01250 873135
Specialities: Fruit Bushes

153 John Redpath

Cupar Road, Tayport, Fife DD6 9PQ
Tel: 01382 552541

154 John Train & Sons

Tarbolton, Tarbolton, Mauchline, Ayrshire KA5 5NT
Tel: 01292 541336
Specialities: Annuals/Biennials/Bedding, Herbaceous Perennials, Hanging Baskets

155 Jordanhill Garden Supplies

425 Shelley Road, Glasgow, Lanarkshire G12 0XW
Tel: 01413 571551

156 Joshua Plumtree

30 Bank Street, Kirriemuir, Angus DD8 4BG
Tel: 01575 572030

157 Just Fuchsia

Ladysbrae Farm, Banff, Aberdeenshire AB45 2JR
Tel: 01261 861259
Specialities: Fuchsia

158 The Kerse Nursey

Kilbirnie Road, Lochwinnoch,

Renfrewshire PA12 4DT
Tel: 01505 503804

159 Keske Nurseries

Clachan Locheport, Isle of North Uist HS6 5HD
Tel: 01876 580333
Specialities: Trees, Plants for Windy Places

160 Kintyre Alpines

Torrisdale, Campbeltown, Argyll PA28 6QT
Tel: 01583 431317
Specialities: Alpines/Rock Plants

161 Kirkcaldy Garden Centre

Thornton Road, Kirkcaldy, Fife KY1 3NN
Tel: 01592 652861

162 Kirkdale Nursery

Daviot, Inverurie, Aberdeenshire AB51 0JL
Tel: 01467 671264

163 Kirktown Garden Centre

Kirktown, Fetteresso, Stonehaven, Kincardineshire AB39 3UP
Tel: 01569 766887

164 Klondyke Garden Centre

Whins of Milton, Stirling, Stirlingshire FK7 8ER
Tel: 01786 816167

165 Klondyke Garden Centre

Campus Roundabout, Livingston, West Lothian EH54 7AW
Tel: 01506 410053

166 Klondyke Garden Centre

Beancross Road, Polmont, Falkirk, Stirlingshire FK2 0XS
Tel: 01324 717035

167 K-Vision Ltd

1 Commercial Street, Port William, Newton Stewart, Dumfries and Galloway DG8 9SJ
Tel: 01988 700334
Specialities: Bedding Plants, Garden Machinery

168 Lady Helen Nurseries

Cardenden, Lochgelly, Fife KY5 0AL
Tel: 01592 720740

169 Laighparks Nurseries

Laighparks Farm, Killearn, Glasgow, Lanarkshire G63 9PT
Tel: 01360 550248

170 Lakeview

Lonmay, Fraserburgh, Aberdeenshire AB43 8SS
Tel: 01346 532929
Specialities: Aquatics, Water Products and Features

171 Lamberton Nursery

No. 3 Lamberton, Nr Berwick-upon-Tweed, Berwickshire TD15 1XB
Tel: 01289 308515
e-mail: ron@lambertonnursery.com
website: www.lambertonnursery.co.uk
Specialities: Rare and Unusual Rock Plants

172 Lilliesleaf Nursery

Linthill Garden Cottage, Melrose, Roxburghshire TD6 9HU
Tel: 01835 870415

173 Linden Garden Centre

Main Street, Glencarse, Perth, Perthshire PH2 7LX
Tel: 01738 860673

174 Lindores Nursery

The Old Post Office, Lindores, Cupar, Fife KY14 6JD
Tel: 01337 840256

175 Linn Botanic Gardens & Nursery

The Linn, Cove, Helensburgh, Dunbartonshire G84 0NR
Tel: 01436 842242

176 The Linns

Sheriffmuir, Dunblane, Perthshire FK15 0LP
Tel: 01491 575075

177 Loch Duich Plants

Foresters Bungalow, Inverinate, Kyle, Ross-Shire IV40 8HE
Tel: 01599 511407

178 Loch Lomond Homes Garden Centre and Nurseries

Gartocharn Road, Balloch, Alexandria, Dunbartonshire G83 8NB
Tel: 01389 752947
Specialities: Restaurant/Coffee Shop, Annuals/Biennials/Bedding, Childrens Play Area

179 Lochend Nurseries

Lochend Rd, Gartcosh, Glasgow, Lanarkshire G69 8BE
Tel: 01236 874683
Specialities: Restaurant/Coffee Shop

180 Loch-Hills Plant Centre

Loch-Hills, Ellon, Aberdeenshire AB41 8DR
Tel: 01358 720694
Specialities: Annuals/Biennials/Bedding, Herbaceous Perennials

181 Lochlands Garden Centre

Dundee Road, Forfar, Angus DD8 1XF
Tel: 01307 463621
Specialities: Annuals/Biennials/Bedding, Herbaceous Perennials, Restaurant/Coffee Shop

182 Logie Steading Farm & Garden Shop

Logie Steading, Forres, Moray IV36 2QN
Tel: 01309 611222 Fax: 01309 611300
e-mail: panny@logie.co.uk
website: www.logie.co.uk

Specialities:

Unusual Hardy Perennials; Shrubs; Ground Cover Plants, Plants for Wet Places

Description:

Many of the plants on sale here are grown in the garden at Logie; all are grown in Scotland. The owners are always sourcing new varieties of hardy perennials, shrubs and ground cover plants to add to their already extensive collections. In the Farm Shop you'll find Longhorn beef, naturally bred and raised at Logie, and hung for at least 3 weeks, along with a huge choice of good wholesome products from Scotland and further afield.

Directions:

Logie Estate lies in the beautiful Findhorn Valley, 6 miles south of Forres. Logie Steading is on the B9007, signposted off the A940.

183 M G M Garden & Leisure

Peace and Plenty, Ayr Road, Kilmarnock, Ayrshire KA1 5LF
Tel: 01563 528911

184 Mackenzie and Cruickshank Ltd

1 Greshop Estate, West Road, Greshop Industrial Estate, Forres, Morayshire IV36 2GW
Tel: 01309 696940

185 Macplants

Berrybank Nursery, Pencaitland, Tranent, East Lothian EH34 5BA
Tel: 01875 341179

186 Mansfield Garden Centre

22 The Avenue, Girvan, Ayrshire KA26 9DS
Tel: 01465 714268
Specialities: Annuals/Biennials/Bedding, Herbaceous Perennials

187 Maybury Gardens

14 North Street, Sandwick, Isle of Lewis, Isle of Lewis HS2 0AD
Tel: 01851 705798
Specialities: Shrubs, Trees

188 Mclaren Nurseries

Lochlibo Road, Uplawmoor, Glasgow, Lanarkshire G78 4DN
Tel: 01505 850666
Specialities: Annuals/Biennials/Bedding, Shrubs, Trees

189 Meadowlands Nursery

Old Mearns Road, Newton Mearns, Glasgow, Lanarkshire G77 5RY
Tel: 01416 392120

190 Meadowside Nursery

Lanark Road, Braidwood, Carluke, Lanarkshire ML8 4PW
Tel: 01555 771292
Specialities: Herbaceous Perennials

191 Meikle Yett

Tongland, Kirkcudbright, Dumfries & Galloway DG6 4NA
Tel: 01557 330798

192 Merryhatton Garden Centre

East Fortune, East Fortune, North Berwick, East Lothian EH39 5JS
Tel: 01620 880278
Specialities: Annuals/Biennials/Bedding, Shrubs

193 Mill Cottage Garden Centre

37 Mill Street, Rothesay, Isle of Bute PA20 0EX
Tel: 01700 505724
Specialities: Garden Furniture, Annuals/Biennials/Bedding

194 Mill Garden Centre

Armadale, Bathgate, West Lothian EH48 3AP
Tel: 01501 732347

195 Moffat Garden Centre

High Street, Moffat, Dumfriesshire DG10 9HF
Tel: 01683 220442

196 Monymusk Walled Garden

Home Farm, Monymusk, Inverurie,
Aberdeenshire AB51 7HL
Tel: 01467 651543

197 Morangie Garden Centre

Morangie Road, Tain, Ross-Shire IV19 1PY
Tel: 01862 893993
Specialities: Annuals/Biennials/Bedding

198 Mossatburn Water Garden

Mossat, Alford, Aberdeenshire AB33 8PL
Tel: 01975 571235
Specialities: Aquatics, Water Products and Features

199 Mr J Coutts

Newtonairds, Dumfries,
Dumfries & Galloway DG2 0JL
Tel: 01387 820203
Specialities: Herbaceous Perennials, Hosta

200 Mugdock Plantaria

The Walled Garden, Mugdock Garden, Milngavie,
Glasgow, Lanarkshire G62 8EL
Tel: 01419 550011

201 Munro's Nurseries

Bogallan, North Kessock, Inverness,
Highland IV1 3XE
Tel: 01463 731210

202 Mutehill Nursery & Gardens Ltd

Kirkcudbright, Dumfries & Galloway DG6 4XE
Tel: 01557 331483

203 Nevis Garden Centre

North Road, Fort William,
Inverness-Shire PH33 6PP
Tel: 01397 700500

204 New Hopetoun Gardens

Dougal Philip Limited, New Hopetoun Gardens,
Hopetoun Wood, by Newton Village,
Broxburn, West Lothian EH52 6QZ
Tel: 01506 834433 Fax: 01506 834444
e-mail: newhopetoungardens@icscotland.net
website: www.newhopetoungardens.co.uk

Specialities:

Largest Range of Garden Plants in Scotland; Beautiful Centre with Lovely Tearoom; 18 Inspirational Themed Gardens

Products and Services:

- Books and Stationery
- Credit Cards Accepted
- Disabled Facilities
- Display Gardens
- Dried and Artificial Flowers
- Garden Design
- Garden Furniture
- Garden Products
- Giftware
- Houseplants
- Information Desk
- Large Car Park
- Paving
- Plant Guarantee
- Plants
- Restaurant/Coffee Shop
- Seeds and Bulbs
- Trees

Directions:

The nursery is located on the A904, 10 miles west of Edinburgh, past the Forth Road Bridge roundabout heading towards Linlithgow

Description:

Dougal and Lesley work full time at the gardens together with their long-serving friendly and experienced staff. They also fit in show garden judging for the RHS and Lesley is a presenter on the BBC Scotland TV programme Beechgrove Garden. The nursery started in the walled garden of a stately home in 1978, expanding in 1998 to a custom-built centre set in 6 acres of woodland. There are 18 themed gardens designed and built by Dougal and Lesley, covering themes such as fragrant, calm and relaxed to conditions like sun/shade and damp/dry. The 3 buildings on site house the Garden Shop, the Orangery Tea Room & Gift Shop, and Dixon & Ramsay's lifestyle goods. Surrounding them are the 18 themed gardens full of ideas for the gardener and, of course, thousands of garden plants for sale. The nursery also hosts demonstrations and master classes on subjects ranging from garden design and plant arranging to Christmas gifts and decorations or living willow weaving.

205 Newarthill Nursery

Legbrannock Road, Motherwell,
Lanarkshire ML1 5SZ
Tel: 01698 733577

The history of **Nurseries Direct** goes back to 1920 when John Gemmell bought a cottage from the Auchinleck Estate. He built a greenhouse to grow tomatoes and other vegetables for local retailers. By 1965 the nursery was supplying markets in Glasgow and Edinburgh in addition to local retailers and the public. Tomatoes remained their main glasshouse crop. Flowers were grown outside.

In 1991 Ian Gemmell's son Colin bought the nursery and started growing bedding plants for Glasgow and Edinburgh markets and local retailers. He renamed the nursery Nurseries Direct to reflect the nature of the business. Over the next few years he started opening to the public and built a retail area and coffee shop. Nurseries Direct grew quickly from 1991 turning more and more towards the retail sector. By 2004 Nurseries Direct was a fully fledged garden centre supplying the public only. The coffee shop was extended to seat 124.

In March 2006 a large canopy was constructed to the rear of the new building. This provided a further 1040sqm of covered area. Within this space plants, shrubs, heathers and bedding plants are displayed throughout the year. And in 2007 the remaining growing greenhouse was brought into the retail domain providing yet more space for plant sales.

The whole place is so much more than a garden centre. It is, in many ways, a tourist attraction, as you could spend the whole day here, in this fascinating cornucopia, not just for the committed gardener, but also for the beginner. Come along and see for yourself.

- GIFT SHOP
  - BOOKS AND STATIONERY
  - CLOTHING
  - DRIED AND ARTIFICIAL FLOWERS
  - FLORISTRY
  - GIFTWARE
- FURNITURE
  - BARBECUES
  - GARDEN FURNITURE
- PLANTS
  - SEASONAL DISPLAYS
  - DISPLAY GARDENS
  - FRUIT TREES AND ORNAMENTALS
  - HOUSEPLANTS
  - SHRUBS AND BEDDING PLANTS
  - TREES
  - SEEDS AND BULBS
  - GARDEN PRODUCTS
- GENERAL
  - RESTAURANT/COFFEE SHOP
  - FOOD HALL
  - CREDIT CARDS ACCEPTED
  - DISABLED FACILITIES
  - LARGE CAR PARK
  - CHILDRENS PLAY AREA

ESTABLISHED SINCE 1991.....

NURSERIES DIRECT GARDEN CENTRE

TRABBOCH ROAD, STAIR, MAUCHLINE, AYRSHIRE KA5 5JD
TEL: 01292 591900 FAX: 01292 591080
E-MAIL: INFO@NURSERIESDIRECT.CO.UK
WEBSITE: WWW.NURSERIESDIRECT.CO.UK

OPENING TIMES
9AM-5.30PM, (MONDAY, WEDNESDAY -SATURDAY)
9.30AM-5.30PM (TUESDAY)
10AM-5.30PM (SUNDAY)

206 Newhailes Nursery

Newhailes Road, Musselburgh,
East Lothian EH21 6RY
Tel: 0131 665 8291 Fax: 01506 834853

Specialities:

Trees & Shrubs, Bedding Plants, Customer Care

Products and Services:

- Credit Cards Accepted
- Disabled Facilities
- Fruit Trees
- Garden Products
- Information Desk
- Large Car Park
- Seeds and Bulbs
- Trees

Directions:

From Edinburgh, take the by-pass on the A1, then take the A6095 to Musselburgh. The Nursery is on this road.

207 Nurseries Direct Garden Centre

Trabboch Road, Stair, Mauchline,
Ayrshire KA5 5JD
Tel: 01292 591900 Fax: 01292 591080
e-mail: info@nurseriesdirect.co.uk
website: www.nurseriesdirect.co.uk

SEE ADVERTISEMENT PAGE 348

Specialities:

Shrubs, Bedding Plants, Trees

Description:

Ayrshire's largest independent garden centre, Nurseries Direct is set in a beautiful rural location with 4000 square metres of retail space. It also has a large gift department with an array of unusual items, a friendly coffee shop seating 147 people and a children's play area.

Directions:

From the Ayr take the A70 through Coylton and Coalhall. Go under the railway bridge then take the 1st left (B730). Then take the 1st right following the signposts for Nurseries Direct (1 mile).

208 Old School Bedding

Dykehead House, Earnock Road, Hamilton,
Lanarkshire ML3 8RH
Tel: 07714 804873
Specialities: Annuals/Biennials/Bedding

209 Organic Trees

Doire Na Mairst, Morvern, Oban, Argyll PA34 5XE
Tel: 01967 421203
Specialities: Organic Trees

210 Orkney Garden Centre

5 Islands View Road, Kirkwall, Orkney KW15 1YP
Tel: 01856 877602

211 Ornamental Garden Products

Windy Neuk, Gight, Ellon, Aberdeenshire AB41 7HX
Tel: 01651 806340
Specialities: Ornamentals

212 Ostel Bay Nursery

Collacha Ardlemont, Kames, Tighnabruaich,
Argyll PA21 2AH
Tel: 01700 811524

213 Outsideinside Aquatics

5a Knox Place, Pencaitland Road, Haddington,
East Lothian EH41 4DY
Tel/Fax: 01620 829299
e-mail: outsideinsidefish@btinternet.com
website: www.outsideinsideaquatics.co.uk

Specialities:

Pond & Aquarium Plants, Pond Supplies, Tropical and Coldwater Fish Specialist

Products and Services:

- Aquatics
- Books (Aquatic)
- Credit Cards Accepted
- Information Desk
- Small Car Park
- Pet Products (Fish)
- Plants (Pond/Bog)

Directions:

From the Edinburgh bypass (A720), take the A1 eastbound until you reach the A6137 to Haddington. Follow the main road into town and turn right at the church onto Knox Place/Pencaitland Road. Outsideinside is approximately 100 yards on the left.

214 Parkhill Nurseries & Garden Centre

Parkhill, Dyce, Aberdeen, Aberdeenshire AB21 7AT
Tel: 01224 722167
Specialities: Annuals/Biennials/Bedding, Herbaceous Perennials, Restaurant/Coffee Shop

215 Pathhead Nurseries

St. Monans, Anstruther, Fife KY10 2DQ
Tel: 01333 730716

216 Pearsons Home & Garden Centre

The Cheeklaw Centre, Station Road, Duns,
Berwickshire TD11 3EL
Tel: 01361 882294

217 Pentland Plants Limited

Pentland Garden Centre, Loanhead,
Midlothian EH20 9QG
Tel: 01314 400895

218 Pinewood Park Nurseries

Countesswells Road, Aberdeen,
Aberdeenshire AB15 7AL
Tel: 01224 318744

219 The Plant Centre

29a Bonnygate, Cupar, Fife KY15 4BU
Tel: 01334 650500

220 Plantiecrub Enterprises

Tingwall, Gott, Shetland, Shetland Islands ZE2 9SH
Tel: 01595 840600

221 Plantiecrub Garden Shop

22-24 North Road, Lerwick, Shetland,
Shetland Islands ZE1 0NQ
Tel: 01595 695444

222 Po'Teak

Edgar Road, Elgin, Morayshire IV30 6YQ
Tel: 01343 544441

223 The Potting Shed

Windy-Yett Farm, Old Glasgow Road, Stewarton,
Kilmarnock, Ayrshire KA3 5JU
Tel: 01560 482138

224 Poyntzfield Herb Nursery

Nr Balblair, Poyntzfield, Dingwall,
Ross-Shire IV7 8LX
Tel: 01381 610352
Specialities: Herbs

225 Quarry View

Crarae, Minard, Inveraray, Argyll PA32 8YB
Tel: 01546 886692

226 Quercus Garden Plants

Rankeilour Gardens, Cupar, Fife KY15 5RE
Tel: 01337 810444

227 Quinish Garden Nursery

Dervaig, Dervaig Tobermory, Isle of Mull,
Argyll PA75 6QL
Tel: 01688 400344
Specialities: Shrubs, Trees, Rhododendron (Inc Azalea)

228 R H Miller

Tweed View, South Parks Estate, Peebles,
Borders EH45 9ED
Tel: 01721 724569
Specialities: Annuals/Biennials/Bedding, Herbaceous Perennials, Shrubs

229 R.J Morrison & Son

Barraston Nursery Torrance, Torrance, Glasgow,
Lanarkshire G64 4DP
Tel: 01360 620354

230 Raemoir Garden Centre

Raemoir Road, Banchory,
Kincardineshire AB31 4ER
Tel: 01330 825059
Specialities: Restaurant/Coffee Shop, Giftware

231 Redcraigs Nursery

Redgraigs Toll, Dunfermline, Fife KY12 9LS
Tel: 01383 734511
Specialities: Herbaceous Perennials

232 Reynard Nursery

Reynard, Carluke, Lanarkshire ML8 5HW
Tel: 01555 770033
Specialities: Annuals/Biennials/Bedding, Hanging Baskets

233 Rhu Garbh Trees & Plants

Shore Cottage, Barcaldine, Oban, Argyll PA37 1SE
Tel: 01631 720577
Specialities: Annuals/Biennials/Bedding, Trees

234 Riccarton Nurseries Ltd

Riccarton Mains Road, Currie, Midlothian EH14 5AA
Tel: 01314 494004
Specialities: Shrubs, Annuals/Biennials/Bedding

235 Richard Aitken Seedsman Ltd

123 Harmony Row, Glasgow, Lanarkshire G51 3NB
Tel: 01414 400033
Specialities: Lawn

236 Ridgeview Nursery

Bathgate, West Lothian EH47 9AB
Tel: 01501 771144

237 Riverbank Nursery

Riverbank Road, Conon Bridge, Dingwall,
Ross-Shire IV7 8BT
Tel: 01349 861720

238 Robert Mcivor & Co

Corramill Farm, Crossford, Carluke,
Lanarkshire ML8 5QP
Tel: 01555 860239

239 Rosebank Plant Centre and Nursery

Rosebank Hall, by Carluke, Lanarkshire ML8 5QB
Tel: 01555 860644

240 Rossiebraes Market Garden

Rossiebraes, Ferryden, Montrose, Angus DD10 9TA
Tel: 01674 674677

Specialities:

Plants & Shrubs, Handmade Garden Furniture, Local Produce

Products and Services:

- Seasonal Displays
- Cut Flowers
- Floristry
- Fruit Trees
- Garden Furniture
- Garden Products
- Giftware
- Greenhouses and Sheds
- Houseplants
- Information Desk
- Large Car Park
- Pet Products
- Plant Guarantee
- Plants
- Seeds and Bulbs
- Trees

Directions:

Rossiebraes Market Garden can be found a half mile south of Montrose on the A92, beside a public layby.

241 Rouken Glen Garden Centre

18 Cauldcoats, Linlithgow, West Lothian EH49 7LY
Tel: 01506 834346
Specialities: Annuals/Biennials/Bedding, Garden Furniture

242 Rouken Glen Garden Centre

Rouken Glen Road, Giffnock, Glasgow,
Lanarkshire G46 7JL
Tel: 01416 200566
Specialities: Giftware, Garden Furniture

243 Rudkins

Denholm Rd, Hawick, Roxburghshire TD9 8SY
Tel: 01450 370555

244 Saigon U.K Ltd

Unit 7, Clydebrae Street, Glasgow,
Lanarkshire G51 2AJ
Tel: 01414 452629

245 Sands Rose Nurseries

Kincardine, Kincardine, Alloa,
Clackmannanshire FK10 4AR
Tel: 01259 730402
Specialities: Rose

246 Sandyholm Garden Centre

Crossford, by Carluke, Lanarkshire ML8 5QF
Tel: 01555 860205

247 Scotplants Ltd.

Brownhills, Burnhouse, Beith, Ayrshire KA15 1HH
Tel: 01505 502717 Fax: 01505 503321
e-mail: enquiries@scotplants.co.uk
website: www.scotplants.co.uk

Specialities:

Shrubs; Heathers; Trees

Description:

Scotplants offers an excellent range of plants from heathers to mature trees. Owner Douglas Boatman has an Honours Degree in Horticulture and during the winter gives demonstrations on a range of horticultural topics. Inspection of the stock at Scotplants is encouraged and there is always someone on hand to give you a conducted tour. The nursery is open daily in summer; closed at weekends from October until the end of February.

Directions:

From Paisley take the A737 to Beith and then turn left onto the B706. At Greenhills turn left and then follow the road to the crossroads. Turn right and go under the railway bridge. The nursery is on the left at Brownhills.

248 Scotsburn Garden Centre

Scotsburn Road, Tain, Ross-Shire IV19 1PP
Tel: 01862 892377
Specialities: Annuals/Biennials/Bedding, Shrubs, Garden Products

249 Scotts Garden & Pet Centre

29 Reform Street, Blairgowrie,
Perthshire PH10 6BD
Tel: 01250 874230

250 Seaforth Nurseries

Liddesdale Rd, Stranraer, Wigtownshire DG9 0AX
Tel: 01776 702911

251 Sheila Northway Auriculas

Craig Lodge Balmaclellan, Balmaclellan, Castle Douglas, Kirkcudbrightshire DG7 3QR
Tel: 01644 420661

252 Shinval Garden Centre

Glengolly, Thurso, Caithness KW14 7XN
Tel: 01847 894306

253 Silver Birch (Scotland) Ltd

6 Antermony Road, Milton of Campsie, Glasgow, Lanarkshire G66 8DB
Tel: 01417 767098

254 Silverbirch Garden Centre

Lanark Road, Crossford, Carluke, Lanarkshire ML8 5QQ
Tel: 01555 860623
Specialities: Giftware, Shrubs, Restaurant/Coffee Shop

255 Silverwells Plant Nursery & Garden Centre

Montrose Road, Arbroath, Angus DD11 5RA
Tel: 01241 875634

256 Simpson & Florence

Millmoss Nursery, Turriff, Aberdeenshire AB53 8BJ
Tel: 01888 563511

257 Simpson Garden Centre

Inshes, Inverness, Highland IV2 5BA
Tel: 01463 250200
Specialities: Restaurant/Coffee Shop, Cut Flowers

258 Skipness Plants

The Gardens, Skipness, Tarbert, Argyll PA29 6XU
Tel: 01880 760201

259 Skye Shrubs Garden Centre

Viewfield Road, Portree, Isle of Skye IV51 9EU
Tel: 01478 613132

260 Smeaton Nursery & Gardens

Smeaton Estate, Preston Road, East Linton, East Lothian EH40 3DT
Nursery Tel: 01620 860501
Tearoom Tel: 01620 861366

Specialities:

Ornamental Trees & Shrubs; Friendly Customer Advice; Victorian Conservatory Tearoom

Products and Services:

- Trees & Shrubs
- Seasonal Bedding Plants
- Seeds & Bulbs
- Pots and Ornaments
- Compost & Garden Sundries
- Gardeners Gift Ideas
- Tearoom Refreshments
- Plant information
- Display gardens
- Credit Cards Accepted
- Local Deliver Service
- HTA Gift Vouchers

Description:

Smeaton Nursery is set in a beautiful old walled-garden in the heart of rural East Lothian. This traditional-style 2 acre nursery is run by a team of enthusiastic plants-people, always available to give knowledgeable and friendly advice. An extensive range of quality shrubs, trees, seasonal-plants and essential gardening products are stocked all-year-round.

The popular Smeaton Tearoom offers an appetising selection of home-made soups, paninis & baguettes, salads & platters and best of all, our ever-popular mouth-watering home-baked cakes and scones. A member of the HTA, Smeaton also sells & accepts HTA gift vouchers.

Directions:

East Linton is 25 miles east of Edinburgh and is situated between the towns of Haddington, North Berwick & Dunbar. The entrance to the Smeaton Estate is on the outskirts of East Linton village, close to the NTS Preston Mill. The walled-gardens are at the top of the drive.

261 Soleburn Garden Centre Ltd

Kirkcolm, Stranraer, Wigtownshire DG9 0PW
Tel: 01776 870664
Specialities: Annuals/Biennials/Bedding, Restaurant/Coffee Shop, Garden Furniture

262 South Logie Nursery

Logiebrae Road, Westfield, Bathgate, West Lothian EH48 3DG
Tel: 01506 631769

263 Speyside Heather Visitor & Garden Centre

Skye of Curr Road, Dulnain Bridge, Morayshire PH26 3PA
Tel: 01479 851359
Specialities: Heathers

264 Spott Nurseries

Stewarton, Kilmarnock, Ayrshire KA3 5HU
Tel: 01560 485063
Specialities: Annuals/Biennials/Bedding, Hanging Baskets

265 St Andrews Garden Centre

199 South Street, St. Andrews, Fife KY16 9EE
Tel: 01334 473342

266 Stanley Nurseries

9 Stanley Avenue, Ardrossan, Ayrshire KA22 8NT
Tel: 01294 461771

267 Sterling Garden Centre Ltd

Moss Road, Tillicoultry, Clackmannanshire FK13 6NS
Tel: 01259 752945
Specialities: Greenhouses and Sheds, Fireworks

268 Stewart Stenhouse Plants

Old Sale Ground, East Linton, East Lothian EH40 3DT
Tel: 01620 860051

269 Stoneymeadow Patio Centre

Stoneymeadow Road, Blantyre, Glasgow, Lanarkshire G72 9UE
Tel: 01355 228217

270 Stornoway Trust Nurseries

The Meadows, Lews Castle Grounds, Isle of Lewis, Isle of Lewis HS2 0XR
Tel: 01851 701655

271 Strawberry Corner Garden Centre

1 Wallyford Toll, Wallyford, Musselburgh, Midlothian EH21 8JU
Tel: 01316 659299

272 Sunnyside Cottage Garden Nurseries

Off Grassyards Road, Kilmarnock, Ayrshire KA3 6HY
Tel: 01560 700678
Specialities: Cottage Garden Plants

273 Sunnyside Nursery (Kilwinning) Ltd

Hazelgrove, Kilwinning, Ayrshire KA13 7JL
Tel: 01294 552113
Specialities: Shrubs, Herbaceous Perennials

274 Sunnyside Perennials

Sunnyside Cottage, Pettinain, Lanark, Lanarkshire ML11 8SS
Tel: 01555 871421
Specialities: Herbaceous Perennials

275 Teviot Water Gardens

Kirkbank House, Kelso, Roxburghshire TD5 8LE
Tel: 01835 520253
Specialities: Aquatics, Water Products and Features

276 Threaplands Nursery and Garden Centre

Threapland, Lhanbryde, Morayshire IV30 8LN
Tel: 01343 842245 Fax: 01343 842242
e-mail: s.higgins@threaplands.com
website: www.threaplands.com

277 Torwood Garden Centre

Bellsdyke Road, Larbert, Stirlingshire FK5 4EG
Tel: 01324 553152

278 The Tree Shop (Ardkinglas Estate Nurseries)

Cairndow, Cairndow, Argyll PA26 8BH
Tel: 01499 600263
Specialities: Trees

279 Trinity Park Garden Centre

Trinity, Brechin, Angus DD9 7PS
Tel: 01356 622487

280 Tynreich Nursery

Ballinluig, Pitlochry, Perthshire PH9 0NA
Tel: 01796 482274
Specialities: Annuals/Biennials/Bedding, Herbaceous Perennials, Rockery Plants

281 Uzumara Orchids

9 Port Henderson, Port Henderson, Gairloch, Ross-Shire IV21 2AS
Tel: 01445 741228
Specialities: Orchids

282 W.G Donaldson

26 High Street, Grantown-on-Spey, Morayshire PH26 3EH
Tel: 01479 872798

283 Walker Home & Garden Centre

361 Argyll Street, Dunoon, Argyll PA23 7RN
Tel: 01369 704003
Specialities: Garden Products

284 The Walled Garden Centre

Murray Royal Hospital, Muirhall Road, Perth, Perthshire PH2 7BH
Tel: 01738 631777

285 Ward Toll Garden Centre

Balfron Station, Glasgow, Lanarkshire G63 0QY
Tel: 01360 850525
Specialities: Giftware

286 WCF Nursery

1 Baxter Lane, Lanark, Lanarkshire ML11 7AF
Tel: 01555 662263

287 Wellpark Nurseries

Mill Street, Willowburn Road, Kirkwall, Orkney KW15 1NE
Tel: 01856 874203

288 West Carlston Garden Centre

Campsie Road, Torrance, Glasgow, Lanarkshire G64 4BN
Tel: 01360 620248

289 Williamson Design Florist

Beechwood Nurseries, Broxburn, West Lothian EH52 6PA
Tel: 01506 811433
Specialities: Floristry

290 Willowglen Garden Centre

Macauley Rd, Stornoway, Isle of Lewis HS1 2HU
Tel: 01851 705656

291 Woodlands Nursery

Easter Tullybannocher, Comrie, Crieff, Perthshire PH6 2JY
Tel: 01764 670038

292 Woodside Cottage Herbs

Woodside Cottage, Longriggend, Airdrie, Lanarkshire ML6 7RU
Tel: 01236 843826
Specialities: Herbs

293 Woodside Garden Centre

Woodside, Castletown, Thurso, Caithness KW14 8TX
Tel: 01847 821567
Specialities: Garden Products, Annuals/Biennials/Bedding

294 Woodside Nursery

Woodside, Jedburgh, Roxburghshire TD8 6TU
Tel: 01835 830315

WALES

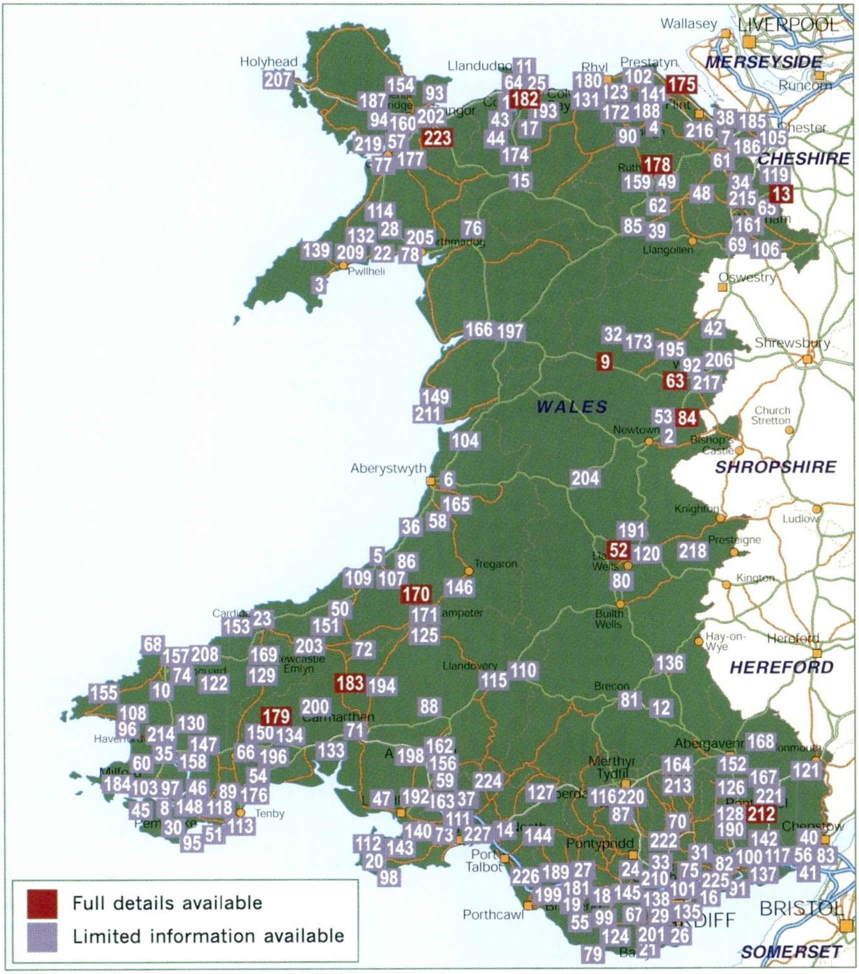

1 Aberconwy Nursery

Graig, Glan Conwy, Colwyn Bay, Clwyd LL28 5TL
Tel: 01492 580875
Specialities: Shrubs, Trees

2 Abermule Nursery Gardens

Kerry Road, Abermule, Montgomery, Powys SY15 6NW
Tel: 01686 630203
Specialities: Annuals/Biennials/Bedding, Shrubs

3 Abersoch Garden Centre

Ty Mawr, Abersoch, Pwllheli, Gwynedd LL53 7UD
Tel: 01758 713098
Specialities: Garden Furniture, Barbeques

4 Aberwheeler Nurseries

Mold Rd, Bodfari, Denbigh, Clwyd LL16 4DP
Tel: 01745 710673
Specialities: Shrubs, Herbaceous Perennials

5 Andrew Bligh

Drefach Organic Nursery, Aberaeron, Dyfed SA46 0JR
Tel: 01545 571344
Specialities: Organic

6 Arthur Newman's Garden Centre and Nursery

Capel Dewi, Aberystwyth, Dyfed SY23 3HS
Tel: 01970 820019
Specialities: Aquatics, Pet Products, Restaurant/ Coffee Shop

7 Aston Hall Farm Nurseries

Lower Aston Hall Lane, Hawarden, Deeside, Clwyd CH5 3EY
Tel: 01244 532465
Specialities: Annuals/Biennials/Bedding, Herbaceous Perennials

8 Bangeston Farm Yard Garden Centre

Cosheston, Pembroke Dock, Dyfed SA72 4RX
Tel: 01646 621566

9 Banwy Valley Nursery

Foel, nr Welshpool, Powys SY21 0PT
Tel: 01938 820281
e-mail: syd@banwnursery.co.uk
website: www.banwnursery.co.uk

Growers of:

Ornamental trees, Fruit trees, Shrubs, Hardy Perennials

Credit Cards Accepted

Directions:

On the A458,

17 miles west of Welshpool, Foel is signposted to the left. The nursery is 1 mile along this minor road.

10 Barrowclift

Letterston, Haverfordwest, Dyfed SA62 5TL
Tel: 01348 840537

11 Battys Nurseries

Penrhyn Old Road, Penrhyn Bay, Llandudno, Gwynedd LL30 3ED
Tel: 01492 549176
Specialities: Annuals/Biennials/Bedding, Shrubs

12 Beacons Nurseries

Llangasty, Brecon, Powys LD3 7PX
Tel: 01874 730169
Specialities: Hanging Baskets, Annuals/Biennials/ Bedding, Herbaceous Perennials

13 Bellis Brothers Ltd.

Wrexham Road Farm, Holt, Wrexham LL13 9YU
Tel: 01829 270304 Fax: 01829 271653
e-mail: info@bellisbrothers.com
website: bellisbrothers.com

Specialities:
Terracotta Pots; Farm Shop; Pick Your Own Fruit

Products and Services:

- Barbeques
- Books and Stationery
- Childrens Play Area
- Seasonal Displays
- Credit Cards Accepted
- Cut Flowers
- Disabled Facilities
- Dried & Artificial Flowers
- Farm Shop
- Floristry
- Fruit Trees
- Garden Furniture
- Garden Products
- Giftware
- Houseplants
- Large Car Park
- Pet Products
- Pick your own
- Plants
- Restaurant/Coffee Shop
- Seeds and Bulbs
- Trees

Directions:
From Wrexham take the A534 east for about 4 miles, then turn left on the B5102 to Holt village. The Garden Centre is 200 metres on the left. Only 9 miles from Chester

Description:
The Bellis Brothers Garden Centre and Farm Shop, located on the outskirts of Holt Village, Wrexham, is a family run business, now in its 5th generation. It was first established in 1876, growing fruit and vegetables for the wholesale markets of North West England. Since then the business has continually expanded. It introduced Pick Your Own Fruit in 1967, opened its first farm shop in 1973 and added a purpose built Garden Centre in 1986. The year 2007 has seen further expansion with the opening of a new Restaurant, sales area and a large car park with spaces for up to 140 cars. In addition to the products and services listed above, the Centre also proclaims that it can supply "Everything for the kitchen and Garden", even logs for your open fire and HTA Garden Vouchers.

14 Betts Nurseries

Cadoxton Road, Neath, West Glamorgan SA10 7AB
Tel: 01639 632746
Specialities: Trees, Shrubs

15 Betws Y Coed Garden & Nursery

Station Approach, Betws-y-Coed,
Gwynedd LL24 0AE
Tel: 01690 710870
Specialities: Annuals/Biennials/Bedding, Shrubs

16 Blooms of Bressingham

Newport Road, St. Mellons, Cardiff,
South Glamorgan CF3 2WJ
Tel: 029 2077 7977

17 Bodnant Garden Nursery

Tal-y-Cafn, Colwyn Bay, Clwyd LL28 5RE
Tel: 01492 650731
Specialities: Annuals/Biennials/Bedding

18 Bordervale Plants

Nantyderi, Ystradowen, Cowbridge,
South Glamorgan CF71 7SX
Tel: 01446 774036
Specialities: Rare and Unusual Plants, Herbaceous Perennials

19 Bridge Farm Nursery

Ewenny, Bridgend, Mid Glamorgan CF35 5AD
Tel: 01656 652154

20 Bright Ideas Plant & Gift Centre

Scurlage Reynoldston, Swansea,
West Glamorgan SA3 1BA
Tel: 01792 390137
Specialities: Annuals/Biennials/Bedding, Herbaceous Perennials, Giftware

21 Broad St Garden Centre

Broad Street, Barry, South Glamorgan CF62 7AE
Tel: 01446 720333

22 Bron Eifion Nursery

Bron Eifion, Criccieth, Gwynedd LL52 0SA
Tel: 01766 522890

23 Brondesbury Park Garden Centre

Aberystwyth Road, Cardigan, Dyfed SA43 1NA
Tel: 01239 615300

24 Brook Farm

Heol Creigiau, Efail Isaf, Pontypridd,
Mid Glamorgan CF38 1BG
Tel: 01443 202318

25 Bryn Euryn Nursery

Dinerth Rd, Rhos On Sea, Colwyn Bay,
Clwyd LL28 4YN
Tel: 01492 546757
Specialities: Annuals/Biennials/Bedding, Shrubs

26 Brynawel Fuchsia & Garden Centre

Sully Road, Penarth, South Glamorgan CF64 2TR
Tel: 029 2070 2660
Specialities: Fuchsia, Giftware

27 Bryncethin Nursery

Pant Hirwaun, Bryncethin, Bridgend,
Mid Glamorgan CF32 9UJ
Tel: 01656 862379

28 Bryncir Garden Centre

Bryncir, Garndolbenmaen, Gwynedd LL51 9LX
Tel: 01766 530685
Specialities: Annuals/Biennials/Bedding, Herbaceous Perennials, Shrubs

29 Brynheulog Nurseries

St. Andrews Road, Wenvoe, Cardiff,
South Glamorgan CF5 6AF
Tel: 029 2059 3375
Specialities: Shrubs, Herbaceous Perennials

30 Bush School Nurseries

Bush Hill, Pembroke, Dyfed SA71 4QU
Tel: 01646 681780

31 Byron Howell & Son

40-42 Commercial Street, Risca, Newport,
Gwent NP11 6AY
Tel: 01633 612781

32 Caecrimon Cottage Garden Plants

Llanwddyn, Nr Oswestry, Shropshire SY10 0LZ
Tel: 01691 870648
Specialities: Annuals/Biennials/Bedding, Herbaceous Perennials

33 Caerphilly Garden Centre

Penrhos, Nantgarw, Cardiff,
South Glamorgan CF15 7UN
Tel: 029 2086 1511
Specialities: Barbeques, Garden Furniture

34 Carlton Garden Centre

Llay Industrial Estate, Llay, Wrexham,
Clwyd LL12 0TL
Tel: 01978 852896
Specialities: Garden Products

35 Cartlett Home & Garden Supplies Ltd

Snowdrop Lane, Haverfordwest, Dyfed SA61 1JD
Tel: 01437 763904
Specialities: Greenhouses and Sheds

36 Castell Bach Nursery

Llanrhystud, Dyfed SY23 5BT
Tel: 01974 202733
Specialities: Annuals/Biennials/Bedding, Shrubs

37 Cefnmyddfain Nursery

Cefnmyddfain Farm, Felindre, Swansea,
West Glamorgan SA5 7NH
Tel: 01792 771932
Specialities: Annuals/Biennials/Bedding, Shrubs

38 Celyn Landscape - Nursery

Sealand Road, Sealand, Deeside, Clwyd CH5 2RH
Tel: 01244 819933
Specialities: Landscape Gardening

39 Celyn Vale Eucalyptus Nurseries

Carrog, Carrog, Corwen, Clwyd LL21 9LD
Tel: 01490 430671
Specialities: Eucalyptus

40 Chepstow Garden Centre

Pwllmeyric, Chepstow, Gwent NP16 6LF
Tel: 01291 626035
Specialities: Trees, Herbaceous Perennials

41 Chepstow Garden Centre

Leechpool Holdings, Portskewett, Caldicot,
Gwent NP26 5UU
Tel: 01291 420767
Specialities: Garden Furniture, Annuals/Biennials/
Bedding, Trees

42 Church House Farm Nursery

Llandysilio, Llanymynech, Powys SY22 6QZ
Tel: 01691 839972

43 Conwy Valley Nurseries

Groesynyd, Tyn-y-Groes, Conwy, Gwynedd LL32 8SS
Tel: 01492 650228
Specialities: Annuals/Biennials/Bedding, Shrubs

44 Conwy Water Gardens

Aquatic Centre & Coarse Fishery, Rowen, Conwy,
Gwynedd LL32 8TP
Tel: 01492 650063
Specialities: Aquatics

45 Corseside Nursery

Angle Telephone Exchange, Angle, Pembroke,
Dyfed SA71 5AA
Tel: 01646 641505
Specialities: Hedging, Shrubs

46 Cosheston Nursery

Mount Pleasant Cross, Cosheston, Pembroke Dock,
Dyfed SA72 4TZ
Tel: 01646 686089

47 Cottage Garden Organic Perennials

Brynsifi, Trimsaran Road, Llanelli, Dyfed SA15 4RE
Tel: 01554 810294
Specialities: Organic, Cottage Garden Plants

48 Cottage Garden Plants

Tudor Farm, Bwlchgwyn, Wrexham,
Clwyd LL11 5YP
Tel: 01978 757165
Specialities: Cottage Garden Plants

49 The Country Garden

Llanrhydd Mill, Llanrhydd, Ruthin, Clwyd LL15 2UT
Tel: 01824 705179

50 Country Lane Nurseries

Plwmp, Llandysul, Dyfed SA44 6HU
Tel: 01239 851015
Specialities: Herbaceous Perennials

51 County Trees

Bier Wood, Manorbier, Tenby, Dyfed SA70 7SW
Tel: 01834 871430
Specialities: Trees

52 Creation Gardens

Swn-y-Gwynt, Nantmel, Llandrindod Wells, Powys LD1 6EL
Tel: 01597 825853 Fax: 01597 822867
e-mail: creationgardens@btconnect.com
websit: www.creationgardens.co.uk

SEE ADVERTISEMENT BELOW

Specialities:

Perennials; Bio-diversity Plants; Environmentally Friendly Products

Description:

Set in the midst of the beautiful mid-Wales countryside, Creation Gardens has a strong environmental and ethical stance to the business, including plants which are good for wildlife. In addition to its extensive range of plants, shrubs and trees, the centre has a garden shop and a Fairtrade tea room.

Directions:

The Centre is on the A44 between Rhayader and Crossgates on the eastern edge of the village of Nantmel.

53 Crimond Nursery

Revel, Revel Berriew, Welshpool, Powys SY21 8QB
Tel: 01686 640475

54 Crofty Nurseries

Llanteg, Narberth, Dyfed SA67 8QE
Tel: 01834 831437
Specialities: Annuals/Biennials/Bedding, Shrubs

55 Cross Farm Nursery

Cross Farm, Pentre Meyrick, Cowbridge, South Glamorgan CF71 7RN
Tel: 01446 774053
Specialities: Annuals/Biennials/Bedding, Shrubs

56 Crown Hill Nursery

Leechpool Holdings, Portskewett, Caldicot, Gwent NP26 5UU
Tel: 01291 420577

57 Crug Farm Plants

Crug Farm, Griffiths Crossing, Caernarfon, Gwynedd LL55 1TU
Tel: 01248 670232
Specialities: Annuals/Biennials/Bedding, Herbaceous Perennials

Creation Gardens is a lovely place to visit and spend some time. This innovative company is set in the midst of the beautiful mid-Wales countryside with a Tea Room overlooking the hillside and a Garden Shop packed full of gifts for gardeners new and old. A speciality at Creation Gardens is to stock plants which are good for wildlife and you will find lots of ideas to encourage wildlife and birds into your garden.

The Garden Centre is stocked with many different perennials from the common to the more unusual, a wide selection of roses, and a good range of plants which are useful for creating a backdrop to any garden, such as shrubs, conifers, bamboos and grasses. Aquatic plants and those suitable for stream or pond sides can also be found here. Trees and fruit trees suitable for growing in mid-Wales and the Marches area can be found in the Garden Centre. Inside the Garden Shop you will find a good selection of gardening tools, seeds and sundries, as well as more unusual garden-related gifts for both adults and children.

The Fairtrade Tea Room is open all day and serves lunches from 12.00 noon until 2.30 pm.

STUNNING VIEWS FROM TERRACE AREA NEXT TO POND AND FROM THE TEAROOM.....

Creation Gardens

Creation Gardens, Swn-y-Gwynt, Nantmel, Llandrindod Wells, Powys LD1 6EL
Tel: 01597 825853 Fax: 01597 822867
e-mail: creationgardens@btconnect.com
website: www.creationgardens.co.uk

Open 9.00 to 5.30
Tuesday to Saturday and Bank Holiday Mondays

58 Cwmclyd Nurseries

Cwmclyd, Rhosygarth, Llanilar, Aberystwyth, Dyfed SY23 4SL
Tel: 01974 241676
Specialities: Annuals/Biennials/Bedding, Shrubs

59 Cwmdu Nursery

Tycroes, Ammanford, Dyfed SA18 3PN
Tel: 01269 593397
Specialities: Annuals/Biennials/Bedding, Shrubs

60 Dale Road Nursery

Dale Rd, Walwyns Castle, Haverfordwest, Dyfed SA62 3AE
Tel: 07773 487485
Specialities: Annuals/Biennials/Bedding, Shrubs

61 Deva Orchids

Littlebrook Farm, Stryt Isa, Penyffordd, Chester, Clwyd CH4 0JY
Tel: 01978 762454
Specialities: Orchids

62 Dibley Nurseries

Llanelidan, Ruthin, Clwyd LL15 2LG
Tel: 01978 790677
Specialities: Annuals/Biennials/Bedding, Herbaceous Perennials

63 Dingle Nursery & Garden

Frochas, nr Welshpool, Powys SY21 9JD
Tel: 01938 555125 Fax: 01938 555778
e-mail: info@dinglenurseries.co.uk
website: www.dinglenurseries.co.uk

Specialities:

Trees; Shrubs; Perennials

Description:

A finalist in the "Nursery of the Year, 2006", Dingle Nursery is a family-run business established in the 1960s and now selling a wide variety of plants, many of which are grown on the wholesale nursery. Alongside it is the famous Dingle Garden, a 4-acre plantsman's garden.

Directions:

From Welshpool take the A490 north. Approximately ¾ mile on, turn left where the Garden Centre is signposted

64 Dolwyd Nurseries

Dolwyd, Colwyn Bay, Clwyd LL28 5HP
Tel: 01492 581089
Specialities: Annuals/Biennials/Bedding, Herbaceous Perennials

65 Dunsdale

Unit 102, Coed Aben Road, Wrexham Industrial Estate, Wrexham, Clwyd LL13 9UH
Tel: 08450 700560

66 Dyffryn Gwyddno Nursery

Dyffryn Farm, Lampeter Velfrey, Narberth, Dyfed SA67 8UN
Tel: 01834 861684
Specialities: Annuals/Biennials/Bedding, Shrubs

67 Dyffryn Nurseries

Home Farm, Dyffryn, Cardiff, South Glamorgan CF5 6SU
Tel: 029 2059 2085
Specialities: Exotic Fruit Trees, Pot Plants

68 Dyffryn Nurseries

Dyffryn, Goodwick, Dyfed SA64 0AE
Tel: 01348 874285
Specialities: Annuals/Biennials/Bedding, Herbaceous Perennials

69 Eastwick Plant Centre

Oswestry Road, Overton, Wrexham, Clwyd LL13 0LG
Tel: 01978 710848
Specialities: Annuals/Biennials/Bedding, Shrubs

70 F G Hilditch

Sunningdale Nurseries, Woodfieldside, Blackwood, Gwent NP12 0PJ
Tel: 01495 228015
Specialities: Annuals/Biennials/Bedding, Herbaceous Perennials

71 Farmyard Nurseries

1 Mansell Street, Carmarthen, Dyfed SA31 1QX
Tel: 01267 220259
Specialities: Herbaceous Perennials

72 Farmyard Nurseries

Farmyard Farm, Dol Llan Road, Llandysul, Dyfed SA44 4RL
Tel: 01559 363389
Specialities: Herbaceous Perennials, Trees, Shrubs

73 Fforest Mill Garden Centre

Pontarddulais Road, Cadle, Swansea,
West Glamorgan SA5 4BA
Tel: 01792 584496
Specialities: Aquatics, Shrubs

74 The Fishguard Garden Centre

Fishguard, Dyfed SA65 9QS
Tel: 01348 874034
Specialities: Trees, Shrubs, Pots

75 Ford Garden Centre

Began Road, Old St. Mellons, Cardiff,
South Glamorgan CF3 6XL
Tel: 029 2077 7292

76 Foundry Nursery

Tanygrisiau, Blaenau Ffestiniog,
Gwynedd LL41 3TW
Tel: 01766 830937
Specialities: Annuals/Biennials/Bedding, Herbaceous Perennials

77 Fron Goch Garden Centre

Fron Goch, Pant Road, Caernarfon,
Gwynedd LL54 5RL
Tel: 01286 672212
Specialities: Annuals/Biennials/Bedding, Trees, Garden Products

78 G Williams

Haulwen, Morfa Bychan, Porthmadog,
Gwynedd LL49 9YD
Tel: 01766 512910

79 The Garden Centre

1 Boverton Road, Boverton, Llantwit Major,
South Glamorgan CF61 1TX
Tel: 01446 792105

80 The Garden Centre

Howey, Howey, Llandrindod Wells, Powys LD1 5PL
Tel: 01597 822890

81 The Garden Centre

53 High Street, Brecon, Powys LD3 7AP
Tel: 01874 625913
Specialities: Pet Products, Flower, Garden Products

82 Gardners Cabin

152a Cromwell Road, Newport, Gwent NP19 0FZ
Tel: 01633 661188

83 Gemini Plants

14 Leechpool Holdings, Portskewett, Caldicot,
Gwent NP26 5TX
Tel: 01291 420495
Specialities: Annuals/Biennials/Bedding, Herbaceous Perennials

84 Glansevern Hall Gardens & Nursery

Berriew, Welshpool, Powys SY21 8AH
Tel: 01686 640644

Specialities:

Herbaceous Plants & Nursery; 20 Acres of Gardens; Tearoom and Gallery Shop

Products and Services:

- Seasonal Displays
- Display Gardens
- Giftware
- Plant Guarantee
- Plants
- Restaurant/Coffee Shop

Directions:

Signposted 4 miles southwest of Powis Castle on the A483

85 Glyndwr Plants

Tafarn Bric, Corwen, Clwyd LL21 9BU
Tel: 01490 413313
Specialities: Annuals/Biennials/Bedding

86 Glynwem Heather Nursery

Cilcennin, Lampeter, Dyfed SA48 8RJ
Tel: 01570 470063

87 Gordon's Nursery

1 Cefnpennar Cottages, Cefn Pennar, Mountain Ash,
Mid Glamorgan CF45 4EE
Tel: 01443 474593
Specialities: Annuals/Biennials/Bedding, Herbaceous Perennials

88 Grafted Walnut Trees

The Manse, Capel Isaac, Llandeilo, Dyfed SA19 7TN
Tel: 01558 669043
Specialities: Trees

89 Grandiflora Plant & Garden Centre

St Florence, Tenby, Dyfed SA70 8LP
Tel: 01834 871778
Specialities: Annuals/Biennials/Bedding, Trees, Shrubs

90 Green Fingers Garden Centre

Rhyl Road, Denbigh, Clwyd LL16 5TQ
Tel: 01745 815279
Specialities: Garden Furniture, Annuals/Biennials/Bedding, Trees

91 Greenmoor Nurseries

Greenmoor Farm, St. Brides Wentlooge, Newport, Gwent NP10 8SQ
Tel: 01633 680572

92 Hardings Sheds and Gardening Supplies

The Garage, Groesllwyd Guilsfield, Welshpool, Powys SY21 9BZ
Tel: 01938 553280
Specialities: Garden Furniture, Greenhouses and Sheds

93 Henllys Lodge Plants

Henllys Lodge, Henllys Lane, Llanfaes, Beaumaris, Gwynedd LL58 8HU
Tel: 01248 810106
Specialities: Herbaceous Perennials, Geranium

94 The Herb Garden & Historical Plant Nursery

Ty Capel Pensarn, Pentre Berw, Gaerwen, Gwynedd LL60 6LG
Tel: 01248 422208
Specialities: Herbs

95 Hill Farm Nurseries

Grove Hill, Pembroke, Dyfed SA71 5PU
Tel: 01646 685971
Specialities: Annuals/Biennials/Bedding, Herbaceous Perennials

96 Hilton Court Gardens & Crafts

Hilton Court Manor, Roch, Haverfordwest, Dyfed SA62 6AE
Tel: 01437 710262
Specialities: Giftware

97 Honeyborough

3 Honeyborough Industrial Estate, Neyland, Milford Haven, Dyfed SA73 1SE
Tel: 01646 601943

98 Horton Nurseries

Horton, Swansea, West Glamorgan SA3 1LQ
Tel: 01792 390322
Specialities: Shrubs, Trees

99 Hurrans Garden Centre

High Street, Cowbridge, South Glamorgan CF71 7YP
Tel: 01446 775053
Specialities: Garden Products

100 Hurrans Garden Centre

Catsash Road, Langstone, Newport, Gwent NP18 2LZ
Tel: 01633 413355

101 J Deen & Son

Maes-Y-Bryn Road, Pontprennau, Cardiff, South Glamorgan CF23 8LZ
Tel: 029 2077 7050

102 Jacksons Nurseries & Garden Centre

Marian, Trelawnyd, Rhyl, Clwyd LL18 6EB
Tel: 01745 570680
Specialities: Garden Products, Herbaceous Perennials

103 John Charles

Elder Cottage, Silverstream, Hubberston, Milford Haven, Dyfed SA73 3PU
Tel: 01646 692253

104 Joseph A&MP Rhoslan Garden Centre

Tre Talybont, Talybont, Dyfed SY24 5EQ
Tel: 01970 832463

105 Kiwi Nurseries

Sealand Road, Sealand, Chester, Clwyd CH1 6BS
Tel: 01244 399299

106 Knolton Nursery

Old Post Office Farm, Overton, Wrexham, Clwyd LL13 0LE
Tel: 01978 710308

107 Ladybird Nursery

Ciliau Aeron, Lampeter, Dyfed SA48 7SG
Tel: 01570 470480
Specialities: Annuals/Biennials/Bedding, Shrubs

108 Little Rhyndaston Nurseries

Hayscastle, Haverfordwest, Dyfed SA62 5PT
Tel: 01437 710656
Specialities: Annuals/Biennials/Bedding, Herbaceous Perennials

109 Llanarth Garden Centre

Llanarth, Dyfed SA47 0NR
Tel: 01545 580271
Specialities: Garden Furniture, Annuals/Biennials/Bedding

110 Llandovery Pet & Garden Centre

Llangadog Road, Brecon Road, Llandovery, Dyfed SA20 0RD
Tel: 01550 720029

111 Longfords Garden Centre

Velindre Road, Llangyfelach, Swansea, West Glamorgan SA5 7LP
Tel: 01792 773386

112 Lovering Gower Nurseries

Llanmadoc, Swansea, West Glamorgan SA3 1DJ
Tel: 01792 386651
Specialities: Annuals/Biennials/Bedding, Shrubs

113 Lydstep Garden Centre

Lydstep, Manorbier, Tenby, Dyfed SA70 7SJ
Tel: 01834 871475
Specialities: Trees, Annuals/Biennials/Bedding

114 Lynn's Nursery

Station House, Pantglas, Garndolbenmaen, Gwynedd LL51 9DX
Tel: 01766 530747

115 M Collins

Llyscoed, Llandovery, Dyfed SA20 0ET
Tel: 01550 720734

116 Maes Newydd Garden Centre

Llewellyn Street, Aberdare, Mid Glamorgan CF44 8HU
Tel: 01685 876843
Specialities: Garden Products

117 Magor Road Nursery

Byford, Magor Road, Langstone, Newport, Gwent NP18 2JX
Tel: 01633 412447
Specialities: Trees, Shrubs

118 Manorbier Garden Centre

Manorbier, Tenby, Dyfed SA70 7SN
Tel: 01834 871206
Specialities: Garden Furniture, Trees

119 Meadow Lane Nurseries

Trevalyn, Rossett, Wrexham, Clwyd LL12 0BT
Tel: 01244 570727
Specialities: Shrubs, Annuals/Biennials/Bedding

120 Midway Nurseries

Penybont, Llandrindod Wells, Powys LD1 5SY
Tel: 01597 851662

121 Millbrook Garden Centre

Mitchel Troy, Monmouth, Gwent NP25 4BD
Tel: 01600 713770
Specialities: Water Products and Features, Garden Furniture, Acer

122 Moorland Cottage Plants

Rhyd-y-Groes, Brynberian, Crymych, Dyfed SA41 3TT
Tel: 01239 891363
Specialities: Cottage Garden Plants

123 Moss Plant Centre

Unit 16, H T M Business Park Abergele Road, Rhuddlan, Rhyl, Clwyd LL18 5UZ
Tel: 01745 592446
Specialities: Annuals/Biennials/Bedding, Herbaceous Perennials, Shrubs

124 Multiple Organics

Lake Farm Barns, St. Athan Road, Cowbridge, South Glamorgan CF71 7HY
Tel: 01446 772964
Specialities: Organic

125 Mynydd Pencarreg Water Plants

Gwarffynnon Farm Cwmann, Cwmann, Lampeter, Dyfed SA48 8ED
Tel: 01570 480452
Specialities: Aquatics

126 Nantyderry Garden Centre

Nantyderry, Nantyderry, Abergavenny, Gwent NP7 9DW
Tel: 01873 880377
Specialities: Annuals/Biennials/Bedding, Herbaceous Perennials

127 Neath Valley Plants

Maes Y Berllan, Glynneath, Neath,
West Glamorgan SA11 5RU
Tel: 01639 722105
Specialities: Annuals/Biennials/Bedding, Herbaceous Perennials

128 New Inn Garden Centre

112 The Highway, Pontypool, Gwent NP4 0PJ
Tel: 01495 762556
Specialities: Trees, Annuals/Biennials/Bedding

129 New Inn Nurseries

Blaenffos, Boncath, Dyfed SA37 0HU
Tel: 01239 841215

130 Newbridge Nursery

Crundale, Crundale, Haverfordwest, Dyfed SA62 4EJ
Tel: 01437 731678
Specialities: Annuals/Biennials/Bedding, Shrubs

131 North Wales Water Gardens

St Asaph Avenue, Kinmel Bay, Rhyl,
Clwyd LL18 5TU
Tel: 01745 338222
Specialities: Aquatics, Water Products and Features

132 Nutwood Tree Nurseries

Rorsedd Bach, Llanarmon, Pwllheli,
Gwynedd LL53 6PU
Tel: 01766 819044
Specialities: Trees

133 Oakgrove Nurseries

Derlwyn, Llangynog, Carmarthen, Dyfed SA33 5HT
Tel: 01267 211878
Specialities: Annuals/Biennials/Bedding, Shrubs

134 Oasis Exotics

Abbey Rd, Whitland, Dyfed SA34 0LG
Tel: 01994 241451
Specialities: Exotics

135 Observatory Bonsai Nursery

Cyncoed Rd, Cardiff, South Glamorgan CF23 5SH
Tel: 029 2048 4892
Specialities: Bonsai

136 Old Railway Line Nursery

Three Cocks, Brecon, Powys LD3 0SG
Tel: 01497 847055
Specialities: Annuals/Biennials/Bedding, Shrubs

137 Olivers Garden Centre

Old Chepstow Road, Langstone, Newport,
Gwent NP18 2ND
Tel: 01633 412386
Specialities: Trees, Annuals/Biennials/Bedding

138 Organic Gardening

156 Whitchurch Road, Cardiff,
South Glamorgan CF14 3NA
Tel: 029 2062 1392
Specialities: Organic

139 Out of the Common Way

Penhyddgan, Boduan, Pwllheli, Gwynedd LL53 8YH
Tel: 01758 721577
Specialities: Aster (Michaelmas Daisy)

140 P & B Fuchsias

Maes y Gwaelod, Penclawdd Road, Penclawdd,
Swansea, West Glamorgan SA4 3RB
Tel: 01792 851669
Specialities: Fuchsia

141 P J Millington

Pen-Yr-Allt, Lloc, Holywell, Clwyd CH8 8SD
Tel: 01745 560204

142 P W Halse & Sons

Sunnyside Nurseries, Chepstow Road, Langstone,
Newport, Gwent NP18 2JN
Tel: 01633 412411
Specialities: Shrubs, Annuals/Biennials/Bedding, Trees

143 Pant Glas Nurseries

Pant Glas, Llanrhidian, Swansea,
West Glamorgan SA3 1EE
Tel: 01792 390935
Specialities: Annuals/Biennials/Bedding, Herbaceous Perennials

144 Pantfechan Nursery

Pentwyn Road, Cynonville, Port Talbot,
West Glamorgan SA13 3HH
Tel: 01639 850553
Specialities: Annuals/Biennials/Bedding, Shrubs

145 Pantycaerau Nursery

Church Road, Pentyrch, Cardiff,
South Glamorgan CF15 9QF
Tel: 029 2089 0697
Specialities: Annuals/Biennials/Bedding, Herbaceous Perennials

146 Pantyfod Garden Nursery

Llandewi Brefi, Llanddewi Brefi, Tregaron, Dyfed SY25 6PE
Tel: 01570 493564

147 Patio & Garden Centre

Quarry Park, Wiston, Haverfordwest, Dyfed SA62 4PH
Tel: 01437 751343
Specialities: Garden Furniture

148 Pembroke Garden Centre

Cosheston, Pembroke Dock, Dyfed SA72 4SX
Tel: 01646 680220
Specialities: Garden Furniture, Annuals/Biennials/Bedding

149 Pendre Garden Centre

Unit 10 Pendre Industrial Estate, Tywyn, Gwynedd LL36 9LW
Tel: 01654 711747

150 Pengawse Hill Nursery

Hafod Forge Llanddewi Velfrey, Llanddewi Velfrey, Narberth, Dyfed SA67 7EH
Tel: 01994 241296
Specialities: Annuals/Biennials/Bedding, Herbaceous Perennials, Shrubs

151 Penlan Perennials

Penlan Farm, Rhydlewis, Llandysul, Dyfed SA44 5QH
Tel: 01239 851244
Specialities: Herbaceous Perennials

152 Penpergwm Plants

Penpergwm Lodge, Penpergwm, Abergavenny, Gwent NP7 9AS
Tel: 01873 840422
Specialities: Annuals/Biennials/Bedding, Herbaceous Perennials

153 Penrallt Garden Centre Limited

Moylegrove, Cardigan, Dyfed SA43 3BX
Tel: 01239 881295
Specialities: Garden Furniture, Trees

154 Pentraeth Nurseries

Pentraeth, Gwynedd LL75 8BZ
Tel: 01248 450269
Specialities: Annuals/Biennials/Bedding, Herbaceous Perennials, Shrubs

155 The Perennial Nursery

Rhosygilwen, St. Davids, Haverfordwest, Dyfed SA62 6DB
Tel: 01437 721954
Specialities: Herbaceous Perennials

156 Petals Garden Centre

Pontardulais Road, Tycroes, Ammanford, Dyfed SA18 3QD
Tel: 01269 596388
Specialities: Garden Furniture, Garden Products

157 Peters Town Garden Centre

25A West Street, Fishguard, Dyfed SA65 9AL
Tel: 01348 873636
Specialities: Annuals/Biennials/Bedding

158 The Picton Castle Trust Nursery

Picton Castle, Rhoshill, Haverfordwest, Dyfed SA62 4AS
Tel: 01437 751326
Specialities: Rhododendron (Inc Azalea), Woodland Plants

159 Pine Direct Wales - Authorised Compton Agent

Log Buildings, Lon Parcwr Industrial Estate, Ruthin, Clwyd LL15 1NJ
Tel: 01824 705390

160 Plas Newydd Nursery

Plas Newydd, Llanfairpwllgwyngyll, Gwynedd LL61 6DQ
Tel: 01248 716386
Specialities: Annuals/Biennials/Bedding, Shrubs

161 Plassey Plants

The Plassey, Eyton, Wrexham, Clwyd LL13 0SP
Tel: 01978 780788
Specialities: Herbaceous Perennials, Shrubs

162 Ponderosa Garden Centre

Srd Caerbryn Ammanford, Penygroes Road, Caerbryn, Ammanford, Dyfed SA18 3EQ
Tel: 01269 850758
Specialities: Garden Furniture

163 Pontarddulais Garden Centre

Allt Y Graban Road, Pontarddulais, Swansea, West Glamorgan SA4 8DS
Tel: 01792 882561
Specialities: Garden Products

164 Prices Garden Centre

1 Church Crescent, Beaufort, Ebbw Vale,
Gwent NP23 5PE
Tel: 01495 303951
Specialities: Garden Products

165 Prime Perennials

Llettymoel, Rhosygarth, Llanilar, Aberystwyth,
Dyfed SY23 4SG
Tel: 01974 241505
Specialities: Herbaceous Perennials

166 R Roberts & Son

The Gardens, Llanelltyd, Dolgellau,
Gwynedd LL40 2SU
Tel: 01341 422358

167 Raglan Garden Centre

Croes Bychan, Abergavenny Road, Raglan, Usk,
Gwent NP15 2BH
Tel: 01291 690751
Specialities: Garden Furniture

168 Redcastle Nurseries

Cross Ash, Abergavenny, Gwent NP7 8PB
Tel: 01873 821232

169 Rhoshill Garden Plants

Tripp Farmhouse, Rhoshill, Cardigan,
Dyfed SA43 2TX
Tel: 01239 841827
Specialities: Trees, Shrubs, Bonsai

170 Rhoslwyn Plants

Silian, Lampeter SA48 8AX
Tel/Fax: 01570 422672
e-mail: janet.arthur5@virgin.net

Specialities:

Perennials; Shrubs; Trees

Products and Services:

- Credit Cards Accepted
- Display Gardens
- Fruit Trees
- Garden Furniture
- Garden Products
- Large Car Park
- Pet Products
- Plants
- Seeds and Bulbs
- Trees

Directions:

From Lampeter, take the A485 north towards Tregaron. After about 2 miles, turn left on minor road signposted "Silian".

Description:

Rhoslwyn Plants was opened in 2005 by Jan and Phil Arthur – Jan grows the plants while Phil assists and makes the coffee! The coffee – and tea – is not

just for them but for anyone visiting the nursery who would like a cup for free. The site is beautifully laid out and visitors can see the plants growing in the gardens, sit on the patios, and admire the wonderful views while listening to the birds and wildlife. Everything you see has been grown here – nothing is imported. The displays include many unusual plants you won't find in larger centres, and many are imaginatively planted to provide inspiration for your own garden. In addition to plants, the nursery stocks shrubs, roses, climbers, ornamental trees, grasses, heathers and seasonal plants, as well as bird tables and other garden accessories. The friendly and knowledgeable staff are happy to provide any advice or information you might need.

171 Robert's Garden Centre

Barley Mow, Lampeter, Dyfed SA48 7BY
Tel: 01570 422756
Specialities: Aquatics, Shrubs

172 Rosehill Nurseries

Bryn Esgob, Glascoed Road, St. Asaph,
Clwyd LL17 0LH
Tel: 01745 582507
Specialities: Annuals/Biennials/Bedding, Shrubs

173 Rosemary's Farmhouse Nursery

Llwyn-y-moel-gau, Llanfihangel, Llanfyllin,
Powys SY22 5JE
Tel: 01691 648196

174 Roualeyn Nurseries

Trefriw, Gwynedd LL27 0SX
Tel: 01492 640548
Specialities: Annuals/Biennials/Bedding, Shrubs

175 Saith Ffynnon Wildlife Plants

Saith Ffynnon Farm, Whitford, Holywell,
Flintshire CH8 9EQ
Tel: 01352 711198 Fax: 01352 716777
e-mail: jan@7wells.org
website: www.7wells.co.uk

Specialities:

Butterfly-Attracting Plants & Garden Design; Wildlife Gardening Plants and Books; Natural Dye Plants mainly by mail order through website.

Description:

Saith Ffynnon specialises in plants for wildlife gardens and is also home to the National Collection of Eupatoriums which flower in August and September. Viewing by appointment, please telephone first.

Directions:

From the A55, take the exit for Holywell. After about 3 miles turn by the Halfway House pub following the sign for Whitford, Trout Farm and Pennant Golf Course. Take the 2nd lane on the right (signposted for Trout Farm). Saith Ffynnon Farm is first on the right.

176 Sandyhill Nurseries

Sandyhill Road, Saundersfoot, Dyfed SA69 9DR
Tel: 01834 813744

177 Seiont Nurseries

Pontrug, Rhosbodrual, Caernarfon,
Gwynedd LL55 2BB
Tel: 01286 672524

178 Seven Oaks Garden Centre

Lon Parcwr Business Park, Ruthin,
Denbighshire LL15 1NJ
Tel: 01824 702567
e-mail: ian@sevnurs.freeserve.co.uk

Specialities:

Japanese Maples; Herbaceous; Shrubs & Climbing Plants

Products and Services:

- Barbeques
- Books and Stationery
- Seasonal Displays Clothing
- Credit Cards Accepted
- Disabled Facilities
- Dried & Artificial Flowers
- Fruit Trees
- Garden Furniture
- Garden Products
- Giftware
- Greenhouses and Sheds
- Houseplants
- Information Desk
- Large Car Park
- Plants
- Seeds and Bulbs
- Trees

Directions:

From the Tesco/Craft Centre roundabout head towards Lon Parcwr Business Park. Turn right to junction, left, then immediately right and the Garden Centre is at the end of this road.

Description:

Ian Forsyth, the owner of the Seven Oaks Garden Centre, started work in horticulture at the age of 15 and went on to study horticulture at Hadlow College. He worked in nurseries in America before returning to Seven Oaks which his family has owned and run since 1988. The nursery has 250 square metres of covered sales area, plus 160 square metres of interior space. Within these areas is displayed a huge range of plants, shrubs, trees, bushes, climbers, herbaceous varieties, glazed and terracotta pots, stoneware and statuary, fruit trees and bushes, herbs and vegetables. The nursery also stocks traditional hand-crafted tools, garden furniture, birdcare items, water features, screening products and a selection of David Austin roses. Seven Oaks is a member of the HTA and also of the National Garden Gift Voucher Scheme. The nursery has good disabled access with hard paths throughout the site.

179 Shipton Bulbs

Y Felin, Henllan Amgoed, Whitland,
Carmarthenshire SA34 0SL
Tel: 01994 240125 Fax: 01994 241180
e-mail: bluebellzoo.co.uk
website: bluebellbulbs.co.uk

Specialities:

Native Bulbs; Native Perennials; Other Bulbs for Naturalising

Description:

Owners John and Alison Shipton have been growing native bulbs on their Welsh smallholding since 1986 and guarantee that none of their bulbs has been taken from the wild. Most of their business is mail order but visitors are welcome – just call ahead to make sure they will be on site.

Directions:

From the A40 at Whitland follow the signs for Henllan Amgoed. In the village, turn right by the school (signposted "Church"). Keep right at the church and turn right at the "Shipton Bulbs" sign.

180 Silver Birch Nurseries

St. Asaph Avenue, Kinmel Bay, Rhyl,
Clwyd LL18 5HA
Tel: 01745 351366
Specialities: Shrubs

181 Singleton Garden Centre

Coychurch, Bridgend, Mid Glamorgan CF35 6AA
Tel: 01656 658201

182 Snowdonia Nurseries & Garden Centre

Llanrwst Road, Glan Conwy, nr Llandudno LL28 5SR
Tel: 01492 580703
e-mail: snowdonianurseries@hotmail.co.uk
website: www.snowdonianurseries.co.uk

Specialities:

Bedding Plants; Garden Furniture; Coffee Shop

Description:

Established more than 20 years ago, Snowdonia Nurseries provides year round plants on its 5-acre site. They offer their own home-grown bedding plants and a full range of shrubs and trees, along with an excellent range of garden furniture. There's also a coffee shop which is open daily.

Directions:

From Conwy cross the river and turn onto the A470, heading southwards. The nursery is on the left hand side as you enter Glan Conwy.

183 Soames Alpines Nursery

Station Yard, Llanpumsaint, nr
Carmarthen SA33 6BU
Tel: 01267 281604

Specialities:

Alpines; Dianthus; Saxifraga

Products and Services:

- Plant Guarantee
- Plants

Directions:

From Carmarthen take the A485 north. About 7 miles along this road, a minor road on the left leads to Llanpumsaint. The nursery occupies the site of the former railway station yard, close to the old railway bridge.

184 St Ishmaels Nurseries & Garden Centre

St. Ishmaels, St. Ishmaels, Haverfordwest,
Dyfed SA62 3SX
Tel: 01646 636343
Specialities: Trees, Shrubs

185 Stone's Garden Centre Ltd

Sealand Road, Sealand, Deeside, Clwyd CH5 2LQ
Tel: 01244 812714
Specialities: Garden Products

186 Strikes Daleside Garden Centre

Gladstone Way, Hawarden, Deeside, Clwyd CH5 3HF
Tel: 01244 532041

187 Strikes Holland Arms Garden Centre

Pentre Berw, Gaerwen, Gwynedd LL60 6LA
Tel: 01248 421655

188 Sundawn Nurseries Garden Centre

A55 Main St Asaph Road, Pen Y Cefn, Caerwys,
Mold, Clwyd CH7 5BL
Tel: 01352 720840
Specialities: Annuals/Biennials/Bedding, Aquatics, Garden Furniture

189 Sunny Hill Nurseries

Langewydd Bridgend, Llangewydd, Bridgend,
Mid Glamorgan CF32 0EU
Tel: 01656 746933
Specialities: Annuals/Biennials/Bedding, Herbaceous Perennials

190 Sunnydale Garden Centre

Treherbert Road, Croesyceiliog, Cwmbran,
Gwent NP44 2BZ
Tel: 01633 484004
Specialities: Trees, Garden Furniture

191 Sunnydale Nursery

Abby-cwm-Hir, Abbeycwmhir, Llandrindod Wells,
Powys LD1 6PG
Tel: 01597 851814
Specialities: Annuals/Biennials/Bedding, Herbaceous Perennials

192 Swiss Valley Garden Centre

Felinfoel, Llanelli, Dyfed SA14 8NJ
Tel: 01554 759712
Specialities: Garden Furniture, Trees, Annuals/Biennials/Bedding

193 Tal Goed Nurseries

Bryn Mynan, Glan Conwy, Colwyn Bay, Clwyd LL28 5TH
Tel: 01492 596095
Specialities: Annuals/Biennials/Bedding, Herbaceous Perennials, Shrubs

194 Tal-y-Bont Nurseries

Alltwalis Road, Pontarsais Sythe, Carmarthen, Dyfed SA32 7DZ
Tel: 01267 253302
Specialities: Herbaceous Perennials

195 Tan-y-Llyn Nurseries

Meifod, Powys SY22 6YB
Tel: 01938 500370
Specialities: Herbaceous Perennials, Alpines/Rock Plants

196 Tavernspite Garden Centre

Tavernspite, Whitland, Dyfed SA34 0NL
Tel: 01834 831671
Specialities: Restaurant/Coffee Shop

197 Tir Mab Cynan Nursery

Brithdir, Dolgellau, Gwynedd LL40 2RW
Tel: 01341 450339
Specialities: Annuals/Biennials/Bedding, Shrubs

198 Tom Jones & Sons

Gorslas, Llanelli, Dyfed SA14 7HP
Tel: 01269 842257

199 Ton Farm Nursery

Merthyr Mawr House, Bridgend, Mid Glamorgan CF32 0LR
Tel: 01656 653636

200 Tradewinds Nursery

Ffynnonwen, Penybont, Carmarthen, Dyfed SA33 6PX
Tel: 01994 484744

201 Trebor Nurseries

Gilbert Lane, Barry, South Glamorgan CF63 1BN
Tel: 029 2051 2195
Specialities: Herbaceous Perennials

202 Treborth Garden Centre

Treborth Road, Bangor, Gwynedd LL57 2RZ
Tel: 01248 371001
Specialities: Garden Products, Garden Furniture

203 Trefhedyn Garden Centre

Bridge Street, Newcastle Emlyn, Dyfed SA38 9DU
Tel: 01239 710292
Specialities: Garden Furniture, Annuals/Biennials/Bedding

204 Trefnant Plants & Herbs

Cwmbelan Tylwch, Cwmbelan, Llanidloes, Powys SY18 6QB
Tel: 01686 413563
Specialities: Annuals/Biennials/Bedding, Herbs

205 Tremadog Garden Centre

High Street, Tremadog, Porthmadog, Gwynedd LL49 9RD
Tel: 01766 513036
Specialities: Garden Furniture, Water Products and Features, Trees

206 Triafon Nursery

Bryn Teg, Trewern, Welshpool, Powys SY21 8EB
Tel: 01938 570166
Specialities: Herbaceous Perennials, Lavender

207 TWR Nurseries

TWR Farm, South Stack Road, Holyhead, Gwynedd LL65 1YL
Tel: 01407 764631

208 Ty Rhos Trees

Ty Rhos, Velindre, Crymych, Dyfed SA41 3XD
Tel: 01239 820701
Specialities: Trees

209 Tyddyn Fachau Nurseries

Arfryn, Y Ffor, Pwllheli, Gwynedd LL53 6UB
Tel: 01758 612626
Specialities: Shrubs, Herbaceous Perennials

210 Tynant Nurseries

Ty Nant Road, Morganstown, Cardiff, South Glamorgan CF15 8LB
Tel: 029 2084 2017
Specialities: Garden Products

211 Tynyrardd Nurseries

Bryn Hyfryd Road, Tywyn, Gwynedd LL36 9HG
Tel: 01654 710283
Specialities: Annuals/Biennials/Bedding, Shrubs

212 Usk Garden Centre

Llanbadoc, Usk, Monmouthshire NP15 1TG
Tel: 01291 673603 Fax: 01291 673360
e-mail: enquiries@morrisusk.demon.co.uk

Specialities:
Plants A-Z; Plants- Specimen Selection; Annual & Perennial Plants

Products and Services:

- Credit Cards Accepted
- Disabled Facilities
- Display Gardens
- Fruit Trees
- Garden Design
- Garden Furniture
- Garden Products
- Giftware
- Houseplants
- Large Car Park
- Pet Products
- Plants
- Restaurant/Coffee Shop
- Seeds and Bulbs
- Trees

Directions:
From the A449, take the Usk exit, drive through the town, cross the river bridge and immediately turn left on minor road to Caerleon. About a mile along this road you will see the Centre on your left.

Description:

The Usk Garden Centre is owned and run by the Morris family with all members involved. They took over here in 1982 when it was just a tomato nursery. In 1990 they opened a small retail outlet and since then the business has expanded year on year. It is now a well laid out traditional garden centre offering an extensive choice of plants, shrubs and trees, including some 200 different varieties of fruit trees. You can be confident of finding almost anything you are looking for, but if it isn't here the nursery will gladly source it for you. The friendly staff are also happy to offer specialist advice, whether it be on soil types or planting schemes. Within the spacious 5-acre site there's also a coffee shop and restaurant, a pet and bird care area, and some live pets such as hamsters, rabbits and guinea pigs.

213 Victoria Garden Centre

Victoria, Ebbw Vale, Gwent NP23 8FP
Tel: 01495 350222
Specialities: Garden Furniture

214 Vincent Davies Garden Centre

Fishguard Road, Haverfordwest, Dyfed SA62 4BT
Tel: 01437 763849

215 Waterways Aquatic & Garden Centre

Llan Y Pwll, Holt Road, Wrexham, Clwyd LL13 8NE
Tel: 01978 660289
Specialities: Aquatics

216 Welsh College of Horticulture

Holywell Road, Northop, Mold, Clwyd CH7 6AA
Tel: 01352 840436

217 Welshpool Garden Centre

2-3 Stanley Street, Welshpool, Powys SY21 7JH
Tel: 01938 553680
Specialities: Organic Compost, Annuals/Biennials/Bedding

218 Whimble Nursery

Kinnerton, Presteigne, Powys LD8 2PD
Tel: 01547 560413

219 White Lodge Nursery

White Lodge, Newborough, Llanfairpwllgwyngyll, Gwynedd LL61 6RS
Tel: 01248 440254

220 Willows Garden Centre

The Willows, Abercanaid Industrial Estate, Merthyr Tydfil, Mid Glamorgan CF48 1YF
Tel: 01685 384415
Specialities: Aquatics, Garden Products

221 Willows of Usk

Baron Street, Usk, Gwent NP15 1AS
Tel: 01291 672039

222 Woodfields Nursery

Twyn Road, Ystrad Mynach, Hengoed, Mid Glamorgan CF82 7EW
Tel: 01443 814241
Specialities: Herbaceous Perennials, Annuals/Biennials/Bedding, Pots

223 World of Ferns

Carreg y Fedwen, Lôn Rallt, Pentir, Bangor, Gwynedd, LL57 4RP
Tel/Fax: 01248 600385
e-mail: info@world-of-ferns.co.uk
website: www.world-of-ferns.co.uk

SEE ADVERTISEMENT ABOVE

Specialities:

Tree Ferns; Hardy Ferns; Tropical Ferns; Architectural and Unusual Plants

Description:

World of Ferns (formerly Rickards Ferns Ltd) is Europe's leading fern specialist nursery offering the widest vatiety of indoor, conservatory, garden and tree ferns from around the world including many exclusive, unusual, specimens and architectural plants.

Directions:

From Exit 11 of the A55, near Bangor, take the A5 south towards Betws-y-Coed. At the first roundabout, take the A4244/B4366 towards Caernarfon. Take the 2nd left (after about 1.5 miles) then left again onto Lôn Rallt, continue up, ignoring 2 right hand turns, until you reach World of Ferns.

224 Wrenvale Nurseries

Heol Y Gors, Cwmgors, Ammanford, Dyfed SA18 1RW
Tel: 01269 826700
Specialities: Annuals/Biennials/Bedding, Shrubs

225 Wyevale Garden Centre (Castleton)

Newport Road, Castleton, Cardiff, South Glamorgan CF3 2UQ
Tel: 01633 680002

226 Wyevale Garden Centre (Pyle)

2 Heol Mostyn, Village Farm Industrial Estate, Pyle, Bridgend, Mid Glamorgan CF33 6BJ
Tel: 01656 741443

227 Wyevale Garden Centre (Swansea)

Siemens Way, Swansea Enterprise Park, Swansea, West Glamorgan SA7 9FT
Tel: 01792 310052

ORDER FORM

To order any of our publications just fill in the payment details below and complete the order form. For orders of less than 4 copies please add £1 per book for postage and packing. Orders over 4 copies are P & P free.

Please Complete Either:

I enclose a cheque for £ ______ *made payable to Travel Publishing Ltd*

Or:

CARD NO: ______ EXPIRY DATE: ______

SIGNATURE: ______

NAME: ______

ADDRESS: ______

TEL NO: ______

Please either send, telephone, fax or e-mail your order to:

Travel Publishing Ltd, 64-66 Ebrington Street, Plymouth, Devon PL4 9AQ
Tel: 01752 276660 Fax: 01752 276699 e-mail: info@travelpublishing.co.uk

	Price	Quantity
Hidden Places Regional Titles		
Cornwall	£8.99	....
Devon	£8.99	....
Dorset, Hants & Isle of Wight	£8.99	....
East Anglia	£8.99	....
Lake District & Cumbria	£8.99	....
Northumberland & Durham	£8.99	....
Peak District and Derbyshire	£8.99	....
Yorkshire	£8.99	....
Hidden Places National Titles		
England	£11.99	....
Ireland	£11.99	....
Scotland	£11.99	....
Wales	£11.99	....
Country Pubs and Inns Titles		
Cornwall	£5.99	....
Devon	£7.99	....
Sussex	£5.99	....
Wales	£8.99	....
Yorkshire	£7.99	....

	Price	Quantity
Country Living Rural Guides		
East Anglia	£10.99	....
Heart of England	£10.99	....
Ireland	£11.99	....
North East of England	£10.99	....
North West of England	£10.99	....
Scotland	£11.99	....
South of England	£10.99	....
South East of England	£10.99	....
Wales	£11.99	....
West Country	£10.99	....
Other Titles		
Off The Motorway	£11.99	....
Garden Centres & Nurseries of Britain	£11.99	....

TOTAL QUANTITY ______

TOTAL VALUE ______

READER REACTION FORM

The **Travel Publishing** *research team would like to receive readers' comments on any Garden Centre or Nursery reviewed in the book and also recommendations for suitable entries to be included in the next edition. This will help ensure that the* **Garden Centres and Nurseries of Britain** *continues to provide its readers with useful information. To provide your comments or recommendations would you please complete the forms below and overleaf as indicated and send to:*

The Research Department,
Travel Publishing Ltd, 64-66 Ebrington Street, Plymouth, Devon PL4 9AQ

YOUR NAME:

YOUR ADDRESS:

YOUR TEL NO:

Please tick as appropriate: **COMMENTS** ☐ **RECOMMENDATION** ☐

ESTABLISHMENT:

ADDRESS:

TEL NO:

CONTACT NAME:

PLEASE COMPLETE FORM OVERLEAF

READER REACTION FORM

COMMENT OR REASON FOR RECOMMENDATION:

READER REACTION FORM

The **Travel Publishing** *research team would like to receive readers' comments on any Garden Centre or Nursery reviewed in the book and also recommendations for suitable entries to be included in the next edition. This will help ensure that the* **Garden Centres and Nurseries of Britain** *continues to provide its readers with useful information. To provide your comments or recommendations would you please complete the forms below and overleaf as indicated and send to:*

The Research Department,
Travel Publishing Ltd, 64-66 Ebrington Street, Plymouth, Devon PL4 9AQ

YOUR NAME:

YOUR ADDRESS:

YOUR TEL NO:

Please tick as appropriate: **COMMENTS** ☐ **RECOMMENDATION** ☐

ESTABLISHMENT:

ADDRESS:

TEL NO:

CONTACT NAME:

PLEASE COMPLETE FORM OVERLEAF

READER REACTION FORM

COMMENT OR REASON FOR RECOMMENDATION:

READER REACTION FORM

The **Travel Publishing** *research team would like to receive readers' comments on any Garden Centre or Nursery reviewed in the book and also recommendations for suitable entries to be included in the next edition. This will help ensure that the* **Garden Centres and Nurseries of Britain** *continues to provide its readers with useful information. To provide your comments or recommendations would you please complete the forms below and overleaf as indicated and send to:*

The Research Department,
Travel Publishing Ltd, 64-66 Ebrington Street, Plymouth, Devon PL4 9AQ

YOUR NAME:

YOUR ADDRESS:

YOUR TEL NO:

Please tick as appropriate: **COMMENTS** ☐ **RECOMMENDATION** ☐

ESTABLISHMENT:

ADDRESS:

TEL NO:

CONTACT NAME:

PLEASE COMPLETE FORM OVERLEAF

READER REACTION FORM

COMMENT OR REASON FOR RECOMMENDATION:

INDEX OF SPECIALITIES

It is virtually impossible to index every single product or service supplied by a garden centre or nursery. As a result the *Country Living Garden Centres & Nurseries of Britain* has not indexed generic products and services (e.g. "plants", "shrubs", "trees" etc) which are supplied by most garden centres and nurseries, but has concentrated on indexing more specific products and services (e.g. "Begonias", "Orchids", "Tulips", "Cornus" etc) which may be more difficult to locate and which have been identified by the garden centres and nurseries themselves. Over 120 specialities are included in the Index of Specialities but if there are specialities which readers would like to be included in the next edition please e-mail us on info@travelpublishing.co.uk

Asters (Michaelmas Daisy)

Bamboos

Begonias

Black Plants a Speciality

Bonsai Plants and Trees

Buddleja (Buddleia)

Exotic Seeds

Fagus (Beech)

Farm Shop

Fencing

Ferns

Floristry

Flower Seeds

Foliage Plants

Fruit Bushes and Trees

Fuchsia

Garden Design

Garden Furniture

Garden Ornaments

Geraniums

Gladiolus

Grasses

Greenhouses and Sheds

Hanging Baskets

Hardy Garden Palms

Hardy Plants

Hardy Woodland Plants

Heathers (Calluna or Erica)

Hedera (Ivy)

Hedgerow Shrubs and Hedging

Hellebores

Hemerocallis (Daylily)

Herbs

Homegrown Plants

Hostas

Hot Tubs and Swimming Pools

Houseplants

Ilex (Holly)

Iris

Japanese Maples

Landscape Gardening

Lavenders

Lilium (Lily)

Seeds

Shade Loving Plants

Specialist Bedding

Specimen Plants and Trees

Syringa (Lilac)

Terracotta Pots, Planters and Containers

Topiary

Tulipa (Tulip)

Turf and Lawn Supplies

Vegetable Plants and Seeds

Viburnum

Vines

Water Products and Features

Wildlife Plants and/or Wild Flowers

Window Boxes

Winter Pansies

Woodland Plants